Microbial Fertilizer Technology for Sustainable Crop Production

Current Advances in Biodiversity, Conservation and Environmental Sciences

Microbial Fertilizer Technology for Sustainable Crop Production

Jeyabalan Sangeetha, PhD
Abdel Rahman Mohammad Al-Tawaha, PhD
Devarajan Thangadurai, PhD
Editors

AAP | APPLE ACADEMIC PRESS

First edition published 2026

Apple Academic Press Inc.
1265 Goldenrod Circle, NE,
Palm Bay, FL 32905 USA

760 Laurentian Drive, Unit 19,
Burlington, ON L7N 0A4, Canada

CRC Press
2385 NW Executive Center Drive,
Suite 320, Boca Raton FL 33431

4 Park Square, Milton Park,
Abingdon, Oxon, OX14 4RN, UK

Library and Archives Canada Cataloguing in Publication

Title: Microbial fertilizer technology for sustainable crop production / Jeyabalan Sangeetha, PhD, Abdel Rahman Mohammad Al-Tawaha, PhD, Devarajan Thangadurai, PhD, editors.
Names: Sangeetha, Jeyabalan, editor | Al-Tawaha, Abdel Rahman Mohammad, editor. | Thangadurai, Devarajan, 1976- editor
Series: Current advances in biodiversity, conservation, and environmental sciences (Series)
Description: First edition. | Series statement: Current advances in biodiversity, conservation, and environmental sciences | Includes bibliographical references and index.
Identifiers: Canadiana (print) 20250303051 | Canadiana (ebook) 2025030306X | ISBN 9781774915684 (hardcover) | ISBN 9781003594826 (ebook)
Subjects: LCSH: Biofertilizers. | LCSH: Crops and soils—Management. | LCSH: Sustainable agriculture.
Classification: LCC S654.5 .M53 2026 | DDC 631.8/6—dc23

Library of Congress Cataloging-in-Publication Data

CIP data on file with US Library of Congress

ISBN: 978-1-774915-68-4 (hbk)
ISBN: 978-1-003594-82-6 (ebk)
DOI: 10.1201/9781003594826

ABOUT THE BOOK SERIES: CURRENT ADVANCES IN BIODIVERSITY, CONSERVATION AND ENVIRONMENTAL SCIENCES

SERIES EDITORS:

Jeyabalan Sangeetha, PhD
Assistant Professor, Central University of Kerala, Kasaragod, India

Devarajan Thangadurai, PhD
Professor, Karnatak University, Dharwad, India

Nature is an entity composed of three main internally interactive components: biodiversity–human–environment. These are triangularly related aspects of life science, and the interdependency of these three reflects the crucial need of maintenance of equilibrium in their interaction. Both human and biodiversity are two interactive phases of environment that provide a necessary platform for interaction. Being part of biodiversity, human life is almost dependent on biodiversity and its products. This is the point where the human–biodiversity interaction fluctuates due to overexploitation of biodiversity, where humans take more than the basic needs of human life. Over the past few centuries, this fluctuation in interaction has caused dramatic depletion of biodiversity and thus a drastic change in environment. This has now boomeranged on human life significantly. This variation in the interaction triangle has already travelled the long path of time. Until now, conservationists and life scientists are thinking of restoring the equilibrium in the interaction triangle by taking innovative steps in conserving biodiversity and protecting the environment.

Conservation biology and environmental science are the hotspots of research activities over the past several decades. This has caused much-needed development in observing, identifying, gathering, storing, and analyzing information, social awareness, planning, and advancement in conservation tools and techniques. Conservation biologists have given more importance to building stronger networks of specialists and experts in biology, agriculture, food technology, medicine, mathematics, engineering, architecture, robotics, information technology, and other branches of science and technology to gain innovative ideas to successfully treat the threats to biodiversity and the environment.

Gathering and monitoring information about biodiversity, conservation, and environment is a crucial need of biologists. Surveys from biological survey agencies, establishment of bio-inventories, modern tools and technologies to monitor the biodiversity and collection of data, and strengthening the scientific networks to make awareness, to generate the data, and for the accumulation of both traditional and scientific knowledge about biodiversity conservation and environmental management have been achieved in the recent decades. Conservationists need technological innovations to resolve the threats to biodiversity and environment that are now within reach.

The *Current Advances in Biodiversity, Conservation and Environmental Sciences* book series publishes titles on a regular basis for a better understanding of biodiversity conservation and environmental management to achieve sustainable development goals for future generations.

TITLES IN THE SERIES TO DATE

- **Algal Farming Systems: From Production to Application for a Sustainable Future**
  Editors: Jeyabalan Sangeetha, PhD, and Devarajan Thangadurai, PhD
- **Algal Genetic Resources: Cosmeceuticals, Nutraceuticals, and Pharmaceuticals from Algae**
  Editors: Jeyabalan Sangeetha, PhD, and Devarajan Thangadurai, PhD
- **Beneficial Microbes for Sustainable Agriculture and Environmental Management**
  Editors: Jeyabalan Sangeetha, PhD, Devarajan Thangadurai, PhD, and Saher Islam, MPhil
- **Biodiversity and Conservation: Characterization and Utilization of Plants, Microbes, and Natural Resources for Sustainable Development and Ecosystem Management**
  Editors: Jeyabalan Sangeetha, PhD, Devarajan Thangadurai, PhD, Hong Ching Goh, PhD, and Saher Islam, MPhil
- **Environmental Phycology: Climate Change Mitigation and Hazardous Pollutant Cleanup using Algal Resources**
  Editors: Jeyabalan Sangeetha, PhD, and Devarajan Thangadurai, PhD
- **Microbial Fertilizer Technology for Sustainable Crop Production**
  Editors: Jeyabalan Sangeetha, PhD,
  Abdel Rahman Mohammad Al-Tawaha, PhD, and
  Devarajan Thangadurai, PhD

- **Mycorrhizal Technology: Managing Plant Stress and Mitigating Climate Change using Mycorrhizae**
  Editors: Jeyabalan Sangeetha, PhD,
  Abdel Rahman Mohammad Al-Tawaha, PhD, and
  Devarajan Thangadurai, PhD
- **Organic Farming for Sustainable Development**
  Editors: Jeyabalan Sangeetha, PhD, Kasem Soytong, PhD,
  Devarajan Thangadurai, PhD, and
  Abdel Rahman Mohammad Al-Tawaha, PhD

ABOUT THE EDITORS

Jeyabalan Sangeetha, PhD
Assistant Professor, Central University of Kerala, Kasaragod, India

Jeyabalan Sangeetha, PhD, is an Assistant Professor at the Central University of Kerala, Kasaragod, India. She has edited and co-edited several books within her research areas, which include environmental toxicology, environmental microbiology, environmental biotechnology, and environmental nanotechnology. She earned her BSc in Microbiology and PhD in Environmental Science from Bharathidasan University, Tiruchirappalli, Tamil Nadu, India. She holds an MSc degree in Environmental Science from Bharathiar University, Coimbatore, Tamil Nadu, India. She is the recipient of a Tamil Nadu Government Scholarship and a Rajiv Gandhi National Fellowship of the University Grants Commission, Government of India for her doctoral studies. She served as Dr. D.S. Kothari Postdoctoral Fellow and UGC Postdoctoral Fellow at Karnatak University, Dharwad, India, from 2012 to 2016, with a funding from the University Grants Commission, Government of India, New Delhi.

Abdel Rahman Mohammad Al-Tawaha, PhD
Professor, Al Hussein Bin Talal University, Jordan

Abdel Rahman Mohammad Al-Tawaha, PhD, is a full Professor of Plant Science, former Head of the Department of Biological Sciences, and former Director of the Planning, Information, and Quality Unit at Al Hussein Bin Talal University, Jordan. He obtained his PhD from McGill University, Montreal, Canada, in 2006. He is the author and coauthor of more than 200 publications in plant science in leading peer-reviewed journals, chapters in edited volumes, and books on a broad range of development issues. He is the Founder and Editor in Chief of *Advances in Environmental Biology*. He has also participated in many international conferences as chair or member of the scientific committee.

Devarajan Thangadurai, PhD
Professor, Karnatak University, Dharwad, India

Devarajan Thangadurai, PhD, is a Professor at Karnatak University, India. He has authored and edited over 45 books with national and international publishers and has visited 24 countries in Asia, Europe, Africa, and the Middle East for academic purposes, scientific meetings, and international collaborations. He received his PhD in Botany from Sri Krishnadevaraya University, India, as a CSIR Senior Research Fellow with funding from the Ministry of Science and Technology, Government of India. He has served as a postdoctoral fellow at the University of Madeira, Portugal, University of Delhi, India, and ICAR National Research Centre for Banana, India. He is a recipient of a Best Young Scientist Award with a Gold Medal from Acharya Nagarjuna University, India, and the VGST-SMYSR Young Scientist Award of the Government of Karnataka, India.

CONTENTS

CONTRIBUTORS

Muhammad Aasim
Department of Plant Protection, Faculty of Agricultural Sciences and Technologies, Sivas University of Science and Technology, Sivas, Turkey

Sabaa Jawad Abdulkadhim
Al Furat Al-Awsat Technical University, Al-Mussaib Technical College, Iraq

Afroza Akhter
Division of Vegetable Science, SKUAST-K, Shalimar, Srinagar, Jammu and Kashmir, India

Taha A. Al-Issa
Department of Plant Production, Faculty of Agriculture, Jordan University of Science and Technology, Irbid, Jordan

Abdel Rahman Mohammad Al-Tawaha
Department of Biological Sciences, Al Hussein bin Talal University, Maan, Jordan

Abdel Razzaq Al-Tawaha
National Agricultural Research Center (NARC), Jordan

Hiba Alatrash
University of Helsinki, Helsinki, Finland

Alla Aleksanyan
Institute of Botany aft. A.L. Takhtajyan NAS RA/Department of Geobotany and Plant Eco-Physiology, Yerevan, Armenia

Iftikhar Ali
Center for Plant Sciences and Biodiversity, University of Swat, Charbagh, Pakistan

Amanullah
Department of Agronomy, The University of Agriculture, Peshawar, Pakistan

Neethu Asokan
Department of Life Sciences, Sri Sathya Sai University for Human Excellence, Kalaburagi, Karnataka, India

Girma Birru
Agroecosystem Management Research Unit, USDA-ARS, NE, USA

Vedavyas Shivanand Chavan
Department of Botany, Karnatak University, Dharwad, Karnataka, India

Pavitra Chippalakatti
Department of Botany, Karnatak University, Dharwad, Karnataka, India

Javeid Ahmed Dar
Division of Vegetable Science, SKUAST-K, Shalimar, Srinagar, Jammu and Kashmir, India

Abhijit Dey
Department of Life Sciences, Presidency University, Kolkata, West Bengal, India

Samar A. Ghanem
Soil and Water Department, Faculty of Agriculture, Tishreen University, Latakia, Syria

Khursheed Hussain
Division of Vegetable Science, Sher-e-Kashmir University of Agricultural Sciences and Technology-Kashmir, Shalimar, Srinagar, India

Imran
Department of Agronomy, The University of Agriculture, Peshawar, Pakistan

Sayed Azrah Indrabi
Division of Vegetable Science, SKUAST-K, Shalimar, Srinagar, Jammu and Kashmir, India

Yaman Jabbour
General Commission for Scientific Agriculture Research, Syria

Edappayil Janeeshma
Assistant Professor, Department of Botany, MES KEVEEYAM College, Valanchery, Kerala, India

Insha Javeed
Division of Vegetable Science, SKUAST-K, Shalimar, Srinagar, Jammu and Kashmir, India

Razzagh Karimirad
Department of Animal Science, Lorestan University, Khorramabad, Iran

Roghayeh Karimirad
Department of Horticultural Sciences, Faculty of Agriculture and Natural Resources, University of Mohaghegh Ardabili, Ardabil, Iran

Shah Khalid
Department of Agronomy, The University of Agriculture, Peshawar, Pakistan

Alaa H. Khallouf
General Commission for Scientific Agricultural Research (GCSAR), Damascus, Syria; Soil and Land Reclamation Department, Faculty of Agriculture, Al-Baath University, Homs, Syria

Samia Khanum
Department of Botany, University of the Punjab, Lahore, Pakistan

Ekaterina Kozuharova
Medical University of Sofia, Faculty of Pharmacy, Department of Pharmacognosy, Sofia, Bulgaria

Aradhna Kumari
Department of Plant Physiology, College of Agriculture, Jawaharlal Nehru Krishi Vishwa Vidyalaya, Ganj Basoda, Madhya Pradesh, India

Poojashree Nagappa Kummur
Department of Botany, Karnatak University, Dharwad, Karnataka, India

Abeer Jubily
Centre for Agricultural Research, General Commission for Scientific Agricultural Research, Syria

Sameena Lone
Division of Vegetable Science, Sher-e-Kashmir University of Agricultural Sciences and Technology of Kashmir, Shalimar, Srinagar, India

Khalid Masoodi
Transcriptomics Laboratory, Division of Plant Biotechnology, Sher-e-Kashmir University of Agricultural Sciences and Technology-Kashmir, Shalimar, Srinagar, India

Sana Mounaimi
Mohammed VI Polytechnic University, Faculty of Agriculture, Morocco

Shahnaz Mufti
Division of Vegetable Science, SKUAST-K, Shalimar, Srinagar, Jammu and Kashmir, India

Kumar Pandey
Department of Biotechnology, School of Bioengineering and Biosciences, Lovely Professional University, Phagwara, Punjab, India

Devendra Kumar Pandey
Department of Biotechnology, School of Bioengineering and Biosciences, Lovely Professional University, Phagwara, Punjab, India

Simran Panigatti
Department of Botany, Karnatak University, Dharwad, Karnataka, India

Siddhartha Pati
NatNov Bioscience Private Limited, Balasore, Odisha, India

Jos Thomas Puthur
Plant Physiology and Biochemistry Division, Department of Botany, University of Calicut, Calicut, Kerala, India

Vishnu D. Rajput
Academy of Biology and Biotechnology, Southern Federal University, Rostov-on-Don, Russia

Najmu Sakib
Division of Vegetable Science, SKUAST-K, Shalimar, Srinagar, Jammu and Kashmir, India

Khaldoun Othman Al Sane
Directorate of Plant Biodiversity and Plant Genetic Resources, National Agricultural Research Center (NARC), Jordan

Jeyabalan Sangeetha
Department of Environmental Science, Central University of Kerala, Kasaragod, Kerala, India

P. Saranraj
Department of Microbiology, Sacred Heart College (Autonomous), Tirupattur, Tamil Nadu, India

R. Z. Sayyed
Department of Microbiology, PSGVP Mandal's Arts, Science and Commerce College, Shahada, Maharashtra, India

Samar S. Shawaqfeh
Department of Plant Production and Protection, College of Agriculture, Jerash University, Jerash, Jordan

Santosh Kumar Singh
Department of Soil Science, Dr. Rajendra Prasad Central Agricultural University, Pusa, Samastipur, Bihar, India

P. Sivasakthivelan
Department of Agricultural Microbiology, Faculty of Agriculture, Annamalai University, Chidambaram, Tamil Nadu, India

Amreena Sultan
Division of Vegetable Science, SKUAST-K, Shalimar, Srinagar, Jammu and Kashmir, India

Devarajan Thangadurai
Department of Botany, Karnatak University, Dharwad, Karnataka, India

Munir Turk
Department of Plant Production, Faculty of Agriculture, Jordan University of Science and Technology, Irbid, Jordan

Puteri E. Megat Wahab
Department of Crop Science, Faculty of Agriculture, University Putra Malaysia, Serdang, Selangor, Malaysia

PREFACE

The constantly growing world population scenario is putting great pressure on crop productivity, escalating the excessive use of chemical pesticides and fertilizers, which may deteriorate the quality of soil and food. For this reason, there is a foremost need to develop alternative fertilizer resources to enhance agriculture productivity and soil sustainability. Biofertilizers comprise microorganisms such as bacteria, fungi, and algae, which have the capacity to enrich the topsoil with valuable rudiments including nitrogen, potassium, phosphorous, and organic matter. Biofertilizers as soil remediators, renewable environmentally friendly bioprotectors, and cost-effective growth boosters facilitate restoration of soil integrity, fertility, and water-holding capacity. For example, plant-growth-promoting rhizobacteria is one of the beneficial microorganisms that emerge as a biofertilizer, biopesticide, and biocontrol agent for sustainable agriculture. In this way, many microbial strains capable of degrading soil compounds are increasingly getting exploited as a potential inoculant for several food crops to sustain global agricultural productivity.

The present book comprises 13 chapters contributed by several distinguished academicians, experts, and researchers who discuss the extensive need to improve soil health through sustainable practices. This book is primarily about utilizing the potential soil microorganisms of diverse genera to overcome issues instigated by usage of chemical pesticides and fertilizers in the long run. Most of the chapters in this book address the formulation, preparation, production, and applications of microbial biofertilizers for agricultural, forestry, industrial, and environmental sectors. Moreover, this book provides extensive information about emerging plant growth promoters, including arbuscular mycorrhiza, rhizobia, endophytes, and cyanobacteria along with their latest case studies, and also presents challenges and potential perspectives for various agricultural crops. In general, the discussion on tremendous progress made in the last 10 years in the biofertilizer industry through the development/bioengineering of stress-tolerant soil microbial strains is a hallmark feature of this book.

With in-depth coverage and strategic emphasis, the book will prove to be immensely useful and informative for microbiologists, biotechnologists, plant biologists, bioengineers, industrialists, and agronomists who deal with production and assembly methods for microbial biofertilizers. Also, it would be an essential guide for agriculture development executives, trainees in the microbial biofertilizer industry, and teachers, research scholars, and postgraduate students who are interested in understanding science behind sustainable biofertilizer production, agroecology and soil chemistry.

Jeyabalan Sangeetha, PhD
Abdel Rahman Mohammad Al-Tawaha, PhD
Devarajan Thangadurai, PhD

CHAPTER 1

ROLE OF PLANT–MICROBE INTERACTION IN DROUGHT STRESS MITIGATION

SAMEENA LONE[1], KHURSHEED HUSSAIN[1], KHALID MASOODI[2], ABDEL RAHMAN MOHAMMAD AL-TAWAHA[3*], HIBA ALATRASH[4], DEVARAJAN THANGADURAI[5], JEYABALAN SANGEETHA[6], EKATERINA KOZUHAROVA[7], and VISHNU D. RAJPUT[8]

[1]Division of Vegetable Science, Sher-e-Kashmir University of Agricultural Sciences and Technology of Kashmir, Shalimar, Srinagar, India

[2]Transcriptomics Laboratory, Division of Plant Biotechnology, Sher-e-Kashmir University of Agricultural Sciences and Technology of Kashmir, Shalimar, Srinagar, India

[3]Department of Biological Sciences, Al Hussein bin Talal University, Maan, Jordan

[4]University of Helsinki, Helsinki, Finland

[5]Department of Botany, Karnatak University, Dharwad, Karnataka, India

[6]Department of Environmental Science, Central University of Kerala, Kasaragod, Kerala, India

[7]Medical University of Sofia, Faculty of Pharmacy, Department of Pharmacognosy, Sofia, Bulgaria

[8]Academy of Biology and Biotechnology, Southern Federal University, Rostov-on-Don, Russia

**Corresponding author.*

Microbial Fertilizer Technology for Sustainable Crop Production. Jeyabalan Sangeetha, Abdel Rahman Mohammad Al-Tawaha, & Devarajan Thangadurai (Eds.)

DOI: 10.1201/9781003594826-1

ABSTRACT

In many parts of the world, the effects of drought stress on plant growth, development, and productivity are significant barriers that must be overcome. While plant-associated bacteria, fungi, and viruses can help plants cope with the negative effects of drought, they can also help them to become more stress resistant. This is accomplished by inducing a variety of mechanisms in the plants that entail biochemical and physiological changes. Abiotic stressors, also known as environmental restrictions, are the most significant and significant limiting factors in the loss of agricultural output throughout the world. Drought is one of the most prevalent and destructive environmental constraints that has a significant impact on the growth and development of crops. Drought is defined as a period of time during which an area or region receives low precipitation, that is, below-normal precipitation, either snow or rain, resulting in a decrease in soil moisture or groundwater, diminished stream flow, crop damage, and a general water shortage in the area. Further research has indicated that increased drought stress resulting from global warming is having a severe impact on plant life, with photosynthetic CO_2 fixation suppression and metabolic impairment leading to a significant reduction of plant growth and, in some cases, final plant death.

1.1 INTRODUCTION

Abiotic stresses or environmental constraints are the chief and major limiting factors for the reduction in agricultural productivity worldwide. One of the dominant and most destructive environmental constraints causing a vast impression on the growth and development of the crop is drought, a stretch of time when an area or region experiences low precipitation, that is, below-normal, either snow or rain, which causes a decline in soil moisture or groundwater, diminished stream flow, crop damage and a general water shortage (National Geographic Society, 2020). According to the United Nations (UN) report in 1990, more than 50% of the region was affected by drought once every 4 years. Drought has intensified over the past decades due to extreme and swift variations in worldwide climatic conditions, posing a threat to successful crop production and thus affecting the world's food security. It may range from short (when the precipitation deficit is equal to 25% to its normal) and moderate (when the precipitation deficit is >25% to its normal) to extremely severe (when the precipitation deficit is >50% to its normal) and prolonged duration, restricting the crop yields (Bottner et al., 1995; Vurukonda et al., 2016). It is expected to cause serious plant growth problems for more than 50% of the arable lands by 2050 (Kasim et al., 2013) and has affected 64% of the global land area (Mittler, 2006; Cramer et al., 2011). According to the Food and Agriculture Organization of the United Nations (FAO) (2007), only 3.5% of worldwide land is unaffected by environmental constraints (Abulencia et al., 2006). Drought will raise the planet's average temperature by 1.4–5.8°C by the century's end. Thus, a global water catastrophe is imminent (Assad et al., 2006).

Climate change-induced drought stress suppresses photosynthetic CO_2 fixation and metabolic function, causing growth reduction and plant death (Oliver et al., 2009). An array of protective mechanisms has been acquired by the plant during the course of evolution to combat adverse drought situations. Such mechanisms cause metabolic reprogramming in the cells to facilitate routine biophysical–chemical processes irrespective of the external situations. One such mechanism that plants adopt to reduce the burden of drought stress is the microbes they inhabit (Table 1.1).

TABLE 1.1 Economic Yield Reduction by Drought Stress in Some Crops.

Crop	Stage of growth	Yield reduction (%)	References
Rice	Grain filling	60	Basnayake et al. (2006)
Maize	Reproductive	63–87	Kamara et al. (2003)
Soybean	Reproductive	46–71	Samarah et al. (2006)
Potato	Flowering	13	Kawakami et al. (2006)
Cowpea	Reproductive	11–60	Ogbonnaya et al. (2003)
Beans	Reproductive	58–87	Martinez et al. (2007)

1.2 CHARACTERISTICS OF DROUGHT

- Drought is distinguished by diminished soil moisture or soil moisture deficit.

- Drought incidence entirely depends upon the amount and distribution pattern of rainfall.
- Drought stress is more common in dry and semi-arid areas than in humid zones.
- Drought damages the chloroplasts and reduces the output of the photosynthates.
- Xerophytic plants are more resistant to drought stress than mesophytes.
- Drought leads to a reduction in the quality of economic products of crop plants.
- Drought resistance is a genetically regulated physiological quality of plants that is linked to morphological and physiological characteristics.

1.3 VARIOUS IMPACTS OF DROUGHT STRESS

1.3.1 PHYSICAL IMPACTS

This includes insufficient renewal of soil moisture, surface run-off, reduction in groundwater level, dry up of water bodies, dry up of soil, and soil erosion.

1.3.2 ENVIRONMENTAL IMPACTS

This includes degradation of landscape quality, forest range fires, degradation of wildlife habitat, degradation of water and air quality, degradation of animal and plant species, loss of biodiversity, loss of wetlands, and loss of vegetation.

1.3.3 SOCIAL IMPACTS

This includes drinking water, food, and fodder, which are in short supply; increase in the rate of inflation; death of livestock; insufficient food intake; widespread malnutrition; hunger; and starvation-induced diseases including diarrhea, dysentery, cholera, and ophthalmia to spread.

1.3.4 ECONOMIC IMPACTS

- A decrease in cultivated land and a drop in agricultural output.
- Employment in the agricultural sector is declining.
- Purchase power dwindles.
- Agony in the livestock sale.
- Anguish about land, jewelry, and personal property sales and mortgages.
- Due to a drop in agricultural production and a drop in purchasing power, secondary and tertiary activities are slowing down.

1.4 PHYSIOLOGICAL IMPACTS OF DROUGHT ON PLANTS

For their optimal growth, development and reproduction, plants need water in sufficient quantity. However, extreme conditions (below or above the optimal levels) of water stress limit plant growth and development by posing a complex set of stress conditions and thus affecting the physiological status of plants in a number of ways such as (1) reduction in leaf water potential and turgor pressure of cells, (2) decrease in seed viability size, and delay in flowering and fruiting, and (3) inhibition of photosynthesis and increased respiration.

1.5 MECHANISMS ADOPTED BY PLANTS TO AVOID DROUGHT STRESS

Drought stress affects the water relations of plants at cellular, tissue, and organ levels, causing specific as well as unspecific reactions, damage, and adaptation reactions. Plants can sense and react to this kind of stress in many ways that favor their sustenance. These include drought escape, avoidance, tolerance, resistance, and microbe-mediated mitigation strategies (Figure 1.1).

1.5.1 DROUGHT ESCAPE

Under drought conditions, the easiest way of sustenance is to escape drought. It involves rapid plant development to enable the plant to flower early and complete its full life cycle within 4–6 weeks prior to a coming drought event, that is, to reduce the vegetative growth period, to produce new seeds before the harsh environment conditions end the life cycle of the plant. Generally, drought occurs in plants grown in desert regions or in the mid or late season. As a result, drought escape is the most typical technique for these plants to avoid drought stress. However, it plays an important function in the growth of several crop plants. Early wheat, sorghum, maize, and rice yields, for example, are less affected by severe drought than late mature species. All of these crops have a predictable growth pattern. Late maturing spring wheat varieties yield more than early maturing varieties, especially when drought strikes early in the season and ends before anthesis (Singh et al., 2015).

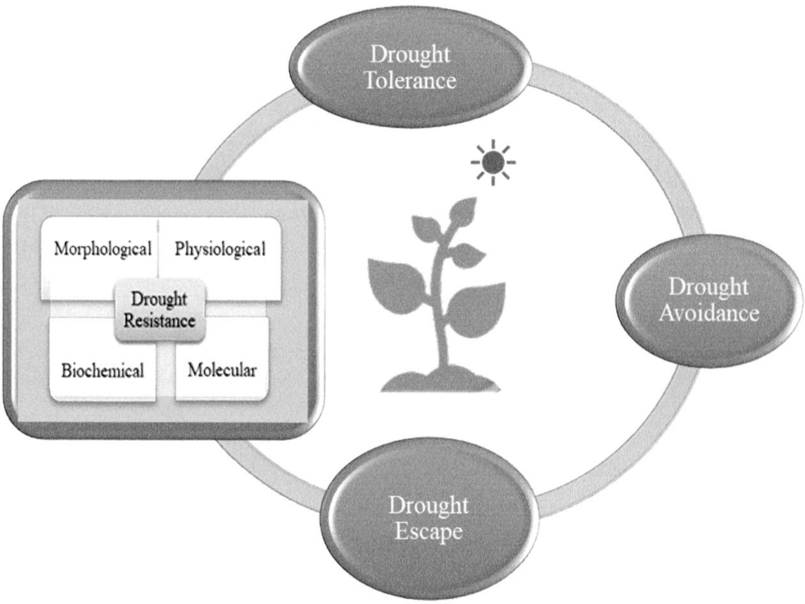

FIGURE 1.1 Mechanisms adopted by plants to avoid drought stress.

1.5.2 DROUGHT AVOIDANCE

In some instances, the plants retain high water contents in their tissues or maintain a favorable internal water balance to avoid the conditions of moisture stress by reducing water loss by controlling the rate of transpiration, increasing water use efficiency, limiting vegetative growth, increasing root growth (roots become thicker and deeper), which permits a more extended growth period in the crop. Therefore, it is one of the most important mechanisms to avoid drought stress than drought tolerance. However, it leads to a reduction in the photosynthetic rate and thereby in the growth of aerial parts of the plant. In cereals, drought avoidance takes place during the vegetative phase of the crop, while tolerance operates during the reproductive phase. They adopt two types of drought avoidance mechanisms:

- By lowering the amount of water lost through perspiration; stomatal traits; as well as the form, size, and orientation of leaves are examples of such characteristics.
- By maintaining water uptake during the drought period (Singh et al., 2015).

1.5.3 DROUGHT TOLERANCE

Drought tolerance refers to the ability of certain plants to combat low moisture content in their tissues under stress conditions by reducing the number and size of leaves to reduce water consumption and avoid yield loss, by improving root growth rate, root density, root spread, and size, and at the same time plants bearing small leaves (phenotypic flexibility), and by improving root growth rate, root density, root spread and size, and at the same time plants bearing small leaves (phenotypic flexibility). It is more desirable since tolerant crops produce higher yields with less water and have greater germination, seedling growth, and photosynthesis. Drought tolerance generally operates during the reproductive phase, in the case of cereals. while, in sorghum, a drought-resistant line exhibited a higher photosynthetic rate at low leaf water potential than a less drought-resistant line. Drought tolerance is distinguished from avoidance in the way that in the case of drought tolerance plants can withstand low water content in their tissues, which is not the case in drought avoidance.

1.5.4 DROUGHT RESISTANCE

Drought resistance refers to a plant's ability to grow, blossom, and generate an economic crop, or to preserve leaf area and growth over an extended vegetative stage, or to enhance cell membrane stability under sub-optimal water supply conditions. The basis of drought resistance is as follows:

1.5.4.1 MORPHOLOGICAL ASPECTS

It is correlated with various morphological characteristics of the plants: (1) earliness and sunken, small size and less number of stomatal openings, (2) cuticular thickness of leaves and waxiness of leaf surface, (3) leaf rolling and hairiness, (4) light green and golden colored leaves in barley, (5) deep, wide, thick, and prolific root system, and (6) indeterminate varieties in case of upland cotton and awn development in wheat and barley.

1.5.4.2 PHYSIOLOGICAL ASPECTS

It is correlated with various physiological features of the plants:

- Osmotic adjustment by which the plant cells reduce osmotic potential, consequently preserving cell turgor by increasing the gradient for water influx-enhanced tissue water status and physiological processes of the plant.
- At high concentrations, the use of osmolytes (compatible solutes) or osmoprotectants helps the plants to survive under water stress conditions.
- Reducing the enhancement of reactive oxygen species (ROS), which are produced as a result of higher leakage of electrons toward O_2 during photosynthesis and respiration processes, by the use of antioxidants.

1.5.4.3 BIOCHEMICAL ASPECTS

It is correlated with various phytohormones produced inside the plants in response to stress conditions. Under drought stress, the production of endogenous auxins is reduced, and the production of abscisic acid and ethylene increases gradually, which are natural growth inhibitors that act as a signal for the initiation of processes involved in adaptation to drought.

Synthesis of abscisic acid causes closure of stomatal openings and reduces water loss through reduced transpiration. Its high levels also lead to changes in plant growth rate, an increase in root-to-shoot dry weight ratio, and a decrease in leaf area.

1.5.4.4 MOLECULAR ASPECTS

Plants' gene expression changes when they are subjected to drought. At the transcriptional level, several genes are stimulated, and the gene products play an important role in drought tolerance. Although it is widely understood that drought tolerance is a complicated mechanism involving the intensive action of numerous genes, gene expression can be induced as a direct result of stress conditions or damage responses. Plants create proteins in reaction to stress in order to survive under drought conditions. The majority of stress proteins are water soluble, and by hydrating the cellular structure, they play a crucial role in stress tolerance. Signal transmission is aided by mitogen-activated protein kinase cascades and interaction between various transcription factors. It also relates external stimuli perception to cellular reactions.

1.6 PLANT–MICROBE INTERACTIONS

Microorganisms, which are found naturally in a wide variety of environments, are thought to be the natural partners that regulate local and systemic mechanisms in plants to offer defense under adverse environmental conditions to mitigate drought stress. They do this by providing a buffer zone for plants against stress, producing a variety of growth-promoting hormones for plants, and increasing the availability of nutrients. Plant–microbe interactions are intricate systems in the plant cellular system and essential to the living ecosystem. The plant rhizosphere, a "hot zone" of microbial activity around the roots, is the focus of plant stress response. Plants interact with a variety of microorganisms, such as bacteria, fungi, algae, and protozoa, which can have positive, negative, or neutral impacts on plants. Positive effects arise from symbiotic relationships of plants with microorganisms where plants provide photosynthetically fixed carbon (root exudates—chemicals and nutrients/litter) to microorganisms, and in return, microorganisms increase abiotic stress tolerance, that is, drought tolerance, improve plant nutrient acquisition, promote growth, stimulate hormone production, help in nutrient cycling and organic matter decomposition and provide defense against pathogens. Microbes can have a number of unfavorable consequences on plants, including resource competition, parasitism, and disease, all of which can have a detrimental influence on plant performance (Figure 1.2).

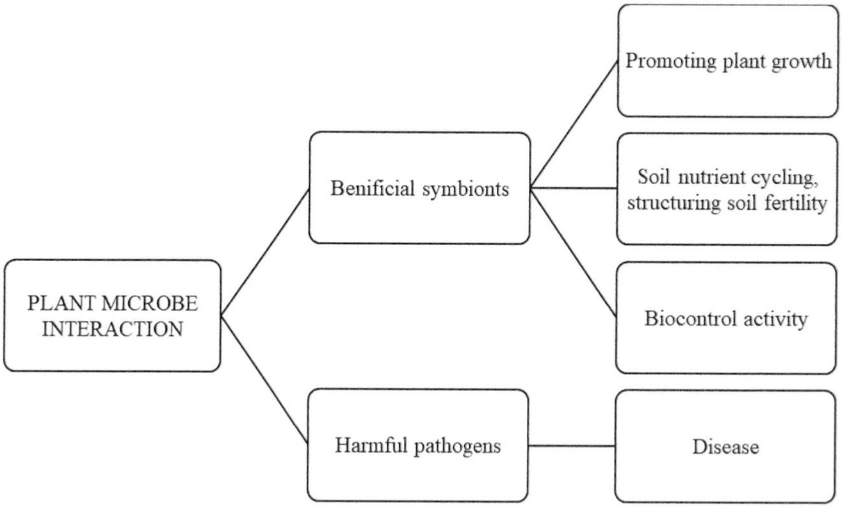

FIGURE 1.2 Plant–microbial interactions.

These plant–microbe interactions can be manipulated to improve plant performance in applications ranging from climate change mitigation to agricultural production. This can be done to enhance different plant physiological metrics including productivity (i.e., yield, growth, photosynthesis), survival, stress tolerance, CO_2 and nutrient uptake, and water use efficiency. Beneficial microbial assemblages are transplanted into host plants with the intention of altering the microbiome in order to improve plant functions. Therefore, efforts have been focused on reprogramming the microbiome to improve a plant function that is of interest during drought and on identifying root-associated microbial communities that can thrive in harsh environments, confer stress tolerance, and potentially be beneficial to the host plant (Table 1.2).

TABLE 1.2 Plant-Microbial Interaction Imparting Drought Tolerance in Plants.

Plant growth-promoting rhizobacteria	Plant species	Effect
Achromobacter piechaudii ARV8	*Solanum lycopersicum* and *Piper nigrum*	Reduced ethylene production and increased fresh and dry weight.
Azospirillum brasilense	*S. lycopersicum*	Nitric oxide as a signaling molecule in IAA-induced pathway, which enhanced lateral root and root hair development.
Azospirillum sp.	*Triticum aestivum*	IAA enhanced root growth, lateral root formation, and increased uptake of water and nutrients under drought stress.
Azospirillum lipoferum	*Zea mays*	Gibberellins increased ABA levels and alleviated drought stress.
ACC deaminase-producing rhizobacteria	*T. aestivum*	Increased root-shoot length, biomass and lateral root number.
Bacillus licheniformis K11	*P. nigrum*	Higher expression of stress genes Cadhn, VA, sHSP and CaPR-10.
Bacillus subtilis	*Platycladus orientalis*	Cytokinin production by PGPR elevated ABA levels in shoots and increased the stomatal conductance.
Bacillus thuringiensis	*Lavandula dentata*	IAA resulted in higher K-content, and proline and decreased the glutathione reductase (GR) and ascorbate peroxidase (APX).
B. thuringiensis AZP2	*T. aestivum*	Reduction of volatile emissions and higher photosynthesis.
Consortia of *Bacillus* isolate 23-B and *Pseudomonas* 6-P with *Mesorhizobium ciceris*	*Cicer arietinum*	Higher proline concentration, improved germination, root and shoot length, and fresh weight of the seedlings.
Pseudomonas putida H-2–3	*Glycine max*	Secretion of gibberellins by *P. putida* improved plant growth.
Pseudomonas fluorescens biotype G (ACC-5)	*Pisum sativum*	Induced longer roots and increased uptake of water.
Phyllobacterium brassicacearum strain STM196	*Arabidopsis thaliana*	Enhanced ABA content resulted in decreased leaf transpiration.

TABLE 1.2 *(Continued)*

Plant growth-promoting rhizobacteria	Plant species	Effect
Rhizobium leguminosarum (LR-30), *M. ciceri* (CR-30 and CR-39), and *Rhizobium phaseoli* (MR-2)	*T. aestivum*	IAA produced by the consortia improved the growth, biomass, and drought tolerance index.
Variovorax paradoxus 5C-2	*P. sativum*	Increase in xylem abscisic acid, higher nodulation, growth, yield, and water-use efficiency.
V. paradoxus 5C-2	*P. sativum*	Increased nodulation, seed yield, seed number, and seed nitrogen content.

Source: Reproduced and modified with permission from Vurukonda et al. (2016), Copyright © Elsevier (2016).

1.7 MICROBE-MODIFIED DROUGHT STRESS

Microbe-mediated drought stress reduction, also known as Induced Systemic Tolerance (IST), is a type of drought stress defense mechanism that plants employ in collaboration with microorganisms. In recent decades, the role of microbes in alleviating abiotic stressors in plants has been a major source of concern (Nadeem et al., 2014; Souza et al., 2015). Microbes help plants cope with abiotic stressors by utilizing their inherent metabolic and genetic capacities (Gopalakrishnan et al., 2015). The role of several rhizospheric occupants belonging to the genera *Pseudomonas* (Sorty et al., 2016), *Azotobacter* (Sahoo et al., 2014a, 2014b), *Azospirillum* (Omar et al., 2009), *Rhizobium* (Sorty et al., 2016), *Pantoea* (Sorty et al., 2016), *Bacillus* (Marulanda et al., 2016; Sorty et al., 2016), *Enterobacter* (Sorty et al., 2016), *Bradyrhizobium* (Panlada et al., 2013), *Methylobacterium* (Meena et al., 2012), *Burkholderia* (Oliveira et al., 2009), *Trichoderma* (Ahmad et al., 2015), and Cyanobacteria (Singh et al., 2011) in plant growth promotion and mitigation of multiple kinds of abiotic stresses especially drought stress, has been documented.

Pandey et al. (2016) have established the involvement of *Trichoderma harzianum* in stress mitigation in rice genotypes by upregulating aquaporin, dehydrin, and malondialdehyde genes, as well as a variety of other physiological indicators. Rhizobacteria-induced drought endurance and resilience (RIDER) has been found for microbe-mediated plant responses, which includes changes in phytohormones, defense-related proteins and enzymes, antioxidants, and epoxy polysaccharide levels. Plants with such techniques are more resistant to abiotic stressors (Kaushal and Wani, 2016). As a result, the selection, screening, and application of stress-tolerant microorganisms could be viable choices for overcoming crop plant productivity limits in stress-prone environments. *T. harzianum* administration improved the uptake of key nutrients, increased the accumulation of antioxidants and osmolytes, and lowered Na^+ uptake in NaCl-affected Indian mustard (*Brassica juncea*) (Ahmad et al., 2015).

Burkholderia phytofirmans strain PsJN has been shown to alleviate drought stress in maize (Naveed et al., 2014b), wheat (Naveed et al., 2014a), and salt stress in *Arabidopsis* (Pinedo et al., 2015). Microorganisms' chemotactic migration toward root exudates acts as a dragging force for microbial populations to colonize the roots. PGPRs may operate as biofertilizers, phytostimulators, or biocontrol agents in the rhizosphere-microenvironment around plant roots, depending on their innate capabilities, mode of interaction, and competing survival conditions. Bacteria that promote plant growth use a variety of direct and indirect strategies to achieve their goals (Braud et al., 2009; Hayat et al., 2010). Direct mechanisms include the production of plant growth regulators, such as iron and zinc sequestration, siderophore production, phosphorus and potassium solubilization, plant hormone production, and atmospheric nitrogen fixation, as well as the synthesis of bacterial compounds that facilitate the uptake of essential nutrients and micronutrients from the soil. Indirect mechanisms, on the other hand, include antagonistic activity against plant pathogenic organisms, synthesis of HCN and antifungal chemicals, and abiotic stress tolerance. Furthermore, bacteria can cause systemic resistance in plants by using their metabolites as extracellular signals, which start a series of internal processes. The translocated signal is eventually detected by distant plant cells, triggering the defensive mechanism to activate. Fungus, particularly mycorrhizal fungi are also essential plant growth boosters. Mycorrhizal fungi and vesicular-arbuscular mycorrhizal (VAM) fungi are the two main types. These fungi generate large networks of very fine hyphae, allowing the roots to absorb more nutrients. By boosting antioxidant levels and enhancing many other features, the root fungal endophyte *Piriformospora indica* produces salt tolerance in barley (Baltruschat et al., 2008) and drought tolerance in Chinese cabbage (Sun et al., 2010; Franken, 2012). As a result, the possibility of microbial interactions with plants plays a multifaceted function. Microbes help plants maintain their growth and development by fixing, mobilizing, and/or producing nutrients, hormones, and organic phytostimulant compounds on the one hand, and by inducing local or systemic stress alleviation response mechanisms in plants to survive under abiotic stress conditions on the other hand. Microorganisms or their communities are strong, feasible, and crucial possibilities for abiotic stress mitigation methods in crop plants because of their varied actions.

Several strategies have been postulated that highlight the involvement of bacteria in abiotic stress reduction. Plant growth has been demonstrated to be enhanced by soil-dwelling microorganisms from the genera *Achromobacter*, *Azospirillum*, *Variovorax*, *Bacillus*, *Enterobacter*, *Azotobacter*, *Aeromonas*, *Klebsiella*, and *Pseudomonas* even in unfavorable environmental conditions (Kaushal and Wani, 2016; Sorty et al., 2016). The importance of bacteria in this field is demonstrated by literature referring to their role in the mitigation of abiotic stresses. Plant growth promoters are all soil bacteria capable of promoting plant development under a variety of physicochemical and environmental circumstances (PGP). Microbes stimulate plant development through a variety of processes. Indole acetic acid (IAA), a plant growth regulator, is primarily generated in the shoot and accumulates in the actively growing sections of roots. IAA and other auxins have a growth-stimulating effect on cell elongation, which leads to the commencement of root growth. Furthermore, these chemicals aid in the formation of lateral roots. Higher auxin concentrations, on the other hand, have been shown to inhibit root development (Jackson, 1991; Sorty et al., 2016). Increased ethylene synthesis can also cause a similar problem (Jackson, 1991). The bacteria that colonize the rhizosphere are said to work in a similar way, producing phytohormones that help plants grow (Belimov et al., 2007). Recent agricultural techniques show that PGPRs not only aid in the mitigation of environmental challenges, but also boost the yield of a variety of crop plants such as rice, maize, barley, and soybeans (Suarez et al., 2015).

Higher expression of antioxidant enzyme machinery in the presence of bacterial inoculant may lead to cell protection in stress circumstances, according to one theory. More research is needed to decipher the molecular pathways underlying plant-bacterial colonizer communication.

1.8 FUTURE PERSPECTIVES

Plants' reaction to drought is complex and growth dependent, involving several dynamic phenotypic alterations that vary with moisture stress duration. The conditional limitations mentioned involve many cellular, physiological, metabolic, and molecular changes. Most of plants' functions, including nutrient uptake and stress response, depend on soil microbial communities in our ecosystem. The root microbiota will change depending on drought duration. Despite the relevance of the root microbiome to plant health and its role in helping plants survive drought stress, few studies have been done on it during drought. Our lack of understanding of the factors that form bacterial communities contributes to this gap. Refinement and consolidation of drought-root microbiome study methodologies will improve understanding. Insight into plant enrichment of microorganisms during drought will reveal taxa involved in plant growth promotion and improve microbial-based soil amendment strategies to reduce drought stress for crops in dry places, strengthening food security against climate change. Thus, microbial inoculation may help drought-stressed crops grow.

KEYWORDS

- **abiotic stresses**
- **crop damage**
- **drought**
- **food security**
- **microbes**
- **photosynthetic**
- **plant growth**

REFERENCES

Abulencia, C. B.; Wyborski, D. L.; Garcia, J. A.; Podar, M.; Chen, W.; Chang, S. H. Environmental Whole-Genome Amplification to Access Microbial Populations in Contaminated Sediments. *Appl. Environ. Microbiol.* **2006,** *72*, 3291–3301.

Acosta-Martínez, V.; Cotton, J.; Gardner, T.; Moore-Kucera, J.; Zak, J.; Wester, D.; et al. Predominant Bacterial and Fungal Assemblages in Agricultural Soils during a Record Drought/Heat Wave and Linkages to Enzyme Activities of Biogeochemical Cycling. *Appl. Soil Ecol.* **2014,** *84*, 69–82.

Ahmad, P.; Hashem, A.; Abd-Allah, E. F.; Alqarawi, A. A.; John, R.; Egamberdieva, D.; et al. Role of *Trichoderma harzianum* in Mitigating NaCl Stress in Indian Mustard (*Brassica juncea* L.) through Antioxidative Defense System. *Front. Plant Sci.* **2015,** *6*, 868.

Albert, K. R.; Ro-Poulsen, H.; Mikkelsen, T. N.; Michelsen, A.; Van Der Linden, L.; Beier, C. Effects of Elevated CO_2, Warming and Drought Episodes on Plant Carbon Uptake in a Temperate Heath Ecosystem are Controlled by Soil Water Status. *Plant Cell Environ.* **2011,** *34*, 1207–1222.

Alster, C. J.; German, D. P.; Lu, Y.; Allison, S. D. Microbial Enzymatic Responses to Drought and to Nitrogen Addition in a Southern California Grassland. *Soil Biol. Biochem.* **2013,** *64*, 68–79.

Anwar, S.; Ali, B.; Sajid, I. Screening of Rhizospheric Actinomycetes for Various in-vitro and in-vivo Plant Growth Promoting (PGP) Traits and for Agroactive Compounds. *Front. Microbiol.* **2016,** *7*, 1334.

Assad, E. D.; Pinto, H. S.; Zullo, J. J.; Avila, A. M. H. Climatic Changes Impact in Agroclimatic Zoning of Coffee in Brazil. *Pesquisa Agropecuaria Brasileira* **2004,** *39* (11), 1057–1064.

Bachar, A.; Al-Ashhab, A.; Soares, M. I. M.; Sklarz, M. Y.; Angel, R.; Ungar, E. D.; et al. Soil Microbial Abundance and Diversity along a Low Precipitation Gradient. *Microb. Ecol.* **2010,** *60*, 453–461.

Badri, D. V.; Chaparro, J. M.; Zhang, R.; Shen, Q.; Vivanco, J. M. Application of Natural Blends of Phytochemicals Derived from the Root Exudates of *Arabidopsis* to the Soil Reveal that Phenolic-Related Compounds Predominantly Modulate the Soil Microbiome. *J. Biol. Chem.* **2013,** *288*, 4502–4512.

Balasooriya, W. K.; Denef, K.; Huygens, D.; Boeckx, P. Translocation and Turnover of Rhizodeposit Carbon within Soil Microbial Communities of an Extensive Grassland Ecosystem. *Plant Soil.* **2014,** *76*, 61–73.

Baltruschat, H.; Fodor, J.; Harrach, B. D.; Niemczyk, E.; Barna, B.; Gullner, G.; et al. Salt Tolerance of Barley Induced by the Root Endophyte *Piriformospora indica* is Associated with a Strong Increase in Anti-Oxidants. *New Phytol.* **2008,** *180*, 501–510.

Barnard, R. L.; Osborne, C. A.; Firestone, M. K. Responses of Soil Bacterial and Fungal Communities to Extreme Desiccation and Rewetting. *ISME J.* **2013,** *7*, 2229–2241.

Bartels, D.; Sunkar, R. Drought and Salt Tolerance in Plants. *Crit. Rev. Plant Sci.* **2005,** *24*, 23–58.

Basnayake, J.; Fukai, S.; Ouk. M. In *Contribution of Potential Yield, Drought Tolerance and Escape to Adaptation of 15 Rice Varieties in Rainfed Lowlands in Cambodia*, Proceedings of the Australian Agronomy Conference, Australian Society of Agronomy, Brisbane, Australia, 2006.

Battisti, D. S.; Naylor, R. L. Historical Warnings of Future Food Insecurity with Unprecedented Seasonal Heat. *Science* **2009,** *323*, 240–244.

Belimov, A. A.; Dodd, I. C.; Hontzeas, N.; Theobald, J. C.; Safronova, V. I.; Davies, W. J. Rhizosphere Bacteria Containing 1-aminocyclopropane-1-carboxylate Deaminase Increase Yield of Plants Grown in Drying Soil via Both Local and Systemic Hormone Signalling. *New Phytol.* **2009,** *181*, 413–423.

Berendsen, R. L.; Pieterse, C. M. J.; Bakker, P. A. H. M. The Rhizosphere Microbiome and Plant Health. *Trends Plant Sci.* **2012,** *17*, 478–486.

Berg, G.; Grube, M.; Schloter, M.; Smalla, K. Unraveling the Plant Microbiome: Looking Back and Future Perspectives. *Front. Microbiol.* **2014,** *5*, 148.

Blazewicz, S. J.; Schwartz, E.; Firestone, M. K. Growth and Death of Bacteria and Fungi Underlie Rainfall-Induced Carbon Dioxide Pulses from Seasonally Dried Soil. *Ecology* **2014,** *95*, 1162–1172.

Bottner, P.; Couteaux, M. M.; Vallejo, V. R. Soil Organic Matter in Mediterranean Type Ecosystems and Global Climatic Changes: A Case Study –The Soils of the Mediterranean Basin. In *Global Change and Mediterranean Type Ecosystems Ecological Studies*; Jose, M., Oechel, W. C., Eds.; Springer-Verlag: New York, 1995; pp 306–325.

Bouasria, A.; Mustafa, T.; De Bello, F.; Zinger, L.; Lemperiere, G.; Geremia, R. A.; et al. Changes in Root-Associated Microbial Communities are Determined by Species-Specific Plant Growth Responses to Stress and Disturbance. *Eur. J. Soil Biol.* **2012,** *52*, 59–66.

Bouffaud, M. L.; Poirier, M. A.; Muller, D.; Moënne-Loccoz, Y. Root Microbiome Relates to Plant Host Evolution in Maize and Other Poaceae. *Environ. Microbiol.* **2014,** *16*, 2804–2814.

Bouskill, N. J.; Lim, H. C.; Borglin, S.; Salve, R.; Wood, T. E.; Silver, W. L.; et al. Pre-Exposure to Drought Increases the Resistance of Tropical Forest Soil Bacterial Communities to Extended Drought. *ISME J.* **2013,** *7*, 384–394.

Bouskill, N. J.; Wood, T. E.; Baran, R.; Hao, Z.; Ye, Z.; Bowen, B. P.; et al. Belowground Response to Drought in a Tropical Forest Soil. II. Change in Microbial Function Impacts Carbon Composition. *Front. Microbiol.* **2016a,** *7*, 323.

Bouskill, N. J.; Wood, T. E.; Baran, R.; Ye, Z.; Bowen, B. P.; Lim, H.; et al. Below Ground Response to Drought in a Tropical Forest Soil. I. Changes in Microbial Functional Potential and Metabolism. *Front. Microbiol.* **2016b,** *7*, 525.

Bradford, M. A.; Fierer, N.; Reynolds, J. F. Soil Carbon Socks in Experimental Mesocosms are Dependent on the Rate of Labile Carbon, Nitrogen and Phosphorus Inputs to Soils. *Funct. Ecol.* **2008,** *22*, 964–974.

Braud, A.; Jezequel, K.; Bazot, S.; Lebeau, T. Enhanced Phytoextraction of an Agricultural Cr-, Hg- and Pb-contaminated Soil by Bioaugmentation with Siderophore Producing Bacteria. *Chemosphere* **2009,** *74*, 280–286.

Bulgarelli, D.; Rott, M.; Schlaeppi, K.; Ver Loren van Themaat, E.; Ahmadinejad, N.; Assenza, F.; et al. Revealing Structure and Assembly Cues for *Arabidopsis* Root-Inhabiting Bacterial Microbiota. *Nature* **2012,** *488*, 91–95.

Calvo, O. C.; Franzaring, J.; Schmid, I.; Müller, M.; Brohon, N.; Fangmeier, A. Atmospheric CO_2 Enrichment and Drought Stress Modify Root Exudation of Barley. *Glob. Change Biol.* **2016,** *23*, 1292–1304.

Canarini, A.; Dijkstra, F. A. Dry-Rewetting Cycles Regulate Wheat Carbon Rhizodeposition, Stabilization and Nitrogen Cycling. *Soil Biol. Biochem.* **2015,** *81*, 195–203.

Canarini, A.; Merchant, A.; Dijkstra, F. A. Drought Effects on *Helianthus annuus* and *Glycine max* Metabolites: From Phloem to Root Exudates. *Rhizosphere* **2016,** *2*, 85–97.

Casanovas, E. M.; Barassi, C. A.; Sueldo, R. J. *Azospirillum* Inoculation Mitigates Water Stress Effects in Maize Seedlings. *Cereal Res. Commun.* **2002,** *30*, 343–350.

Chaparro, J. M.; Badri, D. V.; Vivanco, J. M. Rhizosphere Microbiome Assemblage is Affected by Plant Development. *ISME J.* **2014,** *8*, 790–803.

Chapman, S. C.; Crossa, J.; Edmeades, G. O. Genotype by Environment Effects and Selection for Drought Tolerance in Tropical Maize. Two Mode Pattern Analysis of Yield. *Euphytica* **1997,** *95*, 1–9.

Cherif, H.; Marasco, R.; Rolli, E.; Ferjani, R.; Fusi, M.; Soussi, A.; et al. Oasis Desert Farming Selects Environment-Specific Date Palm Root Endophytic Communities and Cultivable Bacteria that Promote Resistance to Drought: Oasis Palm Endophytes Promote Drought Resistance. *Environ. Microbiol. Rep.* **2015,** *7*, 668–678.

Cho, S. M.; Kang, B. R.; Han, S. H.; Anderson, A. J.; Park, J. Y.; Lee, Y. H.; et al. 2R, 3R-butanediol, A Bacterial Volatile Produced by *Pseudomonas chlororaphis* O6, is Involved in Induction of Systemic Tolerance to Drought in *Arabidopsis thaliana. Mol. Plant Microbe Interact.* **2008,** *21*, 1067–1075.

Cho, S. K.; Kim, J. E.; Park, J. A.; Eom, T. J.; Kim, W. T. Constitutive Expression of Abiotic Stress-Inducible Hot Pepper CaXTH3, which Encodes a Xyloglucan

Endotransglucosylase/Hydrolase Homolog, Improves Drought and Salt Tolerance in Transgenic *Arabidopsis* Plants. *FEBS Lett.* **2006,** *580*, 3136–3144.
Chodak, M.; Gołebiewski, M.; Morawska-Płoskonka, J.; Kuduk, K.; Niklińska, M. Soil Chemical Properties Affect the Reaction of Forest Soil Bacteria to Drought and Rewetting Stress. *Ann. Microbiol.* **2015,** *65*, 1627–1637.
Coleman, D. C.; Reid, C. P. P.; Cole, C. V. Biological Strategies of Nutrient Cycling in Soil Systems. *Adv. Ecol. Res.* **1983,** *13*, 1–55.
Coleman-Derr, D.; Desgarennes, D.; Fonseca-Garcia, C.; Gross, S.; Clingenpeel, S.; Woyke, T.; et al. Plant Compartment and Biogeography Affect Microbiome Composition in Cultivated and Native Agave Species. *New Phytol.* **2016,** *209*, 798–811.
Connon, S. A.; Lester, E. D.; Shafaat, H. S.; Obenhuber, D. C.; Ponce, A. Bacterial Diversity in Hyperarid Atacama Desert Soils. *J. Geophys. Res. Biogeosci.* **2007,** *112*, G04S17.
Cramer, G. R.; Urano, K.; Delrot, S.; Pezzotti, M.; Shinozaki, K. Effects of Abiotic Stress on Plants: A Systems Biology Perspective. *BMC Plant Biol.* **2011,** *11*, 163.
Cruz-Martínez, K.; Suttle, K. B.; Brodie, E. L.; Power, M. E.; Andersen, G. L.; Banfield, J. F. Despite Strong Seasonal Responses, Soil Microbial Consortia are More Resilient to Long-Term Changes in Rainfall than Overlying Grassland. *ISME J.* **2009,** *3*, 738–744.
Csonka, L. N. Physiological and Genetic Responses of Bacteria to Osmotic Stress. *Microbiol. Rev.* **1989,** *53*, 121–147.
da Costa, P. B.; Granada, C. E.; Ambrosini, A.; Moreira, F.; de Souza, R.; dos Passos, J. F. M.; et al. A Model to Explain Plant Growth Promotion Traits: A Multivariate Analysis of 2211 Bacterial Isolates. *PLoS One* **2014,** *9*, e116020.
Delhaize, E.; Ryan, P. R.; Randall, P. J. Aluminum Tolerance in Wheat (*Triticum aestivum* L.) (II. Aluminum-Stimulated Excretion of Malic Acid from Root Apices). *Plant Physiol.* **1993,** *103*, 695–702.
Delmont, T. O.; Prestat, E.; Keegan, K. P.; Faubladier, M.; Robe, P.; Clark, I. M.; et al. Structure, Fluctuation and Magnitude of a Natural Grassland Soil Metagenome. *ISME J.* **2012,** *6*, 1677–1687.
Desgarennes, D.; Garrido, E.; Torres-Gomez, M. J.; Peña-Cabriales, J. J.; Partida-Martinez, L. P. Diazotrophic Potential Among Bacterial Communities Associated with Wild and Cultivated Agave Species. *FEMS Microbiol. Ecol.* **2014,** *90*, 844–857.
Elbl, J.; Plošek, L.; Kintl, A.; Hynšt, J.; Javoreková, S.; Záhora, J.; et al. Effects of Drought on Microbial Activity in Rhizosphere, Soil Hydrophobicity and Leaching of Mineral Nitrogen from Arable Soil Depending on Method of Fertilization. *World Acad. Sci. Eng. Technol.* **2014,** *8*, 741–747.
Fang, Y.; Xiong, L. General Mechanisms of Drought Response and their Application in Drought Resistance Improvement in Plants. *Cell. Mol. Life Sci.* **2015,** *72*, 673–689.
Felsmann, K.; Baudis, M.; Gimbel, K.; Kayler, Z. E.; Ellerbrock, R.; Bruehlheide, H.; et al. Soil Bacterial Community Structure Responses to Precipitation Reduction and Forest Management in Forest Ecosystems Across Germany. *PLoS One* **2015,** *10*, e0122539.
Fenta, B.; Beebe, S.; Kunert, K.; Burridge, J.; Barlow, K.; Lynch, J.; et al. Field Phenotyping of Soybean Roots for Drought Stress Tolerance. *Agronomy* **2014,** *4*, 418–435.
Fierer, N.; Bradford, M. A.; Jackson, R. B. Toward an Ecological Classification of Soil Bacteria. *Ecology* **2007,** *88*, 1354–1364.

Figueiredo, M. V. B.; Burity, H. A.; Martinez, C. R.; Chanway, C. P. Alleviation of Drought Stress in Common Bean (*Phaseolus vulgaris* L.) by Co-inoculation with *Paenibacillus polymyxa* and *Rhizobium tropici*. *Appl. Soil Ecol.* **2008,** *40*, 182–188.

Fonseca-García, C.; Coleman-Derr, D.; Garrido, E.; Visel, A.; Tringe, S. G.; Partida-Martínez, L. P. The Cacti Microbiome: Interplay Between Habitat-Filtering and Host-Specificity. *Front. Microbiol.* **2016,** *7*, 150.

Fontaine, S.; Mariotti, A.; Abbadie, L. The Priming Effect of Organic Matter: A Question of Microbial Competition? *Soil Biol. Biochem.* **2003,** *35*, 837–843.

Ford, D. J.; Cookson, W. R.; Adams, M. A.; Grierson, P. F. Role of Soil Drying in Nitrogen Mineralization and Microbial Community Function in Semi-Arid Grasslands of North-West Australia. *Soil Biol. Biochem.* **2007,** *39*, 1557–1569.

Franken, P. The Plant Strengthening Root Endophyte *Piriformospora indica*: Potential Application and the Biology Behind. *Appl. Microbiol. Biotechnol.* **2012,** *96*, 1455–1464.

Franks, S. J.; Sim, S.; Weis, A. E. Rapid Evolution of Flowering Time by an Annual Plant in Response to a Climate Fluctuation. *Proc. Natl. Acad. Sci. USA* **2007,** *104*, 1278–1282.

Fuchslueger, L.; Bahn, M.; Fritz, K.; Hasibeder, R.; Richter, A. Experimental Drought Reduces the Transfer of Recently Fixed Plant Carbon to Soil Microbes and Alters the Bacterial Community Composition in a Mountain Meadow. *New Phytol.* **2014,** *201*, 916–927.

Fuchslueger, L.; Bahn, M.; Hasibeder, R.; Kienzl, S.; Fritz, K.; Schmitt, M.; et al. Drought History Affects Grassland Plant and Microbial Carbon Turnover During and After a Subsequent Drought Event. *J. Ecol.* **2016,** *104*, 1453–1465.

Gagné-Bourque, F.; Bertrand, A.; Claessens, A.; Aliferis, K. A.; Jabaji, S. Alleviation of Drought Stress and Metabolic Changes in Timothy (*Phleum pratense* L.) Colonized with *Bacillus subtilis* B26. *Front. Plant Sci.* **2016,** *7*, 584.

Gagne-Bourque, F.; Mayer, B. F.; Charron, J. B.; Vali, H.; Bertrand, A.; Jabaji, S. Accelerated Growth Rate and Increased Drought Stress Resilience of the Model Grass *Brachypodium distachyon* Colonized by *Bacillus subtilis* B26. *PLoS One* **2015,** *10*, e0130456.

Gall, H.; Philippe, F.; Domon, J. M.; Gillet, F.; Pelloux, J.; Rayon, C. Cell Wall Metabolism in Response to Abiotic Stress. *Plants* **2015,** *4*, 112–166.

Gassmann, W.; Appel, H. M.; Oliver, M. J. The Interface between Abiotic and Biotic Stress Responses. *J. Exp. Bot.* **2016,** *67*, 2023–2024.

German, D. P.; Chacon, S. S.; Allison, S. D. Substrate Concentration and Enzyme Allocation can Affect Rates of Microbial Decomposition. *Ecology* **2011,** *92*, 1471–1480.

Ghotbi-Ravandi, A. A.; Shahbazi, M.; Shariati, M.; Mulo, P. Effects of Mild and Severe Drought Stress on Photosynthetic Efficiency in Tolerant and Susceptible Barley (*Hordeum vulgare* L.) Genotypes. *J. Agron. Crop Sci.* **2014,** *200*, 403–415.

Glick, B. R. The Enhancement of Plant Growth by Free-Living Bacteria. *Can. J. Microbiol.* **1995,** *41*, 109–117.

Glick, B. R. Bacterial ACC Deaminase and the Alleviation of Plant Stress. *Adv. Appl. Microbiol.* **2004,** *56*, 291–312.

Gopalkrishnan, S.; Arumugam, S.; Vijayabharati, R.; Varshney, R. K.; Gowda, C. L.; Krishnamurthy, L. Plant Growth Promoting Rhizobia: Challenges and Opportunities. *Biotech.* **2015,** *5* (4), 355–377.

Göransson, H.; Godbold, D. L.; Jones, D. L.; Rousk, J. Bacterial Growth and Respiration Responses upon Rewetting Dry Forest Soils: Impact of Drought-Legacy. *Soil Biol. Biochem*. **2013,** *57*, 477–486.

Gornall, J.; Betts, R.; Burke, E.; Clark, R.; Camp, J.; Willett, K.; et al. Implications of Climate Change for Agricultural Productivity in the Early Twenty-First Century. *Philos. Trans. R. Soc. B Biol. Sci*. **2010,** *365*, 2973–2989.

Grönemeyer, J. L.; Burbano, C. S.; Hurek, T.; Reinhold-Hurek, B. Isolation and Characterization of Root-Associated Bacteria from Agricultural Crops in the Kavango Region of Namibia. *Plant Soil* **2012,** *356*, 67–82.

Gunnigle, E.; Frossard, A.; Ramond, J. B.; Guerrero, L.; Seely, M.; Cowan, D. A. Diel-Scale Temporal Dynamics Recorded for Bacterial Groups in Namib Desert Soil. *Sci. Rep*. **2017,** *7*, 40189.

Hartmann, A. A.; Barnard, R. L.; Marhan, S.; Niklaus, P. A. Effects of Drought and N-Fertilization on N Cycling in Two Grassland Soils. *Oecologia* **2013,** *171*, 705–717.

Hartmann, M.; Brunner, I.; Hagedorn, F.; Bardgett, R. D.; Stierli, B.; Herzog, C.; et al. A Decade of Irrigation Transforms the Soil Microbiome of a Semi-Arid Pine Forest. *Mol. Ecol*. **2017,** *26*, 1190–1206.

Hartung, W.; Turner, N. C. Abscisic Acid Relations in Stressed Roots. In *Biology of Root Formation and Development*; Altman, A., Waisel, Y., Eds.; Springer: New York, 1997; pp 125–132.

Hasibeder, R.; Fuchslueger, L.; Richter, A.; Bahn, M. Summer Drought Alters Carbon Allocation to Roots and Root Respiration in Mountain Grassland. *New Phytol*. **2015,** *205*, 1117–1127.

Hayat, R.; Ali, S.; Amara, U.; Khalid, R.; Ahmed, I. Soil Beneficial Bacteria and their Role in Plant Growth Promotion: A Review. *Ann. Microbiol*. **2010,** *60*, 579–598.

Hayden, H. L.; Mele, P. M.; Bougoure, D. S.; Allan, C. Y.; Norng, S.; Piceno, Y. M.; et al. Changes in the Microbial Community Structure of Bacteria, Archaea and Fungi in Response to Elevated CO_2 and Warming in an Australian Native Grassland Soil. *Environ. Microbiol*. **2012,** *14*, 3081–3096.

He, M.; Dijkstra, F. A. Drought Effect on Plant Nitrogen and Phosphorus: A Meta-Analysis. *New Phytol*. **2014,** *204*, 924–931.

Henry, A.; Doucette, W.; Norton, J.; Bugbee, B. Changes in Crested Wheatgrass Root Exudation Caused by Flood, Drought, and Nutrient Stress. *J. Environ. Qual*. **2007,** *36*, 904–912.

Huang, Y. M.; Zou, Y. N.; Wu, Q. S. Alleviation of Drought Stress by Mycorrhizas is Related to Increased Root H_2O_2 Efflux in Trifoliate Orange. *Sci. Rep*. **2017,** *7*, 42335.

Hueso, S.; García, C.; Hernández, T. Severe Drought Conditions Modify the Microbial Community Structure, Size and Activity in Amended and Unamended Soils. *Soil Biol. Biochem*. **2012,** *50*, 167–173.

Hueso, S.; Hernández, T.; García, C. Resistance and Resilience of the Soil Microbial Biomass to Severe Drought in Semiarid Soils: The Importance of Organic Amendments. *Appl. Soil Ecol*. **2011,** *50*, 27–36.

Hutsch, B. W.; Augustin, J.; Merbach, W. Plant Rhizodeposition - An Important Source for Carbon Turnover in Soils. *J. Plant Nutr. Soil Sci*. **2002,** *165*, 397–407.

Jackson, M. B. Regulation of Water Relationships in Flooded Plants by ABA from Leaves, Roots and Xylem Sap. In *Abscisic Acid, Physiology and Biochemistry*; Davis, W. J., Jones, H. G., Eds.; Bios Scientific: Oxford, 1991; pp 217–226.

Jensen, K. D.; Beier, C.; Michelsen, A.; Emmett, B. A. Effects of Experimental Drought on Microbial Processes in Two Temperate Heathlands at Contrasting Water Conditions. *Appl. Soil Ecol.* **2003,** *24*, 165–176.

Jones, L.; McQueen-Mason, S. A Role for Expansins in Dehydration and Rehydration of the Resurrection Plant *Craterostigma plantagineum*. *FEBS Lett.* **2004,** *559*, 61–65.

Kamara, A. Y.; Menkir, A.; Badu, A. B.; Ibikunle, O. The Influence of Drought Stress on Growth, Yield and Yield Components of Selected Maize Genotypes. *J. Agr. Sci.* **2003,** *141*, 43–50.

Kariola, T.; Brader, G.; Helenius, E.; Li, J.; Heino, P.; Palva, E. T. Early Responsive to Dehydration 15, A Negative Regulator of Abscisic Acid Responses in *Arabidopsis*. *Plant Physiol.* **2006,** *142*, 1559–1573.

Karst, J.; Gaster, J.; Wiley, E.; Landhäusser, S. M. Stress Differentially Causes Roots of Tree Seedlings to Exude Carbon. *Tree Physiol.* **2016,** *37*, 154–164.

Kasim, W. A.; Osman, M. E.; Omar, M. N.; Abd El-Daim, I. A.; Bejai, S.; Meijer, J. Control of Drought Stress in Wheat using Plant Growth Promoting Bacteria. *J. Plant Growth Regul.* **2013,** *32*, 122–130.

Kassem, I. I.; Joshi, P.; Sigler, V.; Heckathorn, S.; Wang, Q. Effect of Elevated CO_2 and Drought on Soil Microbial Communities Associated with *Andropogon gerardii*. *J. Integr. Plant Biol.* **2008,** *50*, 1406–1415.

Kaushal, M.; Wani, S. P. Plant-Growth-Promoting Rhizobacteria: Drought Stress Alleviators to Ameliorate Crop Production in Drylands. *Ann. Microbiol.* **2016,** *66*, 35–42.

Kawakami, J.; Iwama, K.; Jitsuyama, Y. Soil Water Stress and the Growth and Yield of Potato Plants Grown from Microtubers and Conventional Seed Tubers. *Field Crop. Res.* **2006,** *95*, 89–96.

Kawasaki, A.; Donn, S.; Ryan, P. R.; Mathesius, U.; Devilla, R.; Jones, A.; et al. Microbiome and Exudates of the Root and Rhizosphere of *Brachypodium distachyon*, A Model for Wheat. *PLoS One* **2016,** *11*, e0164533.

Khan, M. A.; Gemenet, D. C.; Villordon, A. Root System Architecture and Abiotic Stress Tolerance: Current Knowledge in Root and Tuber Crops. *Front. Plant Sci.* **2016,** *7*, 1584.

Khan, Z.; Rho, H.; Firrincieli, A.; Hung, S. H.; Luna, V.; Masciarelli, O.; et al. Growth Enhancement and Drought Tolerance of Hybrid Poplar upon Inoculation with Endophyte Consortia. *Curr. Plant Biol.* **2016,** *6*, 38–47.

Khan, A.; Uddin, S.; Zhao, X. Q.; Javed, M. T.; Khan, K. S.; Bano, A.; et al. *Bacillus pumilus* Enhances Tolerance in Rice (*Oryza sativa* L.) to Combined Stresses of NaCl and High Boron Due to Limited Uptake of Na^+. *Environ. Exp. Bot.* **2016,** *124*, 120–129.

Kim, Y. C.; Glick, B. R.; Bashan, Y.; Ryu, C. M. Enhancement of Plant Drought Tolerance by Microbes. In *Plant Responses to Drought Stress*; Aroca, R., Ed.; Springer: Berlin, 2012; pp 383–413.

Knoth, J. L.; Kim, S. H.; Ettl, G. J.; Doty, S. L. Biological Nitrogen Fixation and Biomass Accumulation within Poplar Clones as a Result of Inoculations with Diazotrophic Endophyte Consortia. *New Phytol.* **2014,** *201*, 599–609.

Kohler, J.; Caravaca, F.; Roldán, A. Effect of Drought on the Stability of Rhizosphere Soil Aggregates of *Lactuca sativa* Grown in a Degraded Soil Inoculated with PGPR and AM Fungi. *Appl. Soil Ecol.* **2009,** *42*, 160–165.

Kramer, S.; Green, D. M. Acid and Alkaline Phosphatase Dynamics and their Relationship to Soil Microclimate in a Semiarid Woodland. *Soil Biol. Biochem.* **2000,** *32*, 179–188.

Kurm, V.; Van der Putten, W. H.; de Boer, W.; Naus-Wiezer, S.; Hol, W. H. Low Abundant Soil Bacteria can be Metabolically Versatile and Fast Growing. *Ecology* **2017,** *98*, 555–564.

Lau, J. A.; Lennon, J. T. Rapid Responses of Soil Microorganisms Improve Plant Fitness in Novel Environments. *Proc. Natl. Acad. Sci. USA* **2012,** *109*, 14058–14062.

Lauber, C. L.; Hamady, M.; Knight, R.; Fierer, N. Pyrosequencing-Based Assessment of Soil pH as a Predictor of Soil Bacterial Community Structure at the Continental Scale. *Appl. Environ. Microbiol.* **2009,** *75*, 5111–5120.

Lee, E. J.; Karoonuthaisiri, N.; Kim, H. S.; Park, J. H.; Cha, C. J.; Kao, C. M.; et al. A Master Regulator σB Governs Osmotic and Oxidative Response as Well as Differentiation Via a Network of Sigma Factors in *Streptomyces coelicolor*. *Mol. Microbiol.* **2005,** *57*, 1252–1264.

Leiros, M.; Alonso, E.; Sanchez, J. A.; Rateb, M. E.; Ebel, R.; Houssen, W. E.; et al. Mitigation of ROS Insults by *Streptomyces* Secondary Metabolites in Primary Cortical Neurons. *ACS Chem. Neurosci.* **2014,** *5*, 71–80.

Lennon, J. T.; Aanderud, Z. T.; Lehmkuhl, B. K.; Schoolmaster, D. R. Mapping the Niche Space of Soil Microorganisms using Taxonomy and Traits. *Ecology* **2012,** *93*, 1867–1879.

Lesk, C.; Rowhani, P.; Ramankutty, N. Influence of Extreme Weather Disasters on Global Crop Production. *Nature* **2016,** *529*, 84–87.

Li, X.; Rui, J.; Xiong, J.; Li, J.; He, Z.; Zhou, J.; et al. Functional Potential of Soil Microbial Communities in the Maize Rhizosphere. *PLoS One* **2014,** *9*, e112609.

Lim, J. H.; Kim, S. D. Induction of Drought Stress Resistance by Multi-Functional PGPR *Bacillus licheniformis* K11 in Pepper. *Plant Pathol. J.* **2013,** *29*, 201–208.

Liptzin, D.; Silver, W. L.; Detto, M. Temporal Dynamics in Soil Oxygen and Greenhouse Gases in Two Humid Tropical Forests. *Ecosystems* **2011,** *14*, 171–182.

Liu, J.; Xia, J.; Fang, Y.; Li, T.; Liu, J. Effects of Salt-Drought Stress on Growth and Physiobiochemical Characteristics of *Tamarix chinensis* Seedlings. *Sci. World J.* **2014,** 765840.

Lundberg, D. S.; Lebeis, S. L.; Paredes, S. H.; Yourstone, S.; Gehring, J.; Malfatti, S.; et al. Defining the Core *Arabidopsis thaliana* Root Microbiome. *Nature* **2012,** *488*, 86–90.

Marasco, R.; Rolli, E.; Ettoumi, B.; Vigani, G.; Mapelli, F.; Borin, S.; et al. A Drought Resistance-Promoting Microbiome is Selected by Root System under Desert Farming. *PLoS One* **2012,** *7*, e48479.

Marasco, R.; Rolli, E.; Vigani, G.; Borin, S.; Sorlini, C.; Ouzari, H.; et al. Are Drought-Resistance Promoting Bacteria Cross-Compatible with Different Plant Models? *Plant Signal. Behav.* **2013,** *8*, e26741.

Marschner, P.; Grierson, P. F.; Rengel, Z. Microbial Community Composition and Functioning in the Rhizosphere of Three *Banksia* Species in Native Woodland in Western Australia. *Appl. Soil Ecol.* **2005,** *28*, 191–201.

Martínez, J. P.; Silva, H.; Ledent, J. F.; Pinto, M. Effect of Drought Stress on the Osmotic Adjustment, Cell Wall Elasticity and Cell Volume of Six Cultivars of Common Beans (*Phaseolus vulgaris* L.). *Eur. J. Agron.* **2007,** *26*, 30–38.

Martiny, J. B.; Martiny, A. C.; Weihe, C.; Lu, Y.; Berlemont, R.; Brodie, E. L.; et al. Microbial Legacies Alter Decomposition in Response to Simulated Global Change. *ISME J.* **2016,** *11*, 490–499.

Marulanda, A.; Azcón, R.; Chaumont, F.; Ruiz-Lozano, J. M.; Aroca, R. Regulation of Plasma Membrane Aquaporins by Inoculation with a *Bacillus megaterium* Strain in Maize (*Zea mays* L.) Plants under Unstressed and Salt-Stressed Conditions. *Planta* **2010,** *232*, 533–543.

Mayak, S.; Tirosh, T.; Glick, B. R. Plant Growth-Promoting Bacteria that Confer Resistance to Water Stress in Tomatoes and Peppers. *Plant Sci.* **2004,** *166*, 525–530.

Meena, K. K.; Sorty, A. M.; Bitla, U. M.; Choudhary, K.; Gupta, P.; Pareek, A.; et al. Abiotic Stress Responses and Microbe-Mediated Mitigation in Plants: The Omics Strategies. *Front. Plant Sci.* **2017,** *8*, 172.

Micallef, S. A.; Shiaris, M. P.; Colon-Carmona, A. Influence of *Arabidopsis thaliana* Accessions on Rhizobacterial Communities and Natural Variation in Root Exudates. *J. Exp. Bot.* **2009,** *60*, 1729–1742.

Mittler, R. Abiotic Stress, The Field Environment and Stress Combination. *Trends Plant Sci.* **2006,** *11*, 15–19.

Mittler, R.; Blumwald, E. Genetic Engineering for Modern Agriculture: Challenges and Perspectives. *Annu. Rev. Plant Biol.* **2010,** *61*, 443–462.

Mohammadipanah, F.; Wink, J. Actinobacteria from Arid and Desert Habitats: Diversity and Biological Activity. *Front. Microbiol.* **2016,** *6*, 1541.

Moura, J. C. M. S.; Bonine, C. A. V.; de Oliveira Fernandes Viana, J.; Dornelas, M. C.; Mazzafera, P. Abiotic and Biotic Stresses and Changes in the Lignin Content and Composition in Plants. *J. Integr. Plant Biol.* **2010,** *52*, 360–376.

Nadeem, S. M.; Ahmad, M.; Zahir, Z. A.; Javaid, A.; Ashraf, M. The Role of Mycorrhizae and Plant Growth Promoting Rhizobacteria (PGPR) in Improving Crop Productivity under Stressful Environments. *Biotechnol. Adv.* **2014,** *32*, 429–448.

Nannipieri, P.; Ascher, J.; Ceccherini, M. T.; Landi, L.; Pietramellara, G.; Renella, G.; et al. Microbial Diversity and Microbial Activity in the Rhizosphere. *Cienc. Suelo.* **2007,** *25*, 89–97.

Nautiyal, C. S.; Dion. P. In *Molecular Mechanisms of Plant and Microbe Coexistence*; Springer: Berlin, 2008.

Naveed, M.; Mitter, B.; Reichenauer, T. G.; Wieczorek, K.; Sessitsch, A. Increased Drought Stress Resilience of Maize through Endophytic Colonization by *Burkholderia phytofirmans* PsJN and *Enterobacter* sp. FD17. *Environ. Exp. Bot.* **2014b,** *97*, 30–39.

Naylor, D.; DeGraaf, S.; Purdom, E.; Coleman-Derr, D. Drought and Host Selection Influence Bacterial Community Dynamics in the Grass Root Microbiome. *ISME J.* **2017,** 11, 2691–2704.

Nessner Kavamura, V.; Taketani, R. G.; Lançoni, M. D.; Andreote, F. D.; Mendes, R.; Soares de Melo, I. Water Regime Influences Bulk Soil and Rhizosphere of *Cereus jamacaru* Bacterial Communities in the Brazilian Caatinga Biome. *PLoS One* **2013,** *8*, e73606.

Ngumbi, E.; Kloepper, J. Bacterial-Mediated Drought Tolerance: Current and Future Prospects. *Appl. Soil Ecol.* **2014,** *105*, 109–125.

Nuccio, E. E.; Anderson-Furgeson, J.; Estera, K. Y.; Pett-Ridge, J.; Valpine, P.; Brodie, E. L.; et al. Climate and Edaphic Controllers Influence Rhizosphere Community Assembly for a Wild Annual Grass. *Ecology* **2016,** *97*, 1307–1318.

Ogbaga, C. C.; Stepien, P.; Johnson, G. N. Sorghum (*Sorghum bicolor*) Varieties Adopt Strongly Contrasting Strategies in Response to Drought. *Physiol. Plant.* **2014,** *152*, 389–401.

Ogbonnaya, C. I.; Sarr, B.; Brou, C.; Diouf, O.; Diop, N. N.; Roy-Macauley, H. Selection of Cowpea Genotypes in Hydroponics, Pots, and Field for Drought Tolerance. *Crop Sci.* **2003,** *43*, 1114–1120.

Oliver, R. J.; Finch, J. W.; Taylor, G. Second Generation Bioenergy Crops and Climate Change: A Review of the Effects of Elevated Atmospheric CO_2 and Drought on Water use and the Implications for Yield. *Global Change Biol. Bioenergy*. **2009,** *1*, 97–114.

Oliveira, C. A.; Alves, V. M. C.; Marriel, I. E.; Gomes, E. A.; Scotti, M. R.; Carneiro, N. P.; et al. Phosphate Solubilizing Microorganisms Isolated from Rhizosphere of Maize Cultivated in an Oxisol of the Brazilian Cerrado Biome. *Soil Biol. Biochem.* **2009,** *41*, 1782–1787.

Omar, M. N. A.; Osman, M. E. H.; Kasim, W. A.; Abd El-Daim, I. A. Improvement of Salt Tolerance Mechanisms of Barley Cultivated Under Salt Stress using *Azospirillum brasilense*. *Tasks Veg. Sci.* **2009,** *44*, 133–147.

Orwin, K. H.; Wardle, D. A. Plant Species Composition Effects on Belowground Properties and the Resistance and Resilience of the Soil Microflora to a Drying Disturbance. *Plant Soil*. **2005,** *278*, 205–221.

Palta, J. A.; Gregory, P. J. Drought Affects the Fluxes of Carbon to Roots and Soil in ^{13}C Pulse-Labelled Plants of Wheat. *Soil Biol. Biochem.* **1997,** *29*, 1395–1403.

Pandey, V.; Ansari, M. W.; Tula, S.; Yadav, S.; Sahoo, R. K.; Shukla, N.; et al. Dose-Dependent Response of *Trichoderma harzianum* in Improving Drought Tolerance in Rice Genotypes. *Planta* **2016,** *243*, 1251–1264.

Panlada, T.; Pongdet, P.; Aphakorn, L.; Rujirek, N. N.; Nantakorn, B.; Neung, T. Alleviation of the Effect of Environmental Stresses using Co-inoculation of Mungbean by *Bradyrhizobium* and Rhizobacteria Containing Stress-Induced ACC Deaminase Enzyme. *Soil Sci. Plant Nut.* **2013,** *59*, 559–571.

Pascault, N.; Ranjard, L.; Kaisermann, A.; Bachar, D.; Christen, R.; Terrat, S.; et al. Stimulation of Different Functional Groups of Bacteria by Various Plant Residues as a Driver of Soil Priming Effect. *Ecosystems* **2013,** *16*, 810–822.

Peaucelle, A.; Braybrook, S.; Höfte, H. Cell Wall Mechanics and Growth Control in Plants: The Role of Pectins Revisited. *Front. Plant Sci.* **2012,** *3*,121.

Peiffer, J. A.; Spor, A.; Koren, O.; Jin, Z.; Tringe, S. G.; Dangl, J. L.; et al. Diversity and Heritability of the Maize Rhizosphere Microbiome Under Field Conditions. *Proc. Natl. Acad. Sci. USA* **2013,** *110*, 6548–6553.

Pinedo, I.; Ledger, T.; Greve, M.; Poupin, M. J. *Burkholderia phytofirmans* PsJN Induces Long-Term Metabolic and Transcriptional Changes Involved in *Arabidopsis thaliana* Salt Tolerance. *Front. Plant Sci.* **2015,** *6*, 466.

Preece, C.; Peñuelas, J. Rhizodeposition Under Drought and Consequences for Soil Communities and Ecosystem Resilience. *Plant Soil* **2016,** *409*, 1–17.

Quiza, L.; St-Arnaud, M.; Yergeau, E. Harnessing Phytomicrobiome Signaling for Rhizosphere Microbiome Engineering. *Front. Plant Sci.* **2015,** *6*, 507.

Reid, C. P. Assimilation, Distribution, and Root Exudation of ^{14}C by Ponderosa Pine Seedlings under Induced Water Stress. *Plant Physiol.* **1974,** *54*, 44–49.

Rolli, E.; Marasco, R.; Vigani, G.; Ettoumi, B.; Mapelli, F.; Deangelis, M. L.; et al. Improved Plant Resistance to Drought is Promoted by the Root-Associated Microbiome as a Water Stress-Dependent Trait: Root Bacteria Protect Plants from Drought. *Environ. Microbiol.* **2015,** *17*, 316–331.

Ruehr, N. K.; Offermann, C. A.; Gessler, A.; Winkler, J. B.; Ferrio, J. P.; Buchmann, N.; et al. Drought Effects on Allocation of Recent Carbon: From Beech Leaves to Soil CO_2 Efflux. *New Phytol.* **2009,** *184*, 950–961.

Sahoo, R. K.; Ansari, M. W.; Dangar, T. K.; Mohanty, S.; Tuteja, N. Phenotypic and Molecular Characterisation of Efficient Nitrogen-Fixing *Azotobacter* Strains from Rice Fields for Crop Improvement. *Protoplasma* **2014a,** *251*, 511–523.

Samarah, N. H.. Effects of Drought Stress on Growth and Yield of Barley. *Agron. Sustain. Dev.* **2005,** *25*, 145–149.

Samarah, N. H.; Mullen, R. E.; Cianzio, S. R.; Scott, P. Dehydrin-Like Proteins in Soybean Seeds in Response to Drought Stress During Seed FIling. *Crop Sci.* **2006,** *46*, 2141–2150.

Sandhya, V.; Ali, S. K. Z.; Grover, M.; Reddy, G.; Venkateswarlu, B. Alleviation of Drought Stress Effects in Sunflower Seedlings by the Exopolysaccharides Producing *Pseudomonas putida* Strain GAP-P45. *Biol. Fertil. Soils* **2009,** *46*, 17–26.

Santos-Medellín, C.; Edwards, J.; Liechty, Z.; Nguyen, B.; Sundaresan, V. Drought Stress Results in a Compartment-Specific Restructuring of the Rice Root-Associated Microbiomes. *mBio*. **2017,** *8*, e00764-17.

Sardans, J.; Peñuelas, J. Increasing Drought Decreases Phosphorus Availability in an Evergreen Mediterranean Forest. *Plant Soil* **2004,** *267*, 367–377.

Sardans, J.; Peñuelas, J. Drought Decreases Soil Enzyme Activity in a Mediterranean *Quercus ilex* L. Forest. *Soil Biol. Biochem.* **2005,** *37*, 455–461.

Sardans, J.; Peñuelas, J.; Estiarte, M. Changes in Soil Enzymes Related to C and N Cycle and in Soil C and N Content Under Prolonged Warming and Drought in a Mediterranean Shrubland. *Appl. Soil Ecol.* **2008,** *39*, 223–235.

Schimel, J.; Balser, T. C.; Wallenstein, M. Microbial Stress-Response Physiology and its Implications for Ecosystem Function. *Ecology* **2007,** *88*, 1386–1394.

Schwartz, E.; Adair, K. L.; Schuur, E. A. Bacterial Community Structure Correlates with Decomposition Parameters Along a Hawaiian Precipitation Gradient. *Soil Biol. Biochem.* **2007,** *39*, 2164–2167.

Sen, S.; Chandrasekhar, C. N. Effect of PGPR on Growth Promotion of Rice (*Oryza sativa* L.) Under Salt Stress. *Asian J. Plant Sci. Res.* **2014,** *4*, 62–67.

Shakya, M.; Gottel, N.; Castro, H.; Yang, Z. K.; Gunter, L.; Labbé, J.; et al. A Multifactor Analysis of Fungal and Bacterial Community Structure in the Root Microbiome of Mature *Populus deltoides* Trees. *PLoS One* **2013,** *8*, e76382.

Shi, S.; Richardson, A. E.; O'Callaghan, M.; DeAngelis, K. M.; Jones, E. E.; Stewart, A.; et al. Effects of Selected Root Exudate Components on Soil Bacterial Communities: Root Exudate Components and Soil Microbial Communities. *FEMS Microbiol. Ecol.* **2011,** *77*, 600–610.

Singh, P. Agrometeorology in the Service of Agriculture. Millenium Lecture Delivered on 21 June 2002 at SV Agricuture College, Tirupati, India, 2002.

Singh, S. A Review on Possible Elicitor Molecules of Cyanobacteria: Their Role in Improving Plant Growth and Providing Tolerance Against Biotic or Abiotic Stress. *J. Appl. Microbiol.* **2014,** *117*, 1221–1244.

Singh, D. P.; Prabha, R.; Yandigeri, M. S.; Arora, D. K. Cyanobacteria-Mediated Phenylpropanoids and Phytohormones in Rice (*Oryza sativa*) Enhance Plant Growth and Stress Tolerance. *Antonie Van Leeuwenhoek* **2011,** *100*, 557–568.

Singh, R.; Singh, P.; Sharma, R. Microorganism as a Tool of Bioremediation Technology for Cleaning Environment: A Review. *Proc. Int. Acad. Ecol. Environ. Sci.* **2014,** *4*, 1–6.

Singh, V.; Van Oosterom, E. J.; Jordan, D. R.; Messina, C. D.; Cooper, M.; Hammer, G. L. Morphological and Architectural Development of Root Systems in Sorghum and Maize. *Plant Soil* **2010,** *333*, 287–299.

Smith, S.; De Smet, I. Root System Architecture: Insights from *Arabidopsis* and Cereal Crops. *Philos. Trans. R. Soc. B Biol. Sci.* **2012,** *367*, 1441–1452.

Song, F.; Han, X.; Zhu, X.; Herbert, S. J. Response to Water Stress of Soil Enzymes and Root Exudates from Drought and Non-Drought Tolerant Corn Hybrids at Different Growth Stages. *Can. J. Soil Sci.* **2012,** *92*, 501–507.

Sorty, A. M.; Meena, K. K.; Choudhary, K.; Bitla, U. M.; Minhas, P. S.; Krishnani, K. K. Effect of Plant Growth Promoting Bacteria Associated with Halophytic Weed (*Psoralea corylifolia* L.) on Germination and Seedling Growth of Wheat under Saline Conditions. *Appl. Biochem. Biotechnol.* **2016,** *180*, 872–882.

Sorty, A. M.; Shaikh, N. R. Novel Co-enrichment Method for Isolation of Magnetotactic Bacteria. *J. Basic Microbiol.* **2015,** *55*, 520–526.

Souza, R. D.; Ambrosini, A.; Passaglia, L. M. P. Plant Growth-Promoting Bacteria as Inoculants in Agricultural Soils. *Genet. Mol. Biol.* **2015,** *38*, 401–419.

Spollen, W. G.; Sharp, R. E. Spatial Distribution of Turgor and Root Growth at Low Water Potentials. *Plant Physiol.* **1991,** *96*, 438–443.

Stark, J. M.; Firestone, M. K. Mechanisms for Soil Moisture Effects on Activity of Nitrifying Bacteria. *Appl. Environ. Microbiol.* **1995,** *61*, 218–221.

Suarez, C.; Cardinale, M.; Ratering, S.; Steffens, D.; Jung, S.; Montoya, A. M. Z.; et al. Plant Growth-Promoting Effects of *Hartmannibacter diazotrophicus* on Summer Barley (*Hordeum vulgare* L.) Under Salt Stress. *Appl. Soil Ecol.* **2015,** *95*, 23–30.

Sun, C.; Johnson, J.; Cai, D.; Sherameti, I.; Oelmüeller, R.; Lou, B. *Piriformospora indica* Confers Drought Tolerance in Chinese Cabbage Leaves by Stimulating Antioxidant Enzymes, the Expression of Drought-Related Genes and the Plastid-Localized CAS Protein. *J. Plant. Physiol.* **2010,** *167*, 1009–1017.

Svenningsson, H.; Sundin, P.; Liljenberg, C. Lipids, Carbohydrates and Amino Acids Exuded from the Axenic Roots of Rape Seedlings Exposed to Water-Deficit Stress. *Plant Cell Environ.* **1990,** *13*, 155–162.

Taketani, R. G.; Lançoni, M. D.; Kavamura, V. N.; Durrer, A.; Andreote, F. D.; Melo, I. S. Dry Season Constrains Bacterial Phylogenetic Diversity in a Semi-Arid Rhizosphere System. *Microb. Ecol.* **2017,** *73*, 153–161.

TerHorst, C. P.; Lennon, J. T.; Lau, J. A. The Relative Importance of Rapid Evolution for Plant–Microbe Interactions Depends on Ecological Context. *Proc. R. Soc. B Biol. Sci.* **2014,** *281*, 20140028.

Thaysen, E. M.; Reinsch, S.; Larsen, K. S.; Ambus, P. Decrease in Heathland Soil Labile Organic Carbon Under Future Atmospheric and Climatic Conditions. *Biogeochemistry* **2017,** *133*, 17–36.

Timm, C. M.; Pelletier, D. A.; Jawdy, S. S.; Gunter, L. E.; Henning, J. A.; Engle, N.; et al. Two Poplar-Associated Bacterial Isolates Induce Additive Favourable Responses in a Constructed Plant–Microbiome System. *Front. Plant Sci.* **2016,** *7*, 497.

Timmusk, S.; Abd El-Daim, I. A.; Copolovici, L.; Tanilas, T.; Kännaste, A.; Behers, L.; et al. Drought-Tolerance of Wheat Improved by Rhizosphere Bacteria from Harsh

Environments: Enhanced Biomass Production and Reduced Emissions of Stress Volatiles. *PLoS One* **2014,** *9*, e96086.
Timmusk, S.; Behers, L.; Muthoni, J.; Muraya, A.; Aronsson, A. C. Perspectives and Challenges of Microbial Application for Crop Improvement. *Front. Plant Sci.* **2017,** *8*, 49.
Timmusk, S.; Kim, S. B.; Nevo, E.; Abd El Daim, I.; Ek, B.; Bergquist, J.; et al. Sfp-Type PPTase Inactivation Promotes Bacterial Biofilm Formation and Ability to Enhance Wheat Drought Tolerance. *Front. Microbiol.* **2015,** *6*, 387.
Timmusk, S.; Paalme, V.; Pavlicek, T.; Bergquist, J.; Vangala, A.; Danilas, T.; et al. Bacterial Distribution in the Rhizosphere of Wild Barley under Contrasting Microclimates. *PLoS One* **2011,** *6*, e17968.
Timmusk, S.; Wagner, E. G. H. The Plant-Growth-Promoting Rhizobacterium *Paenibacillus polymyxa* Induces Changes in *Arabidopsis thaliana* Gene Expression: A Possible Connection between Biotic and Abiotic Stress Responses. *Mol. Plant Microbe Interact.* **1999,** *12*, 951–959.
Tocheva, E. I.; Ortega, D. R.; Jensen, G. J. Sporulation, Bacterial Cell Envelopes and the Origin of Life. *Nat. Rev. Microbiol.* **2016,** *14*, 535–542.
Torres-Cortés, G.; Millán, V.; Fernández-González, A. J.; Aguirre-Garrido, J. F.; Ramírez-Saad, H. C.; Fernández-López, M.; et al. Bacterial Community in the Rhizosphere of the Cactus Species *Mammillaria carnea* During Dry and Rainy Seasons Assessed by Deep Sequencing. *Plant Soil.* **2012,** *357*, 275–288.
Tóth, Z.; Táncsics, A.; Kriszt, B.; Kröel-Dulay, G.; Ónodi, G.; Hornung, E. Extreme Effects of Drought on Composition of the Soil Bacterial Community and Decomposition of Plant Tissue: Bacterial Community and Plant Tissue Decomposition. *Eur. J. Soil Sci.* **2017,** *68*, 504–513.
Treseder, K. K.; Kivlin, S. N.; Hawkes, C. V. Evolutionary Trade-Offs Among Decomposers Determine Responses to Nitrogen Enrichment: Evolutionary Trade-Offs Among Decomposers. *Ecol. Lett.* **2011,** *14*, 933–938.
Tripathi, P.; Rabara, R. C.; Reese, R. N.; Miller, M. A.; Rohila, J. S.; Subramanian, S.; et al. A Toolbox of Genes, Proteins, Metabolites and Promoters for Improving Drought Tolerance in Soybean Includes the Metabolite Coumestrol and Stomatal Development Genes. *BMC Genomics* **2016,** *17*, 102.
UN. What can be Done about Drought? International Decade for Natural Disaster Reduction, IDNDR Secretariat, Geneva, 1990.
Upreti, K. K.; Murti, G. S. R.; Bhatt, R. M. Response of Pea Cultivars to Water Stress: Changes in Morphophysiological Characters, Endogenous Hormones and Yield. *Veg. Sci.* **2000,** *27* (1), 57–61.
Van der Molen, M. K.; Dolman, A. J.; Ciais, P.; Eglin, T.; Gobron, N.; Law, B. E.; et al. Drought and Ecosystem Carbon Cycling. *Agric. For. Meteorol.* **2011,** *151*, 765–773.
Villar-Salvador, P.; Heredia, N.; Millard, P. Remobilization of Acorn Nitrogen for Seedling Growth in Holm Oak (*Quercus ilex*), Cultivated with Contrasting Nutrient Availability. *Tree Physiol.* **2009,** *30* (2), 257–263.
Voothuluru, P.; Sharp, R. E. Apoplastic Hydrogen Peroxide in the Growth Zone of the Maize Primary Root Under Water Stress. I. Increased Levels are Specific to the Apical Region of Growth Maintenance. *J. Exp. Bot.* **2012,** *64*, 1223–1233.

Vurukonda, S. S. K. P.; Vardharajula, S.; Shrivastava, M.; Skz, A. Enhancement of Drought Stress Tolerance in Crops by Plant Growth Promoting Rhizobacteria. *Microbiol. Res.* **2016,** *184*, 13–24.

Wagner, M. R.; Lundberg, D. S.; del Rio, T. G.; Tringe, S. G.; Dangl, J. L.; Mitchell-Olds, T. Host Genotype and Age Shape the Leaf and Root Microbiomes of a Wild Perennial Plant. *Nat. Commun.* **2016,** *7*, 12151.

Wang, S.; Ouyang, L.; Ju, X.; Zhang, L.; Zhang, Q.; Li, Y. Survey of Plant Drought-Resistance Promoting Bacteria from *Populus euphratica* Tree Living in Arid Area. *Indian J. Microbiol.* **2014,** *54*, 419–426.

Wardle, D.; Bardgett, R.; Klironomos, J.; Setälä, H.; van der Putten, W.; Wall, D. Ecological Linkages between Aboveground and Belowground Biota. *Science* **2004,** *304*, 1629–1633.

Welsh, D. T. Ecological Significance of Compatible Solute Accumulation by Micro-organisms: From Single Cells to Global Climate. *FEMS Microbiol. Rev.* **2000,** *24*, 263–290.

Whipps, J. M.; Lynch, J. M. Substrate Flow and Utilization in the Rhizosphere of Cereals. *New Phytol.* **1983,** *95*, 605–623.

Xiong, L.; Wang, R, G.; Mao, G.; Jessica, M. K. Identification of Drought Tolerance Determinants by Genetic Analysis of Root Response to Drought Stress and Abscisic Acid. *Plant Physiol.* **2006,** *142* (3), 1065–1074.

Xu, Y. Envirotyping for Deciphering Environmental Impacts on Crop Plants. *Theor. Appl. Genet.* **2016,** *129*, 653–673.

Yandigeri, M. S.; Meena, K. K.; Singh, D.; Malviya, N.; Singh, D. P.; Solanki, M. K.; et al. Drought-Tolerant Endophytic Actinobacteria Promote Growth of Wheat (*Triticum aestivum*) Under Water Stress Conditions. *Plant Growth Regul.* **2012,** *68*, 411–420.

Zang, U.; Goisser, M.; Häberle, K. H.; Matyssek, R.; Matzner, E.; Borken, W. Effects of Drought Stress on Photosynthesis, Rhizosphere Respiration, and Fine-Root Characteristics of Beech Saplings: A Rhizotron Field Study. *J. Plant Nutr. Soil Sci.* **2014,** *177*, 168–177.

Zhang, M.; Duan, L.; Zhai, Z.; et al. In *Effects of Plant Growth Regulators on Water Deficit-Induced Yield Loss in Soybean*, Proceedings of the 4th International Crop Science Congress, *Brisbane, Australia*, 2004; pp 252–256.

CHAPTER 2

PRODUCTION OF PLANT HORMONES FROM MICROORGANISMS

ABDEL RAHMAN MOHAMMAD AL-TAWAHA[1], ROGHAYEH KARIMIRAD[2*], RAZZAGH KARIMIRAD[3], ABHIJIT DEY[4], SIDDHARTHA PATI[5], ABDEL RAZZAQ AL-TAWAHA[6], ALLA ALEKSANYAN[7], R. Z. SAYYED[8], DEVARAJAN THANGADURAI[9], JEYABALAN SANGEETHA[10], MUHAMMAD AASIM[11], DEVENDRA KUMAR PANDEY[12], IMRAN[13], AMANULLAH[13], SHAH KHALID[13], SAMAR S. SHAWAQFEH[14], and IFTIKHAR ALI[15]

[1]*Department of Biological Sciences, Al Hussein bin Talal University, Maan, Jordan*

[2]*Department of Horticultural Sciences, Faculty of Agriculture and Natural Resources, University of Mohaghegh Ardabili, Ardabil, Iran*

[3]*Department of Animal Science, Lorestan University, Khorramabad, Iran*

[4]*Department of Life Sciences, Presidency University, Kolkata, West Bengal, India*

[5]*NatNov Bioscience Private Limited, Balasore, Odisha, India*

[6]*National Agricultural Research Center (NARC), Baq'a 19381, Jordan*

[7]*Institute of Botany aft. A.L. Takhtajyan NAS RA/Department of Geobotany and Plant Eco-Physiology, Yerevan, Armenia*

[8]*Department of Microbiology, PSGVP Mandal's Arts, Science and Commerce College, Shahada, Maharashtra, India*

Microbial Fertilizer Technology for Sustainable Crop Production. Jeyabalan Sangeetha, Abdel Rahman Mohammad Al-Tawaha, & Devarajan Thangadurai (Eds.)

DOI: 10.1201/9781003594826-2

[9]*Department of Botany, Karnatak University, Dharwad, Karnataka, India*

[10]*Department of Environmental Science, Central University of Kerala, Kasaragod, Kerala, India*

[11]*Department of Plant Protection, Faculty of Agricultural Sciences and Technologies, Sivas University of Science and Technology, Sivas, Turkey*

[12]*Department of Biotechnology, School of Bioengineering and Biosciences, Lovely Professional University, Phagwara, Punjab, India*

[13]*Department of Agronomy, The University of Agriculture, Peshawar, Pakistan*

[14]*Department of Plant Production and Protection, College of Agriculture, Jerash University, Jerash, Jordan*

[15]*Center for Plant Sciences and Biodiversity, University of Swat, Charbagh, Pakistan*

**Corresponding author.*

ABSTRACT

In general, phytohormones are mostly divided into five classes: auxins [indole-3-acetic acid (IAA)], cytokinins (CKs), ethylene, abscisic acid (ABA), and gibberellins (GAs). Apart from phytohormones' role in plant reactions to changes in environmental conditions, phytohormones are the main factor that regulates the expression of the inherent genetic potential of plants. Several biosynthesis pathways have been recognized and the abundance of phytohormones biosynthesis is extensive among plant-associated pathogenesis strains. Phytohormone biosynthesis by microorganisms plays a critical role in regulating plant development by mediating signaling networks involved in plant responses to a broad range of interactions. Interactions between phytohormones-producing micro-organisms and plants cause various results on the plant side, differing from pathogenesis to phytostimulation. For example, there is increasing proof that auxin, is a signaling hormone in microorganisms because auxin influences gene expression in some microorganisms. In this chapter, plant hormone biosynthesis by microorganisms and its role in the association with the plant are discussed.

2.1 INTRODUCTION

Hormones are a class of organic compounds produced in a particular tissue of a plant with low concentrations and regulate different functions associated with physiological, biochemical, and morphological responses in the receptive organ of the plant (Davies, 2010). Phytohormones are the primary factor that balances the expression of the natural genetic capability of plant life. In addition, most phytohormones do not act alone on a physio biochemical process, and therefore, balances between several types of the hormone can be significant. In general, plant hormones are commonly divided into five classes: Phytohormones group [auxins, gibberellins (GAs), cytokinins (CKs), abscisic acid (ABA), ethylene], and with the deepening of investigation, some novel phytohormones such as strigolactones (SLs) salicylates (SAs), jasmonates (JAs), brassinosteroids (BRs), polyamines and nitric oxide have been discovered and recognized. Even though, nitric oxide (NO) is an essential signaling molecule in the tissue of plants, they are now not broadly distinguished as phytohormones in biological science, primarily due to them being an inorganic agent.

These plant hormones have been found to be synthesized not only by higher plants but also in host plant-related bacteria and endophytic fungi. In microorganisms, the plant hormones no longer influence certain hormonal or main physiochemical adjustments. In this case, microbial plant hormone biosynthesized has been associated with modifications in root structure and plant development. The synthesis of phytohormones (IAA, GAs, and CK) by such related pathogens is believed to be one of the primary forms of host-plant pathogens interactions (Mishke, 1988; Kamneva et al., 1999) (Table 2.1). However, the ability to synthesize phytohormones can vary greatly, determined by the plant hormone symbiotic associations and pathogenesis strain (Figure 2.1). According to previous reports, genetic report for the role of numerous plant hormones within the pathogen-plant reaction remains unknown, so questioning the significance of microbial generation. Focused on methods concentrating on the genetic papers for the part of plant hormone along with plant investigation will permit discovering the importance and capacity of this interesting microbial feature.

TABLE 2.1 Efficient PGPR Strains as Phytohormone Producer in Crop Plants.

Hormone produced	Plant growth promoting rhizobacteria	Host plant
IAA	*Aeromonas veronii*	Rice
	Agrobacterium sp.	Lettuce
	Alcaligenes piechaudii	Lettuce
	A. brasilense	Wheat
	Bradyrhizobium sp.	Radish
	Comamonas acidovorans	Lettuce
	Enterobacter cloacae	Rice
	Rhizobium leguminosarum	Radish
CK	*Paenibacillus polymyxa*	Wheat
	Pseudomonas fluorescens	Soybean
	Rhizobium leguminosarum	Rape and lettuce
GAs	*Bacillus* sp.	Alder

Source: Reproduced with permission from Bhattacharyya and Jha (2012), Copyright © Springer Nature (2017).

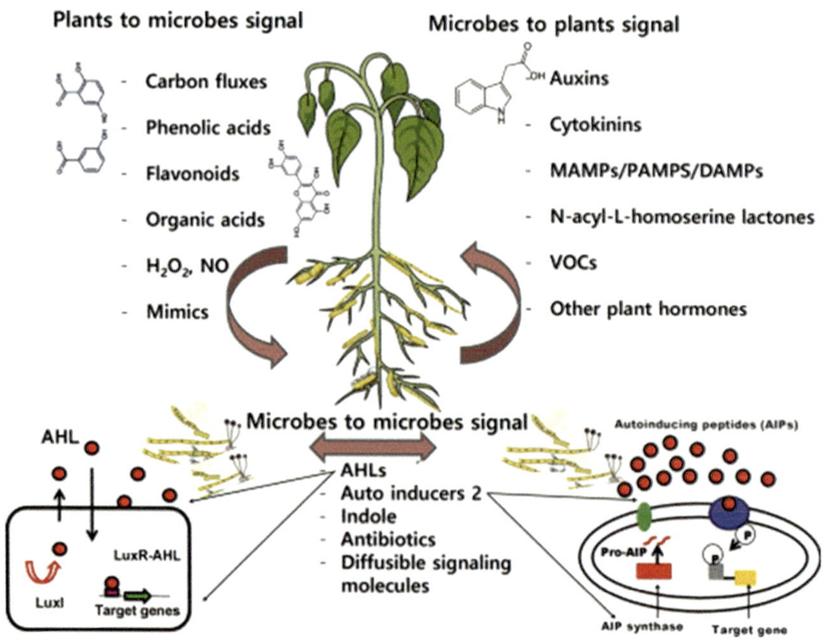

FIGURE 2.1 Signal molecules governing plant-microbe-microbe interactions.

Source: Reproduced with permission from Khare et al. (2020), Copyright © Elsevier (2020).

2.2 INDOLE-3-ACETIC ACID (IAA)

Sergeeva et al. (2007) reported that several varieties of microorganisms, including phototrophic, heterotrophic bacteria, and fungi, possess the functionality to supply IAA. Among auxins, IAA is a common plant hormone that plays a critical part in overall physiological effects such as division, elongation, expansion, and differentiation of cells and tissue of plants (Donati et al., 2013). Donati et al. (2013) demonstrated that these pathogens influence plant physiology and change for their advantage. Numerous research studies have shown the significant proven interaction is that between microorganisms and plants, which can restrain plant development by disturbing the IAA adjustment in plant tissue and cause many diseases (Jameson, 2000; Mole et al., 2007). Further, a few investigations have revealed that auxin is a signaling molecule in microbes since it induces gene expression in a few microbes (Prusty et al., 2004; Yuan et al., 2008). Research by Kim et al. (2011a) revealed that IAA has significant effects on bacterial adaptation to stress conditions, so IAA biosynthetics can present vast compatibility from an evolutionary viewpoint.

Spaepen et al. (2007a) indicated that microorganism biosynthetic consists of complex pathways, with most paths being recommended based on the existence of metabolic intermediates in the growth medium. In many ways, there is no proof the document is accessible, and so the presence and significance of these paths require to be revealed. Duca et al. (2014) reported that for IAA synthesis, tryptophan (Trp) and aromatic amino acid (AAA) are utilized as a precursor in bacteria. According to previous reports, there are two predominant paths based on genetic documents: (1) indole-3-acetamide (IAM) and (2) indole-3-pyruvate (IPyA). In the intermediate IAM pathway, Trp is first converted by a Trp monooxygenase to the intermediate IAM ($C_{10}H_{10}N_2O$), which is then catalyzed to IAA by an intermediate IAM hydrolase. The pathway has been well distinguished in the rhizobia and many microorganisms that induce plant cell growth (Patten et al., 2013). In the intermediate IPyA (pathway plentiful in the useful plant-correlated pathogen, Trp is in first step transaminases to intermediate IPyA (by an aromatic aminotransferase which is then catalyzed to indole-3-acetaldehyde by an, the intermediate IPyA decarboxylase (IPDC) (encoded by the *ipdC gene*). In the second, IPyA is converted to indole-3-acetaldehyde by a chemical reaction (decarboxylation) catalyzed by IPDCs. Then, indole-3-acetaldehyde

($C_{10}H_{10}N_2O$) is changed into IAA. IAAld dehydrogenase gene, *dhaS*, has been recognized in the genome of *Bacillus amyloliquefaciens* (Shao et al., 2015). Other than the two pathways referred to above, different paths of microbe for auxin syntheses have been suggested, but many of these paths prove the document is lacking. In the tryptamine ($C_{10}H_{12}N_2$) paths, Trp by a Trp decarboxylase (TDC1) is first decarboxylated and eventually by an amine oxidase (AOs) change to indole-3-acetaldehyde. Like many plants, a pathway via 3-indoleacetonitrile ($C_{10}H_8N_2$) has been demonstrated based on the shift of $C_{10}H_8N_2$ by nitrilase enzymes to IAA directly or via intermediate IAM, respectively.

However, it is not evident whether genetic examination of a few of these biochemical paths has contributed to our comprehension of the characteristics of pathogen-synthesized auxin during interaction with plants. Therefore, we will concentrate on some well-evidenced high points, the most regulatory parameters, and the mechanisms of IAA synthesis. Plant-associated bacteria that synthesize auxin include *Azospirillum*, *Pseudomonas*, *Agrobacterium*, *Enterobacter*, *Alcaligenes*, *Azotobacter*, *Acetobacter*, *Rhizobium*, *Flavobacterium*, *Rhodococcus*, *Acinetobacter*, *Corynebacterium*, *Aminobacter*, *Bacillus*, *Arthrobacter*, *Xanthomonas*, *Klebsiella*, *Paracoccus*, *Methylovorus*, *Herbaspirillum*, *Achromobacter*, *Bradyrhizobium*, and *Methylobacterium* synthesized IAA (Mishke, 1988; Belimov, 1999; Kamneva, 1999; Ivanova, 2001), *Xanthomonas campestris*, *Pantoea agglomerans*, *Dickeya* spp., *P. savastanoi* (Barash and Manulis-Sasson, 2009), and *Pseudomonas syringae* pathovars (McClerklin et al., 2018). *Pseudomonas*, *Bacillus*, and *Xanthomonas* are broadly recognized in the natural condition, and have been extracted from the root zone of Australian orchids. Previous research demonstrated the ability of plant-associated bacteria to synthesize plant hormones and the impact of such microbes on plant growth under natural conditions of cultivation. At this moment, we brief what is well recognized about the chemical reaction applied by the plant-associated microbes to produce auxin. *AldA*, the gene, has been identified in the genome of *P. syringae* (*Pto*-DC3000). McClerklin et al. (2018) found that mutation of this gene *in Pto*-DC3000 developed in a 70–80% decrease in auxin synthesis in a culture medium when supplemented with indole-3-acetaldehyde. The production of an *AldA* auxin synthesis pathway (mutant in *Pto*DC3000), a model microbial strain, supplies an excellent instrument for exploring the critical role of microbe-isolated auxin (Xin and He, 2013).

Research by Howden et al. (2009) revealed that a gene encodes a nitrilase enzyme in *P. syringae* pv. *syringae* has been purified and recognized (*strainPs*B728a), even though it is not apparent how much of this enzyme activity is associated with converting IAN to auxin by *P. syringae* pv. *syringae* strain. They also predicted that the strain of *Ps*B728a encodes an indole-3-acetaldoxime (IAOx) dehydratase (DH) for producing IAN from IAOx, which the microbes likely acquire from their host plant. Jameson (2000) found that the DNA region is converted to the plant upon contamination, and the Trp is applied to synthesize a high level of auxin, causing the formation of gall in combination with a significant amount of CK. In plant pathology, expression is generally associated with other parameters such as type III secretion system (TTSS). In plant pathogenic bacterium (*Ralstonia solanacearum*), plant hormone production is decreased in the TTSS mutant; on the contrary, in *Erwinia chrysanthemi*, the expression of these genes is reduced in an auxin synthesis pathway mutant (Spaepen et al., 2007b).

Previous research revealed that the fungus *Colletotrichum gloeosporioides* is also capable of using exogenous Trp through the IAM pathway to produce auxin (Maor et al., 2004). Previous research showed that genetically modified *Fusarium* strains (*Orobanche* spp.) overexpressing bacterial Trp-2-monooxygenase (*iaaM*) and indole acetamide hydrolase (*iaaH*) revealed improved auxin synthesize amounts (Kulkarni et al., 2011). *Ustilago maydis* is a fungal pathogen synthesized IAA significantly from Trp. Transgenic *U. maydis* lacking auxin synthesis showed a reduction in host auxin amount upon incubation, while tumor formation was not compromised. Reineke et al. (2008) indicated that, IAA synthesized in fungal mainly associated with auxin levels in the incubation medium, but this is not a single part for induction host tumor formation.

In many mechanisms, the microorganisms themselves synthesize auxin. In this relation, IAA can be considered as a virulence parameter, but still, in other reactions, the microorganisms-induced IAA in the host via the function of virulence gene expression that has developed to regulate host IAA biology. An improvement in IAA levels in incubated host plants created many various activities related to pathogenesis, containing plant epiphytic microorganism colonization, prohibition of host defenses, promotion of host cell division, and microorganism growth in plant tissue (McClerklin et al., 2018). In many plant interactions with pathogens, IAA balances the expression of genes assumed to stimulate interactions with plants. These observations suggest that, IAA induces great level alteration in the transcriptome of *Azospirillum brasilense*, a plant growth-promoting rhizobacteria (PGPR). This contains upregulation of genes involved in auxin production, leading to a significant impact that supports IAA responsiveness (Vande Broek et al., 2005). Van Puyvelde et al. (2011) reported for most genes, changes in expression are involved in the metabolic process, transport, and respiration. These phenomena revealed that auxin stimulated metabolic adaptation for growth in the root area. Additionally, auxin produced the virulence genes in *Dickeya didantii*, a pathogen that is capable of decomposition of cell wall structures. It should be noted that an *iaaM* mutant showed decreased expression of *T3SS*-linked genes, indicating that auxin promotes TTSS. Previous research revealed that auxin stimulated the synthesis of pectate lyase, that responsible for the soft rotting of tissue. Therefore, two groups of virulence expression are significantly adjusted by auxin in *D. didantii* (Yang et al., 2007). In other research, exogenously, IAA amounts improve in host plants incubated with *Puccinia graminis* (Yin et al., 2014), as well as some bacterial *R. solanacearum* (Denance et al., 2013), *P. syringae* (Chen et al., 2007). Auxin in plants after colonization by arbuscular mycorrhizal fungi was evaluated (Meixner et al., 2005). IAA and Indole-3-butyric acid (IBA) stimulate the evolution of lateral roots, which are the fevered contamination area for the arbuscular mycorrhizal fungi (Kaldorf et al., 2000; Yao et al., 2005). Similarly, transgenic tobacco plants expressing the glucuronidase (GUS) reporter gene combined with an IAA-inducible promoter indicated improved IAA amounts in roots incubated by arbuscular mycorrhizal fungi (Jentschel et al., 2007). The reaction of many plants to microbial-produced auxin can differ from advantageous to harmful effects, determined by the level of IAA in the root area. This phenomenon suggests the amounts of auxin delivered by microbes are determined by microbial growth, metabolic activity, species of population, and the expression of genes encoding enzymes in the auxin synthesis paths.

2.3 GIBBERELLINS (GAs)

GAs are a group of plant hormones that promote the development and various growth processes during all stages of plant life, as well as cell elongation and division (from seed germination to formation of fruits and senescence). GAs were first isolated from the *Fusarium fujikuroi* (Sawada, 1917). *F. fujikuroi* played an essential role in researching the construction and biological chemistry pathway of GAs. Bomke and Tudzynski (2009) found that *F. fujikuroi*, which is used to significantly produce GA and GA_3 biosynthesis (GA_{12}-aldehyde as a precursor), has separately developed from the biosynthetic paths of plants and varies mainly during the period in which the 13-hydroxylation and 3β-hydroxylation occurs. Biosynthetic genes in *F. fujikuroi*, are collected along with the genome. In the previous research, GAs synthesizing fungi were mostly plant microbes isolated by the shoot extension sign of the disease associated with GAs. In addition, orchid-incubated *Fusarium proliferatum,* more endophyte species mainly identified from root areas of seemingly healthy tissue, are being found as GA promoters (Tsavkelova et al., 2008). Choi et al. (2005) showed that, *Penicillium commune* and 10 endophytic fungal strains indefinite until now were extracted from the rhizosphere of *Sesamum indicum*. Khan et al. (2008, 2009) demonstrated that microbial GAs' production stimulates the plants *Carex kobomugi* and *Ixeris repens* but mainly has been indicated for fungal pathogens such as *Penicillium citrinum* and *Arthrinium phaeospermum*. The growth capacity of the aforementioned pathogens can be a benefit for many plants under different natural conditions such as those developed in the coastal areas. Although GAs' paths and the related genes in some fungi are recognized, genetic evidence in bacteria species is still unknown. The potential of bacteria to synthesize GAs has been shown for a long time, having been determined for many PGPR, including *Azospirillum lipoferum*, *Rhizobium phaseoli*, *Herbaspirillum seropedicae*, *Acetobacter diazotrophicus*, *Bacillus pumilus*, and *Bacillus licheniformis* (Gutierrez-Maero et al., 2001; MacMillan, 2002; Bottini et al., 2004). Although these microbes are not as it were able to synthesize GAs, they can also promote inactive 3-deoxy GAs available in the rhizosphere by 3β-hydroxylation to produce GAs (GA_1, GA_3, and GA_4) (Cassn et al., 2001a, 2001b). Cassan et al. (2014) reported that diterpene $C_{20}H_{32}$, in *Bradyrhizobium japonicum* supplied some documents for the critical role of operons in GAs synthesis. In *A. lipoferum* GC-MS research, they could recognize the various synthesized GAs, suggesting a supposed path for biosynthesis. Morrone et al. (2009) showed the two supposed diterpene synthases in *B. japonicum* and recognized that they are unrelated to ent-kaurene and ent-copalyl diphosphate synthases. Thus, they supply the first document that this operon might, in fact, be complicated in the GA synthesis pathway. They propose that microbial terpene cyclases and plant enzymes play a role in the gibberellin (GA) synthesis pathway, as well as in the production of other diterpenoids, sharing common precursor sources (Figure 2.2).

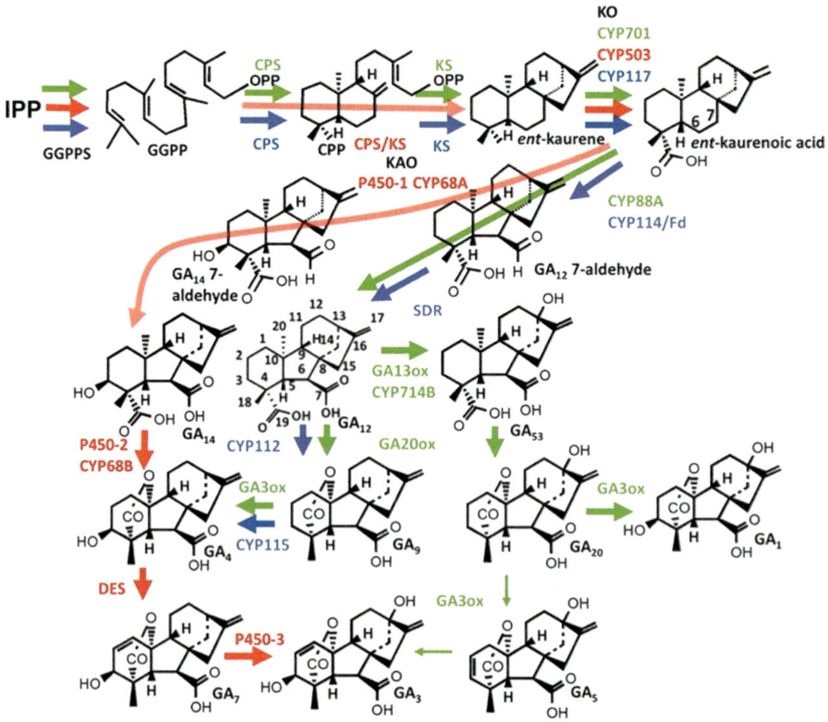

FIGURE 2.2 An overview of GA biosynthesis, comparing pathways for higher plants (in green), the fungus *F. fujikuroi* (red) and bacteria (blue) to biological active products. Enzymes are indicated in the respective colors. C-atom numbers are given for the C_{20}-GA, GA_{12}, GA_{13}ox, GA 13-oxidase; SDR, short-chain dehydrogenase/reductase.

Source: Adapted from Peter (2020), https://creativecommons.org/licenses/by/4.0/

2.4 CYTOKININS (CKs)

CKs are a class of phytohormones that, along with other phytohormones, stimulate cytokinesis and influence many other features of plant growth.

Naturally, CKs are adenine derivatives, a group of most preserved low molecular mass molecules in different microorganisms (fungi, nematodes, algae, bacteria, and amoebae) and humans. However, the constant progress role of CKs as a part of tRNA serves primarily to enhance the translation performance into the more critical function of signaling molecules in plant link microbes, which frequently utilize CKs as a vehicle for manipulating the plant host's evolution (Spichal, 2012). Crozier et al. (2001) showed that 5′-Adenosine monophosphate was a precursor of isopentenyl adenosine 5′-phosphate in *Dictyostelium discoideum*. Dimethylallyl diphosphate: 5′-Adenosine monophosphate transferase promotes this process, and this enzyme was also recognized from tobacco and callus maize kernels medium that converts CK-autonomous. Many bacteria related to the various genera, including *Bacillus*, *Rhizobium*, *Pseudomonas*, and *Azospirillum* can synthesize CKs, which influence the extension of plants and improve tolerance to natural conditions, increase disease resistance (Grobkinsky et al., 2016; Numan et al., 2018). CKs are also synthesized by diverse cyanobacteria (Hussain et al., 2010; Frebortova et al., 2015; Zizkova et al., 2017). Much research is based on the indicated secretion of CKs into the habitat without investigating communicated proteins and genes. According to genetic evidence, the limited presence of adenylate isopentenyl transferase, and the ubiquitous existence of tRNA in microorganisms, it can be supposed that most bacterially synthesized CKs derive from tRNA degradation. Additionally, CK synthesis was also indicated in *Mycobacterium tuberculosis*. In mutants of *M. tuberculosis* with disabled proteasome decomposition, protein (LOG) can collect, leading to incredible amounts of active CKs. Protein amounts are tightly managed by the proteasome, recommending that *M. tuberculosis* could well-tune CKs synthesize. It was assumed that CKs could either be utilized as signaling molecules to interface between mycobacteria or function on the host to promote contamination (Samanovic et al., 2015).

2.5 ETHYLENE

Ethylene (C_2H_4) exists in the natural world at a low level, which promotes the regulation of various biological processes during all stages of the plant, as well as germination, cell expansion, fruit ripening, and senescence.

Ethylene (C_2H_4) can be produced following three various pathways (De Paepe et al., 2005; Ladygina et al., 2006):

i) In plants, methionine is shifted into S-adenosylmethionine (SAM) by SAM synthetase. The further reaction SAM transformed into 1-amino-cyclopane-1-carboxylic acid by 1-aminocyclopane-1-carboxylic acid synthase, followed by the conversion of 1-aminocyclopane-1-carboxylic acid into ethylene (C_2H_4) by 1-aminocyclopane-1-carboxylic acid oxidase. Many microbes synthesize ethylene from two possible pathways.
ii) The critical pathway for fungi and bacteria synthesis, including methionine and 2-oxo-4-methylthiobutyric acid (KMBA1) in which ethylene is produced from KMBA by chemical process (Billington et al., 1979) an α-ketoglutaric acid (KGA) pathway in which KGA is produced from glucose (Fukuda et al., 1993). The diffusion of ethylene (C_2H_4) via the KMBA paths is related to hyphal growth in comparison to the reproduction of *Botrytis cinerea* (Cristescu et al., 2002). Billington et al. (1979) found that these paths can be utilized as an ethylene precursor in *Penicillium digitatum*. However, only low concentrations of this hormone are synthesized via this pathway.
iii) Ethylene synthesized from the 2-oxoglutarate paths, which appears as the principal biosynthetic pathway in *Pseudomonas*. In previous literature reported in *P. syringae* and *P. digitatum* 2-oxoglutarate pathway synthesizes the maximum ethylene amounts (De Paepe et al., 2005; Eckert et al., 2014).

Besides this relevant pathway, four ethylene production pathways have been studied in bacteria (phototrophic), in which this hormone may be produced by the methionine salvage pathway (MSP) (North et al., 2017). Plant hormone (C_2H_4) production by microbes has been reported in some fungal and bacterial species, such as *Botrytis* sp., *Fusarium* sp., *Aspergillus* sp., *Verticillium* sp. (Chague, 2010), *Pseudomonas*, *Erwinia*, and *Xanthomonas* (Tsavkelova et al., 2006). The critical role of this phytohormone in pathogens' host relationship is varying, determined by the exchange. This phytohormone, may be an important part of host resistance to invading microorganisms (Van der Ent and Pieterse, 2012). The diffusion of ethylene through the KMBA pathway is associated with hyphal growth rather than the germination of *B. cinerea* conidia

(Cristescu et al., 2002). Araki et al. (2000) reported that tobacco plants (transgenic), including the gene of *P. syringae* made an improvement in hormone formation for some plants.

2.6 ABSCISIC ACID (ABA)

ABA is a signaling molecule that plays a critical part in some physiological events in plants' cycles. ABA is vital for the control of many processes during the inhibition of seed progression and promotes stomatal closure and fruit ripening. In addition, ABA is essential for reacting to natural stress conditions such as drought, salt, cold, and metal toxicity (Davies, 2010). These phytohormones are produced in plants, and several pathogenic fungi, such as *Cercospora rosicola* (Assante et al., 1977), *B. cinerea* (Marumo et al., 1982), and *Rhizopus nigricans* (Crocoll et al., 1991). The ABA biosynthesis is through the mevalonic acid (MVA) pathway in fungi, whereas primarily by the 2-*C*-methyl-D-erythritol-4-phosphate (MEP) pathway in plants (Hauser et al., 2011). Besides the central ABA produce pathway, precursor acetyl-CoA, genes encoding polycarbohydrates hydrolases, and glucose transporters are related to the ABA biosynthesis in pathogenic fungi (Ding et al., 2016). Phytopathogens may synthesize ABA to inhibit the plant-resistant reaction. It is demonstrated this plant hormone, during inter-species interaction, may function as a signaling molecule. However, whether the host plant ABA would influence the development and modification of fungi is until now unknown. Previous studies reported that ABA might stimulate pathogenic extension and have adverse effects on resistance to plantnpathogens (Kim et al., 2011b; Spence et al., 2015). Bacterial synthesis of ABA has been shown for *B. japonicum* strains, but the production pathways of this hormone are unrevealed. As ABA suppresses CK synthesis, bacterial ABA synthesis can participate with the CK amounts in plants. Under stressful natural conditions, the production of ABA by the bacteria might maintain the internal hormone level in plants and reduce the harmful effects of the burden of stress. This phytohormone in mammals adjusts the resistant modulation and sugar homeostasis, which might have some pharmaceutical applications. Previous research demonstrated that this phytohormone in immune cells might function as a plant-microbe effector, which is controlled by the pathogen; it might act as an interspecies interaction indicator in ecosystems (Bruzzone et al., 2007; Nagamune et al., 2008) (Table 2.2).

TABLE 2.2 Phytohormones Produced by Plants and Microorganisms, and their Effect on Plant Morphology and Development.

Phytohormone		**Plant endogenous production or causative agent**	**Observed effect**
Auxin	Plant	*Zea mays*, *Arabidopsis thaliana*	Cell enlargement, root initiation, vascular differentiation, apical dominance.
	Fungus	*Pisolithus tinctorius*	Plant growth promotion.
	Bacteria	*Azospirillum*, *Rhizobium*, *Bradyrhizobium*	Decrease of root length, increase of root hair development.
		Klebsiella	Increase in root branching and root surface.
		Azospirillum,	Corn seedling inoculated showed an increase on free active IAA and IBA.
		Gluconacetobacter,	
		Herbaspirillum	
		Pseudomonas syringae pv *savastanoi*, *Agrobacterium*,	Induction of gall and tumor formation.
		Erwinia herbicola pv *gypsophilae*	
	Cyanobacteria	*Nostoc*	Symbiotic tissue of *Gunnera.*
GAs	Plant	*Arabidopsis thaliana*, *Oryza sativa*, *Zea mays*, *Pisum sativum*	Seed germination, development and reproduction of plants, floral development.
	Fungus	*Gibberella fujikuroi*	Bakane effect in maize, rice and other plants.
	Bacteria	*A. brasilense*, *A. lipoferum*	Reversion of dwarfism in maize and rice.
		A. brasilense	Promotion of shoot elongation, growth, and root hair density.

TABLE 2.2 *(Continued)*

Phytohormone		**Plant endogenous production or causative agent**	**Observed effect**
CK	Plant	*Arabidopsis thaliana*	Cell division, chloroplast differentiation, photosynthesis, senescence, and nutrient metabolism.
	Bacteria	*Azospirillum*	Plant growth promotion.
		P. syringae pv *savastanoi*, *Agrobacterium tumefaciens*, *Erwinia herbicola*	Induction of gall and tumor formation.

Source: Reproduced with permission and slightly modified from Baca and Elmerich (2007), Copyright © Springer Nature (2011).

2.7 CONCLUSION

With the rapid increase in population, phytohormones are playing a significant role in the agricultural field of the world. However, the high price of the pathogen-derived phytohormones still restricted their wide-scale application. Until now, there is no adequate evidence for their production in natural conditions. Consequently, in the future, much attempt is still needed to make the existing microbial process more effective and economical. With the help of a number of novel investigation tools and more integrative information based upon research on genomics, metabolomics, proteomics, and systems biology, microbial strains would not only attain higher titers of phytohormones but would also obtain higher phytohormones productivities. As both the microorganism and the plant produce IAA, GA, and CKs, it is hard to focus on the share of one specific phytohormone as responsible for the effects noticed. In the future, the increasing quantity of genomic information will further broaden our understanding of the capacity to participate in phytohormone signaling among organisms.

KEYWORDS

- **auxin**
- **biosynthesis**
- **microbial pathways**
- **phytohormones**
- **plant-microbe interactions**

REFERENCES

Araki, S.; Matsuoka, M.; Tanaka, M.; Ogawa, T. Ethylene Formation and Phenotypic Analysis of Transgenic Tobacco Plants Expressing a Bacterial Ethylene-Forming Enzyme. *Plant Cell Physiol.* **2000,** *41*, 327–334.

Assante, G.; Merlini, L.; Nasini, G. (+)-Abscisic Acid, A Metabolite of the Fungus *Cercospora rosicola*. *Experientia* **1977,** *33*, 1556–1557.

Baca, B.; Elmerich, C. Microbial Production of Plant Hormones. In *Associative and Endophytic Nitrogen-Fixing Bacteria and Cyanobacterial Associations. Nitrogen Fixation: Origins, Applications, and Research Progress*; Elmerich, C., Newton, W. E., Eds.; Springer: Dordrecht, 2007. https://doi.org/10.1007/1-4020-3546-2_6.

Barash, I.; Manulis-Sasson, S. Recent Evolution of Bacterial Pathogens: The Gall-Forming *pantoea agglomerans* Case. *Annu. Rev. Phytopathol.* **2009,** *47*, 133–152.

Belimov, A. A.; Ivanchikov, Y. A.; Yudkin, L. V.; Khamova, O. F.; Postavskaya, S. M.; Popolzukhina, P. V.; Shmakova, A. A.; Kozlova, Y. G. New Strains of Associative Growth-Stimulating Bacteria Dominating the Rhizoplane of Barley Seedlings: Characterization and Introduction. *Mikrobiologiya* **1999,** *68*, 392–397.

Bhattacharyya, P. N.; Jha, D. K. Plant Growth-Promoting Rhizobacteria (PGPR): Emergence in Agriculture. *World J. Microbiol. Biotechnol.* **2012,** *28* (4), 1327–50. https://doi.org/10.1007/s11274-011-0979-9.

Billington, D. C.; Golding, B. T.; Primrose, S. B. Biosynthesis of Ethylene from Methionine. Isolation of the Putative Intermediate 4-methylthio-2-Oxobutanoate from Culture Fluids of Bacteria and Fungi. *Biochem. J.* **1979,** *182*, 827–836.

Bomke, C.; Tudzynski, B. Diversity, Regulation, and Evolution of the Gibberellin Biosynthetic Pathway in Fungi Compared to Plants and Bacteria. *Phytochemistry* **2009,** *70*, 1876–1893.

Bottini, R.; Cassan, F.; Piccoli, P. Gibberellin Production by Bacteria and its Involvement in Plant Growth Promotion and Yield Increase. *Appl. Microbiol. Biotechnol.* **2004,** *65*, 497–503.

Bruzzone, S.; Moreschi, I.; Usai, C.; Guida, L.; Damonte, G.; Salis, A.; Scarfi, S.; Millo, E.; De Flora, A.; Zocchi, E. Abscisic Acid is an Endogenous Cytokine in Human Granulocytes with Cyclic ADP-Ribose as Second Messenger. *Proc. Natl. Acad. Sci. USA* **2007,** *104*, 5759–5764.

Cassan, F.; Vanderleyden, J.; Spaepen, S. Physiological and Agronomical Aspects of Phytohormone Production by Model Plant-Growth Promoting Rhizobacteria (PGPR) Belonging to the Genus *Azospirillum*. *J. Plant Growth Regul.* **2014,** *33*, 440–459.

Cassn, F.; Bottini, R.; Schneider, G. *Azospirillum brasilense* and *Azospirillum lipoferum* Hydrolyze Conjugates of GA_{20} and Metabolize the Resultant Aglycones to GA_1 in Seedlings of Rice Dwarf Mutants. *Plant Physiol.* **2001a,** *125*, 2053–2058.

Cassn, F.; Lucangeli, C.; Bottini, R. *Azospirillum* spp. Metabolize [17,17-2H2] Gibberellin A20 to [17,17-2H2] Gibberellin A1 in vivo in dy Rice Mutant Seedlings. *Plant Cell Physiol.* **2001b,** *42*, 763–767.

Chague, V. Ethylene Production by Fungi: Biological Questions and Future Developments towards a Sustainable Polymers Industry. In *Handbook of Hydrocarbon and Lipid Microbiology*; Timmis, K. N., Ed.; Springer: Berlin/Heidelberg, Germany, 2010; pp 3011–3020.

Chen, Z.; Agnew, J. L.; Cohen, J. D.; Kunkel, B. N. *Pseudomonas syringae* Type III Effector AvrRpt2 Alters *Arabidopsis thaliana* Auxin Physiology. *Proc. Natl. Acad. Sci. USA* **2007,** *104*, 20131–20136.

Choi, W. Y.; Rim, S. O.; Lee, J. H.; Lee, J. M.; Lee, I. J.; Cho, K. J.; Rhee, I. K.; Kwon, J. B.; Kim, J. G. Isolation of Gibberellins Producing Fungi from the Root of Several *Sesamum indicum* Plants. *J. Microbiol. Biotechnol.* **2005,** *15*, 22–28.

Cristescu, S. M.; De Martinis, D.; Te Lintel Hekkert, S.; Parker, D. H.; Harren, F. J. M. Ethylene Production by *Botrytis cinerea* in vitro and in Tomatoes. *Appl. Environ. Microbiol.* **2002,** *68*, 5342–5350.

Crocoll, C.; Kettner, J.; Dorffling, K. Abscisic Acid in Saprophytic and Parasitic Species of Fungi. *Phytochemistry* **1991,** *30*, 1059−1060.

Crozier, A.; Kamiya, Y.; Bishop, G.; Yokota, T. Biosynthesis of Hormones and Elicitors Molecules. In *Biochemistry and Molecular Biology of Plants*; Buchanan, B. B., Grussem, W., Jones, R. L., Eds.; American Society of Plant Biologists: Rockville, MD, 2001; pp 850–900.

Davies, P. J. In *Plant Hormones: Biosynthesis, Signal Transduction, Action*. Springer: Dordrecht, 2010.

De Paepe, A.; Van Der Straeten, D. Ethylene Biosynthesis and Signaling: An Overview. In *Vitamins and Hormones*; Litwack, G., Ed.; Academic Press: New York, USA, 2005; vol *72*, pp 399–430.

Denance, N.; Ranocha, P.; Oria, N.; Barlet, X.; Riviere, M. P.; Yadeta, K. A.; Hoffmann, L.; Perreau, F.; Clement, G.; Maia-Grondard, A.; Van den Berg, G. C. M.; Savelli, B.; Fournier, S.; Aubert, Y.; Pelletier, S.; Thomma, B. P. H. J.; Molina, A.; Jouanin, L.; Marco, Y.; Goffner, D. *Arabidopsis* wat1 (walls are thin1)-Mediated Resistance to the Bacterial Vascular Pathogen, *Ralstonia solanacearum*, is Accompanied by Cross-Regulation of Salicylic Acid and Tryptophan Metabolism. *Plant J.* **2013,** *73*, 225–239.

Ding, Z. H.; Zhang, Z. H.; Zhong, J.; Luo, D.; Zhou, J.; Yang, J.; Xiao, L.; Shu, D.; Tan, H. Comparative Transcriptome Analysis between an Evolved Abscisic Acid-Overproducing Mutant *Botrytis cinerea* TBC-A and its Ancestral Strain *Botrytis cinerea* TBC-6. *Sci Rep.* **2016,** *6*, 37487.

Donati, A. J.; Lee, H. I.; Leveau, J. H.; Chang, W. S. Effects of Indole-3-Acetic Acid on the Transcriptional Activities and Stress Tolerance of *Bradyrhizobium japonicum*. *PLoS One* **2013,** *8*, 76559.

Duca, D.; Lorv, J.; Patten, C. L.; Rose, D. Indole-3-Acetic Acid in Plant-Microbe Interactions. *Antonie Van Leeuwenhoek* **2014,** *106*, 85–125.

Eckert, C.; Xu, W.; Xiong, W. Ethylene-Forming Enzyme and Bioethylene Production. *Biotechnol. Biofuels* **2014,** *7*, 33.

Frebortova, J.; Greplova, M.; Seidl, M. F.; Heyl, A.; Frebort, I. Biochemical Characterization of Putative Adenylate Dimethylallyltransferase and Cytokinin Dehydrogenase from *Nostoc* sp. PCC 7120. *PLoS One* **2015,** *10*, e0138468.

Fukuda, H.; Ogawa, T.; Tanase, S. Ethylene Production by Micro-Organisms. In *Advances in Microbial Physiology*; Rose, A., Ed.; Academic Press Inc.: London, UK, 1993; pp 275–306.

Grobkinsky, D. K.; Tafner, R.; Moreno, M. V.; Stenglein, S. A.; Garcia de Salamone, I. E.; Nelson, L. M.; Novak, O.; Strnad, M.; Graaff, E. V. D.; Roitsch, T. Cytokinin Production by *Pseudomonas fluorescens* G20-18 Determines Biocontrol Activity against *Pseudomonas syringae* in *Arabidopsis*. *Sci. Rep.* **2016,** *6*, 23310.

Gutierrez-Maero, F.; Ramos-Solano, B.; Probanza, A.; Mehouachi, J.; Tadeo, F. R.; Talon, M. The Plant-Growth Promoting Rhizobacteria *Bacillus pumilus* and *Bacillus licheniformis* Produce High Amounts of Physiologically Active Gibberellins. *Physiol. Plant.* **2001,** *111*, 206–211.

Hauser, F.; Waadt, R.; Schroeder, J. I. Evolution of Abscisic Acid Synthesis and Signaling Mechanisms. *Curr Biol.* **2011,** *21*, 346–355.

Howden, A. J.; Rico, A.; Mentlak, T.; Miguet, L.; Preston, G. M. *Pseudomonas syringae* pv. syringae B728a Hydrolyses Indole-3- Acetonitrile to the Plant Hormone Indole-3-Acetic Acid. *Mol. Plant Pathol.* **2009,** *10*, 857–865.

Hussain, A.; Krischke, M.; Roitsch, T.; Hasnain, S. Rapid Determination of Cytokinins and Auxin in Cyanobacteria. *Curr. Microbiol.* **2010,** *61*, 361–369.

Ivanova, E. G.; Doronina, N. V.; Trotsenko, Y. A. Aerobic Methylobacteria are Capable of Synthesizing Auxins. *Mikrobiologiya* **2001,** *70*, 452–458.

Jameson, P. Cytokinins and Auxins in Plant-Pathogens Interactions - An Overview. *Plant Growth Regul.* **2000,** *32*, 369–380.

Jentschel, K.; Thiel, D.; Rehn, F. Arbuscular Mycorrhiza Enhances Auxin Levels and Alters Auxin Biosynthesis in *Tropaeolum majus* During Early Stages of Colonization. *Physiol. Plant.* **2007,** *129*, 320–333.

Kaldorf, M.; Ludwig-Muller, J. AM Fungi Might Affect the Root Morphology of Maize by Increasing Indole-3-Butyric Acid Biosynthesis. *Physiol. Plant.* **2000,** *109*, 58–67.

Kamneva, S. V.; Muronets, E. M. Genetic Control of the Processes of Interaction of Bacteria with Plants in Associations. *Genetika* **1999,** *35*, 1480–1494.

Khan, S. A.; Hamayun, M.; Kim, H. Y. A New Strain of *Arthrinium phaeospermum* Isolated from *Carex kobomugi* Ohwi is Capable of Gibberellin Production. *Biotechnol. Lett.* **2009,** *31*, 283–287.

Khan, S. A.; Hamayun, M.; Yoon, H. Plant Growth Promotion and *Penicillium citrinum*. *BMC Microbiol.* **2008,** *8*, 231.

Khare, E.; Tyagi, S.; Patil, K. S. Language of Plant-Microbe-Microbe Interactions in Rhizospheric Ecosystems. In *Molecular Aspects of Plant Beneficial Microbes in Agriculture*; Vivek, S., Richa, S., Laith, K., Tawfeeq, A. A., Eds; Academic Press, 2020; pp 59–76. https://doi.org/10.1016/B978-0-12-818469-1.00005-5.

Kim, T. H.; Hauser, F.; Ha, T.; Xue, S. H.; Bohmer, M.; Nishimura, N.; Munemasa, S. H.; Hubbard, K.; Peine, N.; Lee, B. H.; Lee, S.; Robert, N.; Parker, J. E.; Schroeder, J. I. Chemical Genetics Reveals Negative Regulation of Abscisic Acid Signaling by a Plant Immune Response Pathway. *Curr. Biol.* **2011b,** *21*, 990–997.

Kim, Y. C.; Leveau, J.; McSpadden Gardener, B. B.; Pierson, E. A.; Pierson, L. S.; Ryu, C. M. The Multifactorial Basis for Plant Health Promotion by Plant Associated Bacteria. *Appl. Environ. Microbiol.* **2011a,** *77*, 1548–1555.

Kulkarni, G. B.; Sajjan, S. S.; Karegoudar, T. Pathogenicity of Indole-3-Acetic Acid Producing Fungus *Fusarium delphinoides* Strain GPK Towards Chickpea and Pigeon Pea. *Eur. J. Plant Pathol.* **2011,** *131*, 355–369.

Ladygina, N.; Dedyukhina, E. G.; Vainshtein, M. B. A Review on Microbial Synthesis of Hydrocarbons. *Process Biochem.* **2006,** *41*, 1001–1014.

MacMillan, J. Occurrence of Gibberellins in Vascular Plants, Fungi, and Bacteria. *J. Plant Growth Regul.* **2002,** *20*, 387–442.

Maor, R.; Haskin, S.; Levi-Kedmi, H. In Planta Production of Indole-3-Acetic Acid by *Colletotrichum gloeosporioides* f. sp. *aeschynomene*. *Appl. Environ. Microbiol.* **2004,** *70*, 1852–1854.

Marumo, S.; Katayama, M.; Komori, E. Microbial Production of Abscisic Acid by *Botrytis cinerea*. *Agric. Biol. Chem.* **1982,** *46*, 1967–1968.

McClerklin, S.; Lee, S. G.; Harper, C. P.; Nwumeh, R.; Jez, J. M.; Kunkel, B. N. *Pseudomonas syringae* DC3000-Derived Auxin Contributes to Virulence on *Arabidopsis*. *PLoS Path* **2018,** *14*, e1006811.

Meixner, C.; Ludwig-Muller, J.; Miersch, O. Lack of mycorrhizal Autoregulation and Phytohormonal Changes in the Supernodulating Soybean Mutant nts1007. *Planta* **2005,** *222*, 709–715.

Mishke, I. V. *Mikrobnye fitogormony* v rastenievodstve (Microbial Phytohormones in Agriculture), Zinatne: Riga, 1988.

Mole, B. M.; Baltrus, D. A.; Dangl, J. L.; Grant, S. R. Global Virulence Regulation Networks in Phytopathogenic Bacteria. *Trends Microbiol.* **2007,** *15*, 363–371.

Morrone, D.; Chambers, J.; Lowry, L.; Kim, J.; Anterola, A.; Bender, K.; Peters, R. J. Gibberellin Biosynthesis in Bacteria: Aeparate Ent-copalyl Diphosphate and Ent-kaurene Synthases in *Bradyrhizobium japonicum*. *FEBS Lett.* **2009,** *583*, 475–480.

Nagamune, K.; Hicks, L. M.; Fux, B.; Brossier, F.; Chini, E. N.; Sibley, L. D. Abscisic Acid Controls Calcium-Dependent Egress and Development in *Toxoplasma gondii*. *Nature* **2008,** *451*, 207–210.

North, J. A.; Miller, A. R.; Wildenthal, J. A.; Robert Tabita, F. Microbial Pathway for Anaerobic 50-methylthioadenosine Metabolism Coupled to Ethylene Formation. *Proc. Natl. Acad. Sci. USA* **2017,** *114*, 10455–10464.

Numan, M.; Bashira, S.; Khan, Y.; Mumtaz, R.; Khan Shinwari, Z.; LatifKhan, A.; Khan, A.; AL-Harrasi, A. Plant Growth Promoting Bacteria as an Alternative Strategy for Salt Tolerance in Plants: A Review. *Microbiol. Res.* **2018,** *209*, 21–32.

Patten, C. J.; Blakney, A. J. C.; Coulson, T. J. D. Activity, Distribution and Function of Indole–3-Acetic Acid Biosynthetic Pathways in Bacteria. *Crit. Rev. Microbiol.* **2013,** *39*, 395–415.

Peter, H. The Current Status of Research on Gibberellin Biosynthesis. *Plant Cell Physiol.* **2020,** *61* (11), 1832–1849. https://doi.org/10.1093/pcp/pcaa092.

Prusty, R.; Grisafi, P.; Fink, G. R. The Plant Hormone Indole Acetic Acid Induces Invasive Growth in *Saccharomyces cerevisiae*. *Proc. Natl. Acad. Sci. USA* **2004,** *101*, 4153–4157.

Reineke, G.; Heinze, B.; Schirawski, J.; Buettner, H.; Kahmann, R.; W Basse, C. Indole 3 Acetic Acid (IAA) Bosynthesis in the Smut Fungus *Ustilago maydis* and its Relevance for Increased IAA Levels in Infected Tissue and Host Tumour Formation. *Mol. Plant Pathol.* 2008, *9*, 339–355.

Samanovic, M. I.; Tu, S.; Novak, O.; Iyer, L. M.; McAllister, F. E.; Aravind, L.; Gygi, S. P.; Hubbard, S. R.; Strnad, M.; Darwin, K. H. Proteasomal Control of Cytokinin Synthesis Protects *Mycobacterium tuberculosis* Against Nitric Oxide. *Mol. Cell.* **2015,** *19*, 984–994.

Sawada, K. Contributions on Formosan Fungi. *Trans. Nat. Hist. Soc. Formosa.* **1917,** *7*, 131–133.

Sergeeva, E.; Hirkala, D. M.; Nelson, L. Production of Indole-3-Acetic Acid, Aromatic Amino Acid Aminotransferase Activities and Plant Growth Promotion by *Pantoea agglomerans* Rhizosphere Isolates. *Plant Soil* **2007,** *297*, 1–13.

Shao, J.; Li, S.; Zhang, N.; Cui, X.; Zhou, X.; Zhang, G.; Shen, Q.; Zhang, R. Analysis and Cloning of the Synthetic Pathway of the Phytohormone Indole-3-Acetic Acid in the Plant-Beneficial *Bacillus amyloliquefaciens* SQR9. *Microb. Cell Factories.* **2015,** *14*, 130.

Spaepen, S.; Vanderleyden, J.; Remans, R. Indole-3-Acetic Acid in Microbial and Microorganism-Plant Signaling. *FEMS Microbiol. Rev.* **2007b,** *31*, 425–448.

Spaepen, S.; Versees, W.; Gocke, D.; Pohl, M.; Steyaert, J.; Vanderleyden, J. Characterization of Phenylpyruvate Decarboxylase, Involved in Auxin Production of *Azospirillum brasilense*. *J. Bacteriol.* **2007a,** *189*, 7626–7633.

Spence, C. A.; Lakshmanan, V.; Donofrio, N.; Bais, H. P. Crucial Roles of Abscisic Acid Biogenesis in Virulence of Rice Blast Fungus *Magnaporthe oryzae*. *Front. Plant Sci.* **2015,** *6*, 1082.

Spichal, L. Cytokinins–Recent News and Views of Evolutionally Old Molecules. *Funct. Plant Biol.* **2012,** *39*, 267–284.

Tsavkelova, E. A.; Bomke, C.; Netrusov, A. I.; Weiner, J.; Tudzynski, B. Production of Gibberellic Acids by an Orchid-Associated *Fusarium proliferatum* Strain. *Fungal Genet. Biol.* **2008,** *45*, 1393–1403.

Tsavkelova, E. A.; Klimova, S. Y.; Cherdyntseva, T. A.; Netrusov, A. I. Hormones and Hormone-Like Substances of Microorganisms: A Review. *Appl. Biochem. Microbiol.* **2006,** *42*, 229–235.

Van der Ent, S.; Pieterse, C. M. J. Ethylene: Multi-tasker in Plant–Attacker Interactions. In *Annual Plant Reviews*; American Cancer Society: Atlanta, GA, USA, 2012; pp 343–377.

Van Puyvelde, S.; Cloots, L.; Engelen, K.; Das, F.; Marchal, K.; Vanderleyden, J.; Spaepen, S. Transcriptome Analysis of the Rhizosphere Bacterium *Azospirillum brasilense* Reveals an Extensive Auxin Response. *Microbial. Ecol.* **2011,** *61*, 723–728.

Vande Broek, A.; Gysegom, P.; Ona, O.; Hendrickx, N.; Prinsen, E.; Van Impe, J.; Vanderleyden, J. Transcriptional Analysis of the *Azospirillum brasilense* Indole-3-pyruvate Decarboxylase Gene and Identification of a Cis-acting Sequence Involved in Auxin Responsive Expression. *Mol. Plant Microb. Interact.* **2005,** *18*, 311–323.

Xin, X. F.; He, S. Y. *Pseudomonas syringae* pv. Tomato DC3000: A Model Pathogen for Probing Disease Susceptibility and Hormone Signaling in Plants. *Annu. Rev. Phytopathol.* **2013,** *51*, 473–498.

Yang, S.; Zhang, Q.; Guo, J.; Charkowski, A.O.; Glick, B. R.; Mark Ibekwe, A.; Cooksey, D. A.; Yang, C. H. Global Effect of Indole-3-Acetic Acid Biosynthesis on Multiple Virulence Factors of *Erwinia chrysanthemi* 3937. *Appl. Environ. Microbiol.* **2007,** *73*, 1079–1088.

Yao, Q.; Zhu, H.; Chen, J. Growth Responses and Endogenous IAA and iPAs Changes of Litchi (*Litchi chinensis* Sonn.) Seedlings Induced by Arbuscular mycorrhizal Fungal Inoculation. *Sci. Hort.* **2005,** *105*, 145–151.

Yin, C.; Park, J. J.; Gang, D. R.; Hulbert, S. H. Characterization of a Tryptophan 2 Monooxygenase Gene from *Puccinia graminis* f. sp. *tritici* Involved in Auxin Biosynthesis and Rust Pathogenicity. *Mol. Plant Microb. Interact.* **2014,** *27*, 227–235.

Yuan, Z. C.; Liu, P.; Saenkham, P.; Kerr, K.; W Nester, E. Transcriptome Profiling and Functional Analysis of *Agrobacterium tumefaciens* Reveals a General Conserved Response to Acidic Conditions (pH 5.5) and a Complex Acid-Mediated Signaling Involved in Agrobacterium – Plant Interactions. *J. Bacteriol.* **2008,** *190*, 494–507.

Zizkova, E.; Kubes, M.; Dobrev, P. I. Control of Cytokinin and Auxin Homeostasis in Cyanobacteria and Algae. *Ann. Bot.* **2017,** *119*, 151–166.

CHAPTER 3

MICROALGAL BIORESOURCES FOR SUSTAINABLE AGRICULTURE

SANA MOUNAIMI[1*], ABDEL RAHMAN MOHAMMAD AL-TAWAHA[2], PAVITRA CHIPPALAKATTI[3], JEYABALAN SANGEETHA[4], and DEVARAJAN THANGADURAI[3]

[1]Mohammed VI Polytechnic University, Faculty of Agriculture, Morocco

[2]Department of Biological Sciences, Al Hussein bin Talal University, Maan, Jordan

[3]Department of Botany, Karnatak University, Dharwad, Karnataka, India

[4]Department of Environmental Science, Central University of Kerala, Kasaragod, Kerala, India

**Corresponding author.*

ABSTRACT

Microalgae have recently drawn tremendous worldwide interest due to their vast potential application, notably in the agriculture industry. Microalgae are organic, sustainable, and economical sources of biostimulants, biofertilizers, and biocontrol products. So many microalgae species have been explored for their ability as value-added items of outstanding biological performance. Microalgae biomass produces massive quantities of bioactive compounds, such as phytohormones, polysaccharides, polyamines, and betaines. In agriculture, microalgae have been used in various applications, including amendment, foliar, and/or root application and seed priming. As plant biostimulants, biofertilizers, and biocontrol, microalgae

Microbial Fertilizer Technology for Sustainable Crop Production. Jeyabalan Sangeetha, Abdel Rahman Mohammad Al-Tawaha, & Devarajan Thangadurai (Eds.)

DOI: 10.1201/9781003594826-3

could be used as an appropriate eco-friendly alternative to reduce or manage chemical fertilizers, phytosanitary products (pesticides), and plant growth regulators, generating multiple benefits in terms of enhancing rooting, higher crop yields, and quality and tolerance to biotic and abiotic stress as well. Biostimulants have been used as bioremediants to increase levels of adenosine triphosphate (ATP) and phosphatase and urease activity, increasing the rate of soil xenobiotics degradation and improving beneficial microbial soil communities. Biostimulants can also lead to reducing the amount of hazardous agrochemicals, particularly decreasing fertilizer and pesticide consumption for sustainable agriculture. Furthermore, the demand for products based on microalgae is growing at a considerable rate. The most challenging and crucial issues are the advancement of new technologies to facilitate the economic feasibility of algae and biomass products and appropriate market and economic data to support industrial and institutional strategy. This review outlines the benefits, the economic prospects, and the biological activities of microalgae use for plant growth, development, nutrient uptake, and protection against biotic and abiotic stress, as well as to achieve sustainable agriculture.

3.1 INTRODUCTION

Since individual and collective conscious awareness of environmental issues, multiple strategies are being considered that should ensure greater management of our environment. One of the major ecological challenges facing our societies today is the global warming derailment, partly due to the increased release of greenhouse gases into the atmosphere. Moreover, the increase in temperature on Earth's surface is predicted to cause global climate change. That being said, such changes may lead to perturbations in the quantities and distribution of precipitation, the development of food resources, and intensified melting of glaciers, thereby raising sea levels or accelerating the extinction of some species.

Carbon dioxide (CO_2) is the main greenhouse gas. It is therefore necessary to provide solutions to combat its increase in the atmosphere. Biological capture can be mentioned as one of the approaches proposed to reduce the concentration of CO_2 in the atmosphere. In general, CO_2 sequestration methods must provide long-term sequestration of CO_2 to be effective. One of the solutions suggested will be the use of photosynthetic microorganisms to capture CO_2 at industrial sites. For this purpose, the use of microalgae for the capture and storage of CO_2 is currently an emerging area of research (Figure 3.1).

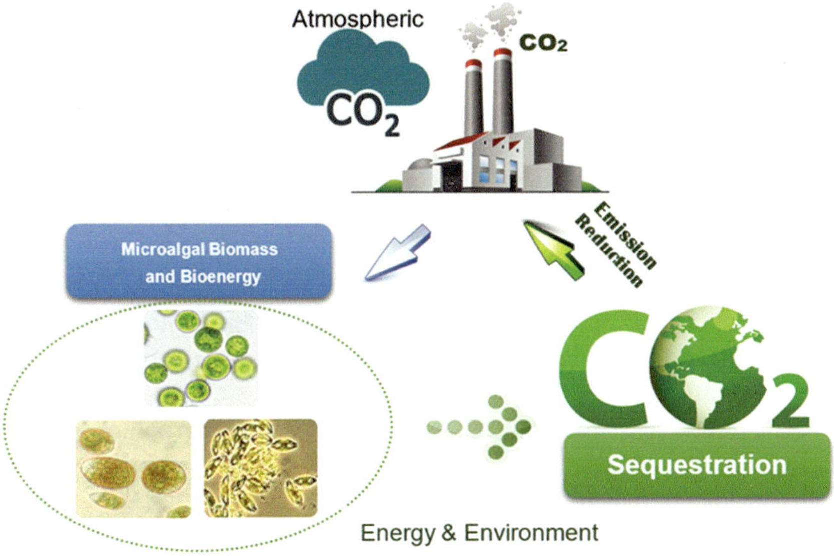

FIGURE 3.1 Energy and environmental relationship based on microalgal-CO_2 sequestration.
Source: Reproduced with permission from Bingtao and Yaxin (2020), Copyright © Elsevier (2020).

Furthermore, microalgae have inhabited either marine, freshwater, or even terrestrial habitats, making them suitable for a wide range of environmental conditions. It is, therefore, present in all climates, from the hottest to the coldest. This broad diversity reveals many possibilities in their use. There are many applications for microalgae nowadays: animal and human food, pigment processing, and molecules of interest to the food industry or even to agriculture (Leung et al., 2014).

Currently, given the increasing worldwide population, a crucial challenge among agricultural issues is meeting food demands and enhancing agricultural sustainability. Therefore, farmers are called upon to increase agricultural production alongside climate change concerns, and researchers must introduce creative products and technologies capable of increasing crop yields and quality.

Microalgae can produce biomass, which can be used in various sectors, such as crop production. Microalgae contain high levels of macronutrients and micronutrients, which are vital for the optimal growth and development of plants. In addition, microalgae have been shown to have potential applications as biostimulants and biofertilizers. Moreover, it attracts the interest of farmers to improve the sustainable production of crops.

Biostimulants and biofertilizers are considered eco-friendly and cost-effective alternatives to reduce and manage the use of synthetic products, such as fertilizers, crop protection products, and regulators for plant growth. First, biostimulants act on the physiology of plants through various pathways, enhancing crop growth, yields, quality, nutrient uptake, abiotic stress tolerance, and the longevity of harvested products. Second, biofertilizers are products that contain living microorganisms or natural substances that can enhance the chemical and biological properties of the soil, stimulate plant growth, and restore fertility in the soil (Ronga et al., 2019).

This report aims to study the feasibility of using microalgae biotechnology to ensure sustainable agriculture and to understand the different mechanisms by which microalgae affect plants, both in terms of growth and development and in terms of protection against all kinds of stress.

Microalgae are unicellular or multicellular photosynthetic eukaryotic microscopic algae, appearing over three billion years ago and inhabiting highly diverse habitats (seas, oceans, fresh, and brackish waters), ranging from microns to so many hundred microns and have a much greater diversity than terrestrial plants (Wolff, 2012). With sunlight, CO_2 is converted into organic complexes, compounds that are ultimately stored within or removed from the cell; a variety of molecules, such as carbohydrates, proteins, and essential amino acids, are produced (Abla, 2015).

3.2 MICROALGAE STRUCTURE

It is in the chloroplasts, and more precisely in the thylakoid membranes, that the oxidation-reduction reaction takes place, which reduces carbon dioxide from the hydrogen in the water to simple sugars and oxygen.

The plasma membrane is a structure composed of polysaccharides and proteins that are more or less complex and in variable proportions depending on the species. It is negatively charged, which is due to the composition of the functional groups associated with it. It confers to the cells a specific resistance to potentially toxic metal ions (El Hifnawy, 2016) (Figure 3.2).

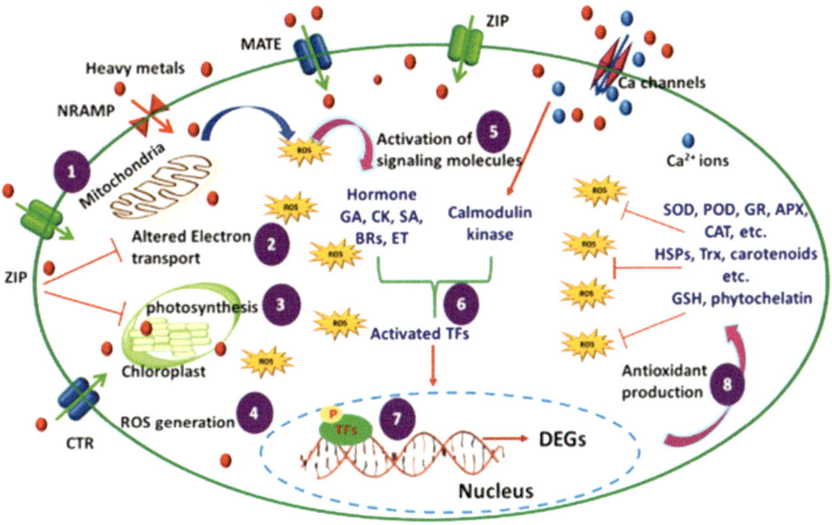

FIGURE 3.2 Sequential stages of cellular events to detoxify heavy metal induced ROS by microalgae: (1) Metal transporters facilitate entry of heavy metal ions inside the cells; disruption of the electron transport chain by heavy metals in mitochondria (2), and photosynthesis in chloroplast (3) via displacing the crucial metal ions, which eventually lead to generation of excessive ROS (4); activation of harmone and calmodulin kinase pathway by generated ROS molecules, along with calcium ions that act as secondary messengers (5); Eventual activation of certain transcription factors (6), and enhancement of antioxidant gene(s) expression (7, 8) to combat excessive ROS, Abbreviations: Zrt-, Irt-like proteins (ZIP); multidrug and toxic compound extrusion (MATE); natural resistance-associated macrophage proteins (NRAMP); Cu transporter (CTR); metal tolerance protein (MTP); arsenite efflux transporter (ACR3); ATP-binding cassette transporters (ABC transporters); CE (cation exchange); differentially expressed genes (DEGs).

Source: Reproduced with permission from Shweta and Krishna (2021). Copyright © Elsevier (2021).

3.3 DIVERSITY AND CLASSIFICATION

Algae are grouped with fungi in the division of Thallophytes (Mahfoud, 2014). The species are classified according to an International Code of Nomenclature Botany. Because of the controversies due to the dynamic nature of research and discussions of the phylogeny, the classification schemes should take into consideration the traditional approaches, such as pigmentation, chemical nature of the photosynthetic products, thylakoid organization, and other chloroplast ultrastructures (Rmiki, 1999).

3.4 MICROALGAE CELL PHYSIOLOGY

3.4.1 PHOTOSYNTHESIS

Microalgae have a similar photosynthetic cycle to the one of higher plants. This photosynthesis occurs on the thylakoid membrane level. Thylakoids are saccule-shaped organelles present in cyanobacteria and the chloroplasts of eukaryotic cells. The internal space, delimited by the thylakoid membranes, is called the lumen. Outer space is the cytoplasm in cyanobacteria or the stroma of chloroplasts in photosynthetic eukaryotic.

3.4.2 REPRODUCTION

The proliferation of microalgae occurs mainly by asexual reproduction or vegetative multiplication: a mother cell then divides into two genetically identical daughter cells. However, like most eukaryotes, microalgae can also multiply by sexual reproduction. This phase of the life cycle is generally triggered by particular environmental conditions that are often multifactorial (Chevanton, 2013).

3.5 BASIC NEEDS AND NUTRITION PATTERNS

3.5.1 NUTRIENTS

Different microalgae species can differ in their nutritional needs; however, the basic requirements for all species are the same. Nitrogen, phosphorus, and carbon serve as the backbone of microalgae and are defined as macronutrients for algal production. Some marine microalgae species also require silicone as a macronutrient. O_2 and H_2 are absorbed by Microalgae from water (Khan et al., 2018). Besides that, micronutrients Mo, K, Co, Fe, Mg, Mn, B, and Zn are only needed in trace amounts but have a significant effect on microalgae growth as they influence several enzymatic activities in algal cells (Khan et al., 2018) (Table 3.1).

3.5.2 NUTRITION MODES

Microalgae have several modes of nutrition; they are generally photo-autotrophic organisms. Autotrophy is a mode of nutrition of microalgae allowing them to use the sunlight to synthesize their energy. They also use an inorganic carbon source, such as CO_2 and hydrogen carbonate HCO_3^- for the synthesis of organic carbon. This organic carbon is essential for the synthesis of all the organic components necessary for their survival. On the contrary, several microalgae have a heterotrophic metabolism of nutrition and do not need solar energy. Instead, they use an organic carbon source for energy production and organic components (El Hifnawy, 2016). Moreover, mixotrophic metabolism microalgae can feed either autotrophically or heterotrophically. Indeed, in the absence of light energy, when a source of organic carbon is available, the development of chloroplasts is inhibited, and these microalgae metabolize their energy in heterotrophic mode.

3.5.3 GROWING CONDITIONS

If all nutrients are present in the environment, the ecophysiological factors must be favorable to allow the growth of microalgae. These are mainly light, temperature, pH, and salinity factors. Each species selected and adapted to its environment has developed during the evolution of the preferences concerning these parameters. Considering an isolated factor, each species develops optimally for a given parameter value (Chevanton, 2013).

3.6 SELECTION

3.6.1 SELECTION FACTORS

There are many strains available, and their choice depends on several criteria. This section lists and discusses these criteria for different applications. The selection and cultivation of a microalgae strain that provides

TABLE 3.1 Nutrients Essential for Autotrophic Microalgae and the Elementary Composition of Algal Cells.

Element	Component added to culture medium	Concentration in culture medium	Cell composition (mg g^{-1} dry weight)
C	CO_2, HCO_3^-, CO_3^{2-}	g L^{-1}	175–650
O	O_2, H_2O	g L^{-1}	205–330
H	H_2O	g L^{-1}	29–100
N	NH_4^+, NO_3^-, NO_2^-, urea	g L^{-1}	10–140
Na	Inorganic salts, i.e., NaCl, Na_2SO_4, Na_3PO_4	g L^{-1}	0.4–47
K	Inorganic salts, i.e., KCl, K_2SO_4, K_3PO_4	g L^{-1}	1–75
Ca	Inorganic salts, i.e., $CaCl_2$, $CaCO_3$	g L^{-1}	0.0–80
P	Inorganic salts, i.e., Na or K phosphates	g L^{-1}	0.5–33
S	Inorganic salts, i.e., $MgSO_4{\cdot}7H_2O$, or amino acids	g L^{-1}	1.5–16
Mg	Inorganic salts, i.e., Mg sulfates or chlorides	g L^{-1}	0.5–75
Cl	As Na^+, K^+, Ca^{2+}, or NH_4^+ salts	g L^{-1}	[a]
Fe	In complex with metal ion buffer (e.g., EDTA)	mg L^{-1}	0.2–34
Zn	Inorganic salts, i.e., $ZnSO_4$, $ZnCl_2$	mg L^{-1}	0.005–1.0
Mn	Inorganic salts, i.e., $MnSO_4$, $MnCl_2$	mg L^{-1}	0.02–0.24
Br	As Na^+, K^+, Ca^{2+} or NH^{4+} salts	mg L^{-1}	[a]
Si	$Na_3SiO_3{\cdot}9H_2O$	mg L^{-1}	0–230
B	H_3BO_3	mg L^{-1}	0.001–0.25
Mo	Na^+ or NH_4^+ molybdate salts	µg L^{-1}	0.0002–0.001
V	$Na_3VO_4{\cdot}16H_2O$	µg L^{-1}	[a]
Sr	As sulfates or chlorides	µg L^{-1}	[a]

TABLE 3.1 *(Continued)*

Element	Component added to culture medium	Concentration in culture medium	Cell composition (mg g^{-1} dry weight)
Al	As sulfates or chlorides	μg L^{-1}	[a]
Rb	As sulfates or chlorides	μg L^{-1}	[a]
Li	As sulfates or chlorides	μg L^{-1}	[a]
Cu	As sulfates or chlorides	μg L^{-1}	0.006–0.3
Co	Vitamin B_{12}, sulfates or chlorides	μg L^{-1}	0.0001–0.2
I	As Na^+, K^+, Ca^{2+} or NH_4^+ salts	μg L^{-1}	[a]
Se	SeO_3^{2-}, SeO_4^{2-}	ng L^{-1}	0–0.9

[a] *Data not available.*

Source: Reproduced with permission from Procházková et al. (2014), Copyright © Springer Nature 2013.

both a satisfactory growth rate and a market-relevant content are essential criteria (Belair et al., 2012). The selected strain should have, as far as possible, the following characteristics:

1. High biomass productivity.
2. High productivity of the value-added product.
3. High CO_2 fixation capacity in autotrophy.
4. High light utilization capacity in autotrophy.
5. Ability to use a low-cost and widely available glucose source in heterotrophy.
6. Ability to dominate native species in an open system.
7. Tolerance to a wide range of temperatures, according to diurnal and seasonal variations (Belair et al., 2012).

3.6.2 PHOTOSYNTHESIS EFFICIENCY

The efficiency of photosynthesis is defined as the fraction of light energy converted into organic energy under photoautotrophic conditions. Only wavelengths between 400 and 740 nm, representing 42.3% of the total energy of sunlight, can be used in photosynthesis (El Hifnawy, 2016).

3.6.3 GROWTH RATE

The choice of species is made first by selecting species with a rapid cell growth rate to achieve high biomass productivity (Belair et al., 2012). To be able to grow faster, the algae must be able to capture a large amount of CO_2 because the ability to capture a lot of CO_2 is combined with a high photosynthesis efficiency; thus, the more an algae produces photosynthesis, the more it will be able to grow (Jérémie, 2019).

3.6.4 ROBUSTNESS

The algae's robustness can be combined with low nutrient consumption, as low nutrient consumption reduces production costs (Jérémie, 2019). That is why the strain must be resistant to the imposed growing conditions and contamination and capable of synthesizing an appreciable quantity of the desired product (Belair et al., 2012).

3.6.5 CULTIVATION FACILITY

First, algae must be capable of surviving in photobioreactors: on the one hand, they should be robust, and on the other hand, they should be able to live in challenging conditions, including when voluntary deficiencies are found. Second, algae in open ponds must be strong enough to withstand and dominate wild species (Jérémie, 2019).

3.6.6 PRODUCTIVITY

The reproducibility of the algae must be as quickly and efficiently as possible to provide a better yield (Jérémie, 2019).

3.7 MICROALGAE REQUIREMENTS

3.7.1 FACTORS AFFECTING THE MICROALGAE GROWTH

Like higher plants, microalgae are affected by various factors, including physical/chemical, biotic, and environmental factors (Figure 3.3).

3.7.2 ABIOTIC FACTORS

3.7.2.1 LUMINOSITY

Light intensity is one of the main determining factors in the production of microalgae. Light duration and intensity have a direct impact on microalgae photosynthesis and also affect the biochemical composition of microalgae and biomass productivity. Moreover, the growth rate and the productivity of biomass are predicted as a function of light (Khan et al., 2018).

Only solar radiation with a wavelength between 400 and 700 nm can be absorbed by microalgae for photosynthesis. Therefore, the effect of luminosity varies with temperature, as some species tolerate a higher intensity when the temperature is raised (El Hifnawy, 2016).

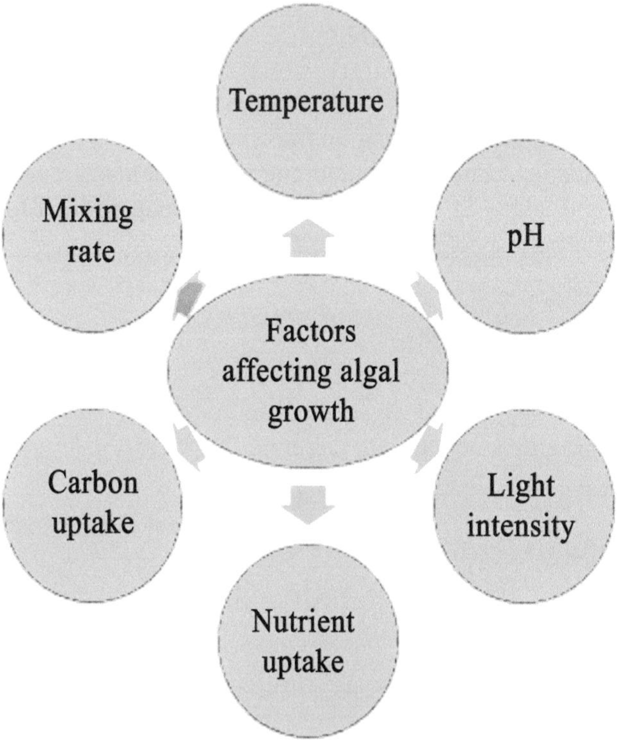

FIGURE 3.3 Factors affecting microalgal growth.

Source: Adapted from Arpita (2017).

3.7.2.2 PHYSICAL/CHEMICAL FACTORS

Temperature is another significant factor contributing to microalgae growth and significantly impacts biochemical processes, including photosynthesis, in the algal cell factory. Thus, every species has its optimum growth temperature. Increasing the temperature to the optimum range exponentially increases algal production but reduces or decreases the temperature above the optimum point delays or stops (El Hifnawy, 2016).

Low temperatures affect photosynthesis by decreasing carbon assimilation activity, although too-high temperatures affect photosynthesis by inactivating photosynthetic proteins and disrupting the energy balance in the cell. Higher temperature also reduces cell size and respiration. The decline in photosynthesis results in a decrease in the growth rate.

The pH of the culture media is another significant factor influencing the development of microalgae. Microalgae species have different pH needs. The large majority grow well throughout the pH range from 6 to 8.76. However, increased pH will also increase the salinity of the culture media, which is very harmful to algae cells (El Hifnawy, 2016).

3.7.3 BIOTIC FACTORS

Besides physical and chemical factors, there are biotic factors affecting the growth of microalgae, such as competition between microalgae organisms. This is among the biotic factors limiting algal growth. For example, cyanobacteria release toxins that have inhibitory effects on the growth of several other microalgae organisms. In the same way, there are several species of microalgae that can release antibacterial substances (El Hifnawy, 2016).

3.8 MOLECULES OF INTEREST PRODUCED BY MICROALGAE

The primary interest of microalgae culture is the production of molecules with high added value. Microalgae are an essential source of polysaccharides, proteins, and pigments (El Hifnawy, 2016).

3.8.1 SUGARS

Commonly the dominant simple sugars are polyols: D-mannitol and D-sorbitol. They are, therefore, cell compounds that are osmo-compatible, resulting from adaptation to the saline environment. Mannitol has some chelating features that can be utilized for the nutrition of plant minerals (Mahfoud, 2014).

3.8.2 PROTEINS AND ENZYMES

Multiple microalgae species produce higher amounts of different essential amino acids (including proline, serine, glycine, and valine) and rich protein sources that can be used to protect plants against various diseases. Moreover, microalgae generally produce 2.5–7.5 tons/ha/year of proteins, in addition to a nonnegligible amount of enzymes such as superoxide dismutase (SOD), that protects plants from oxidative stress, without ignoring the fundamental enzyme carbonic anhydrase, that further plays a central role in converting CO_2 into carbonic acid and bicarbonate (Khan et al., 2018).

3.8.3 LIPIDS

Lipids are indispensable constituents of microalgae cells. They are found at the structural level (phospholipids that make up the membranes) and energetic level (reserve molecules). Globally, microalgae can comprise lipids up to 16–75% of their dry mass, depending on the species. Microalgae store lipids in the form of triglycerols or triglycerides (El Hifnawy, 2016).

3.8.4 PIGMENTS AND ANTIOXIDANTS

Microalgae produce pigments that are essential for photosynthesis. Moreover, pigments are defined as a complex of molecules that have the ability to change the color of reflected or transmitted light according to the absorption of specific wavelengths (El Hifnawy, 2016). Besides, antioxidants are very essential compounds used by plants to shield itself from the harmful effects of free radicals. Considering that biomolecules such as DNA, proteins, and membrane lipids are attacked by reactive oxygen species (ROS) and nitrogen reactive species (NOS) (Khan et al., 2018).

3.8.5 PHYTOHORMONES

Several algae-derived phytohormones were detected for auxins, cytokinins, abscissic acid, and gibberellins (Mahfoud, 2014). Besides that, microalgae phytohormones may serve as regulators of plant species' exogenous growth, affect crop yields and tolerance of abiotic and biotic stress factors, but also work as endogenous components of microalgae, which typically impact the process of biosynthesis of pigment, etc. (Romanenko et al., 2015) (Table 3.2).

3.8.6 TOXIC METABOLITES

Most species of microalgae offer a range of toxic substances, which are commonly called mycotoxins. They are of huge importance as environmental hazards due to the chemistry of their toxicology. They are widely divided into neurotoxins, hepatotoxins, and chemical structures such as lipopolysaccharides. These toxic substances also carry out significant antibacterial and antifungal activities (Khan et al., 2018).

3.8.7 POLYSACCHARIDES

Microalgae biomass production provides a wide variety of algal polysaccharides/exopolysaccharides, which can stimulate plant growth and plant defense responses by regulating various biological processes such as photosynthesis, nitrogen and sulfur assimilation and plant defense mechanisms (Chanda et al., 2019).

3.9 CULTURE OF MICROALGAE

3.9.1 TYPE OF MICROALGAE CULTURE

Large-scale microalgae cultivation can be conducted in two modes, either using high-yield open ponds (the raceway) or in a closed transparent enclosure using natural or artificial light (the photo-bioreactor). Microalgae with an autotrophic metabolism are mainly cultivated in outdoor ponds, photobioreactors, or hybrid models that combine the first two approaches. Microalgae of heterotrophic metabolism are mainly cultivated in closed bioreactors called fermenters in the absence of sunlight (Abla, 2015).

TABLE 3.2 A Summary of Phytohormones, their Biosynthesis, Physiological Actions in Plants and Key Algal Sources.

Phytohormones	Biosynthesis	Physiological action	Key algae sources
Auxins	Via tryptamine or indole-3-pyruvic acid pathway in leaf primordia, young leaves, and fruits.	Cell elongation (via activation of the plasma membrane H^+- [adenosine triphosphatases (ATPase)], differentiation of phloem, apical dominance, tropisms, initiation of root formation, and stress tolerance (drought, salinity, and heat).	Macroalgae: *Ascophyllum*, *Ecklonia*, *Ectocarpus*, *Fucus*, *Laminaria*, *Macrocystis*, *Nereocystis*, *Sargassum*, *Undaria*, *Klebsormidium nitens*, *Caulerpa*, *Cladophora*, *Ulva*, *Polysiphonia*, *Botryocladia*, *Pyropia*, *Prionitislanceolata*, *Porphyra*, *Gelidium*, *Gracilaria*, *Gracilariopsis*, *Chondracanthus*, and *Hypnea*. Microalgae: *Chlorella*, *Coenochloris*, *Acutodesmus*, and *Scenedesmus*; Cyanophyta: *Synechocystis*, *Chroococcidiopsis*, *Anabaena*, *Phormidium*, *Oscillatoria*, *Nostoc*, *Chlorogloeopsis*, and *Oscillatoria*.
Cytokinins	Via biosynthesis of N6-isopentenyladenosine monophosphate from adenosine monophosphate and pyrophosphate (catalyzed by isopentenyl-transferase) in root tips, leaves, and developing seeds or changes in the structure of tRNA comprising cis-zeatin. The latter pathway is involved in algae.	Control cell division, promotion of protein and chlorophyll synthesis, chloroplast and vascular tissue development, bud development, shoot growth, fruit and flower development, apical dominance, development of the leaf blade, and senescence and stress tolerance (drought and heat).	Macroalgae: *Ascophyllum*, *Cystoseira*, *Ecklonia*, *Fucus*, *Macrocystis*, *Sargassum*, *Undaria*, *Dictyota*, *Laminaria*, *Pyropia*, *Sarcothalia*, *Amphiroa*, *Corallina*, *Porphyras*, *Gelidium*, *Gracilaria*, *Gracilariopsis*, *Chondracanthus*, *Hypnea*, *Gigartinaclathrata*, *Hypnea*, *Ulva*, *Cladophora* and *Halimeda*; Microalgae: *Chlorella*, *Chlamydomonas*, *Desmococcus*, *Euglena*, *Myrmecia*, *Stigeoclonium*, *Mychonastes*, *Scenedesmus*, *Nannochloropsis*, *Nautococcus* and *Protococcus*. Cyanophyta: *Arthronema*, *Synechocystis*, *Chroococcidiopsis*, *Anabaena*, *Phormidium*, *Oscillatoria*, *Calothrix*, *Chlorogloeopsis*, and *Rhodospirillum*.

TABLE 3.2 *(Continued)*

Phytohormones	Biosynthesis	Physiological action	Key algae sources
Gibberellins	Via pathways of isoprenoid synthesis (though methylerythritol phosphate pathway) in plastids.	Associated with stem elongation, initiation of seed germination via activation of α-amylase, floral organ development influence protein biosynthesis, and initiation of flowering.	Macroalgae: *Cystoseira*, *Ecklonia*, *Fucus*, *Petalonia* and *Sargassum*, *Caulerpa*, *Ulva*, *Pyropia*, and *Hypnea*. Microalgae: *Scenedesmus*, *Stigeoclonium*, *Gyoerffyana*, *Myrmecia*, *Nautococcus*, *Chorella*, *Scotiellopsis*, *Chlorococcum*, *Chlamydomonas*, and *Nannochloropsis*; *Cyanophyta*: *Arthrospira*, *Anabaenopsis*, *Cylindrospermum*, and *Phormidium*.
Brassinosteroids	From campesterol via 2 pathways, either early C6-oxidation pathway or late C6-oxidation pathway.	Controls division, elongation, and differentiation of vascular system, promotion of ethylene production, and stress tolerance.	Macroalgae: *Ecklonia* and *Hydrodictyon*. Microalgae: *Klebsormidium*.
Ethylene	From methionine in stressed tissues and fruits via 1-aminocyclopropane-1-carboxylate synthase or 1-aminocyclopropane-1-carboxylate oxidase.	Regulates developmental processes such as fruit ripening, opening of flowers, cell division, cell elongation, senescence of vegetative tissues, and biotic and abiotic stress tolerance.	Macroalgae: *Ecklonia*, *Padina*, *Klebsormidium*, *Porphyra*, and *Pyropia*. Microalgae: species of *Chlorella* and *Scenedesmus*; Cyanophyta: species of *Arthrospira*, *Synechococcus*, *Anabaena*, *Nostoc*, *Calothrix*, *Scytonema*, and *Cylindrospermum*.
Abscisic acid	From carotenoids in roots and expanded leaves.	Induces stomatal closure, shoots growth inhibition, and protein storage in seeds with seed dormancy.	Macroalgae: species of *Ascophyllum*, *Ecklonia*, *Saccharina*, *Laminaria*, *Ulva*, *Porphyra*, *Gelidium*, *Gracilaria*, *Gracilariopsis*, *Chondracanthus*, and *Hypnea*; Microalgae: species of *Scenedesmus*, *Chlorella*, *Dunaliella*, *Haematococcus*, *Chlamydomonas* and *Nannochloropsis*. Cyanophyta: species of *Synechococcus*, *Nostocmuscorum*, *Trichormusvariabilis*, and *Anabaena*.

TABLE 3.2 *(Continued)*

Phytohormones	Biosynthesis	Physiological action	Key algae sources
JA	In two steps in chloroplast and peroxisomes from α-linolenic acid.	Regulation of plant defense responses, synthesis of proteinase inhibitors, promotion of tuber formation and senescence.	Macroalgae: species of *Fucus*, *Gelidium*, and *Lithothamnion*. Microalgae: species of *Chlorella*, *Scenedesmus*, *Dunaliella*, *Euglena*, and *Gelidium*. Cyanophyta: species of *Arthrospira*.
SA	From chorismate via either isochorismate synthase (ICS) or PAL pathway.	Regulation of plant defense responses during pathogenesis.	Microalgae: species of *Scenedesmus*. Cyanophyta: species of *Arthrospira*.
Polyamines	Via decarboxylation of arginine or ornithine.	Stress response regulation in plants and metabolic processes such as cell proliferation, differentiation, and growth.	Macroalgae: species of *Gelidium*, *Grateloupia*, *Dictyota*, and *Ulva*. Microalgae: species of *Chlorella*, *Cyanidium*, and *Euglena*. Cyanophyta: species of *Arthrospira*.
Betaines	Via choline and glycine. Glycinebetaine is synthesized from choline via a 2-step oxidation of choline via the toxic intermediate betaine aldehyde.	Acts as an osmolyte and regulates defense mechanisms against abiotic stresses such as salinity, frost, and drought.	Macroalgae: species of *Ascophyllum*, *Fucus*, *Laminaria*, *Laurencia*, *Mastocarpus*, and *Chaetomorpha*.
Rhodomorphin	–	In morphogenesis, induction of fusion of fragments, and restoration of filament integrity.	Macroalgae: species of *Griffithsia*.

Source: Reproduced with permission from Kapoore et al. (2021), Copyright © Elsevier (2021).

3.9.2 IN PHOTOAUTOTROPHIC CULTURE

3.9.2.1 OPEN SYSTEMS (RACEWAYS)

Open culture systems, which are closed-loop recirculation ponds with a depth of a few tens of centimeters, are simple technologies with little initial investment. Including roller tracks made of poured cement or dug directly into the ground and covered with a plastic layer to prevent the penetration of the culture medium into the soil. The mixing and the circulation of the medium are possible thanks to a paddle wheel and a bubbling allows a CO_2 supply. The temperature of the environment fluctuates according to diurnal (night) and seasonal cycles (Abla, 2015).

3.9.2.2 CLOSED SYSTEMS PBR

The use of photobioreactors is more expensive than open pond systems; however, they require less light and space for the cultivation of microalgae. The agitation and turbulence conditions must be compatible with the species' biology because some cells have a particularly fragile flagellum and are extremely sensitive to high shear stresses. Two geometries for closed photobioreactors are preferred for large-scale production: tubular reactors and flat reactors (Abla, 2015).

3.9.2.3 COMPARISON BETWEEN OPEN AND CLOSED SYSTEMS

A comparison of the advantages and disadvantages of either the closed or the open system cultivation of microalgae proves that such methodology could possibly result in significantly higher productivity or even an improvement in the algal industry.

3.9.3 IN HETEROTROPHIC CULTURE

The fundamental culture medium composition for heterotrophic cultures is analogous to the auto-trophic culture, with the exception of the addition of organic carbon. However, limited to a few microalgae organisms, using their heterotrophic growth ability in the absence of light will substitute the fixation of atmospheric CO_2 of autotrophic cultures with organic carbon sources dissolved in the culture media (Perez-Garcia et al., 2010).

3.10 CULTIVATION METHODS

There are two different modes used for the cultivation of microalgae, each of which has its own qualities as well as its own defects.

3.10.1 HARVESTING SYSTEM

All harvesting techniques are designed to eliminate as much culture media from microalgae biomass as possible to enable more downstream processing, such as the extraction of bioactive compounds. Various harvesting techniques are often used to extract biomass, such as filtration, centrifugation, flocculation, and flotation. Sometimes, a combination of two or more methods is used to further maximize harvesting efficiency (Tan et al., 2020).

3.11 APPLICATIONS

A variety of commercial uses have been discovered for microalgae, including food formulation, feed, cosmetics, health products, wastewater treatment, as well as biofertilizers and biostimulants.

3.12 FOOD AND FEED

Thanks to their nutritional properties, microalgae can be considered a desirable food or feed product. Microalgal biomass contains a broad range of biologically active compounds such as polyunsaturated fatty acids, pigments, vitamins, and minerals, or oligosaccharides as extracellular compounds; they may, therefore, serve as nutraceutical compounds (Camacho et al., 2019).

3.13 PHARMACEUTICALS AND COSMETICS

Numerous studies have declared beneficial health effects from microalgal compounds such as carotenoids bearing anticancer, anti-inflammatory, antioxidant, antimicrobial, and antiobesity capacities, in addition to hypocholesterolemic properties (Sialve and Steyer, 2013).

3.14 ENVIRONMENT AND ENERGY

3.14.1 *LIQUID EFFLUENTS TREATMENT*

Since the growth conditions and metabolic activity require water and nutrients (macro- and microelements), microalgae can grow in various effluents. The scientific literature reports a large number of deposits that have been successfully confronted with microalgae: urban wastewater, livestock effluents, and industrial effluents, such as digestate or leachate from landfill (Sialve and Steyer, 2013).

3.14.2 *METAL TREATMENT*

Microalgae are characterized by a certain tolerance to the presence of metals and high yields in fixation. For this reason, these organisms are used as ecotoxicological indicators of water quality, especially in relation to heavy metals (Sialve and Steyer, 2013).

3.14.3 *GASEOUS EFFLUENTS TREATMENT*

The use of photosynthesis to repair anthropogenic CO_2 responds to an environmental challenge from the point of view of industrial activities that are particularly carbon-intensive and to an economic challenge on the mass culture side of microalgae (Sialve and Steyer, 2013).

3.14.4 *ENERGY PRODUCTION*

Thanks to the similarity between higher plants and microalgae in terms of metabolic phenotypic expression, research on microalgae energy recovery primarily concerns the same industries (e.g., gaseous biofuel, crude oil, and methane or ethanol production) (Sialve and Steyer, 2013).

3.15 AGRICULTURE

Algae are widely used as an organic improvement and, in particular, as foliar nutrition due to their remarkable benefits, such as improved germination rate, increased yields, increased resistance to colds, increased absorption of mineral elements from the soil or the shelf life of fruit (Mahfoud, 2014).

3.16 MICROALGAE FOR SUSTAINABLE AGRICULTURE

As the worldwide population rises, coupled with the environmental impacts of climate change, agriculture is continually confronted with constraints such as abiotic stress, which horribly inhibits the growth and development of plants. Alternative strategies, including the irrigation, mechanization, and application of nitrogen, phosphorus, and potassium (NPK) fertilizers and biocides, have, indeed, been implemented to ensure maximum crop productivity and optimize plant growth conditions. Nevertheless, modern agriculture is forced to follow novel eco-friendly strategies due to the negative impact of these intensifications on the environment and the ecosystem. The introduction of innovative technologies based on bioresources, such as microalgae-based products, has proven to be effective strategy for improving crop performance (Mutale-Joan et al., 2020).

3.17 THEIR EFFECT ON PLANT GROWTH AND DEVELOPMENT

Microalgae and seaweed extracts tested both under open-field and under greenhouse conditions have been shown to stimulate germination and growth of seedlings, shoots, and root biomass in plants by modulating different physiological and biochemical processes (Chanda et al., 2019).

There are multiple groups of high-value compounds derived from algae with extremely high application potential-carotenoids and fatty acids. The previous participates in crucial physiological processes, such as breathing and photosynthesis, as well as affecting plant physiology at the enzyme- and hormone-dependent control level, including stress mechanisms and pathogen attacks. The latter, which is typically connected with storage content, contains basic components of the biological membranes' lipid bilayer. As such, fatty acids that are implicated in the coordination of most crucial cellular functions and responses that influence the smooth functioning of the plant body (Chojnacka et al., 2018).

3.17.1 EFFECT ON PLANT GROWTH

The impacts of the application of algal extracts on plant growth include enhanced ROS scavenging, increased ribulose-1,5-bisphosphate carboxylase/oxygenase (RuBisCo) synthesis, increased PSII activity, and enhanced photosynthesis. For example, polysaccharides extracted from *S. platensis* have shown a significant increase in plant growth in *Capsicum annuum* and *Solanum lycopersicum*, as demonstrated by plant weight, plant size, and leaf size/number. Similarly, seaweed polysaccharides have demonstrated a number of biological activities related to plant growth (Mutale-Joan et al., 2020). Moreover, it has been shown that the foliar application of aqueous cell extracts of *A. dimorphus* was found to significantly boost plant growth compared to the control group when using an optimal concentration (Garcia-Gonzalez and Sommerfeld, 2016).

It has been hypothesized that oligoalginate and oligo-carrageenan can interact with plasma membrane receptors that use a co-receptor involved in signal transduction to activate plant growth simultaneously. Furthermore, using algal polysaccharides could boost plant growth and resistance by targeting mechanisms on both the shooting and root systems (Chanda et al., 2019). Besides that, microalgae polysaccharides, such as carrageenan, mediate plant growth, regulation of photosynthesis and carbon fixation in plants, enhancement of RuBisCo activity, cell division, purine and pyrimidine synthetic pathways and metabolic pathways involved in nitrogen, and sulfur assimilation (Shukla et al., 2016).

From another study, tomato seedlings treated with dry *A. dimorphous* biomass (biofertilizer) 22 days prior to transplantation resulted in the highest plant production, the highest number of branching and floral development and early fruit settings, which could probably result in higher agricultural productivity (Garcia-Gonzalez and Sommerfeld, 2016).

3.17.2 EFFECT ON PLANT ROOT DEVELOPMENT

Generally, using algal extracts such as polysaccharides could boost plant growth by targeting mechanisms on root systems. The root target includes root development and metal chelation, which leads to an increase in root surface and access to nutrients (Chanda et al., 2020). Interestingly, microalgal treatment improved seed germination percentage (Garcia-Gonzalez and Sommerfeld, 2016) and growth with significant improvement of soluble carbohydrate, soluble protein, pigment content, and total free amino acids, without ignoring increased seedlings' fresh and dry weight. This is due to the involvement of root-promoting hormones such as auxins, cytokinins, and gibberellic acid, which in turn can promote both soil microbial growth and crop production (Ronga et al., 2019).

It was noted that the use of optimal extract concentration resulted in a significantly higher percentage of germination and the highest growth of plants and also that increasing concentration of extracts had a detrimental effect on seed germination (Garcia-Gonzalez and Sommerfeld, 2016).

3.18 EFFECTS OF MICROALGAE EXTRACT ON NUTRIENT ABSORPTION AND ASSIMILATION

Microalgae are regarded as an organic fertilizer with the potential to prevent loss of nutrients by gradual release of N, P, and K (Ronga et al., 2019). Besides, several studies have revealed that algae are a good source of macro- and micronutrients. These are also well-known for their high content of bioactive compounds. Nevertheless, when researching plant treatment, there are so many metabolites of particular interest: plant hormones (phytohormones), hormone-like compounds, poly- and oligosaccharides, and phenolic compounds (Chojnacka et al. 2018).

Moreover, the use of microalgae as biofertilizers showed a correlation between increased nutrient uptake, increased accumulation of biomass, and increased crop yields. So, at the molecular level, certain genes corresponding to different biological pathways and processes were upregulated by microalgal extracts, particularly primary and secondary metabolisms, and intracellular transports, notably those involved in nutrient uptake related to root traits (Ronga et al., 2019). Moreover, it has now been documented that polysaccharides are rich in functional groups that have the potential to bind to certain microelements with significant nutritional value, thereby enhancing the nutrient quality of the roots (Chanda et al., 2020).

The foliar spraying of biochemical, organic substances is among the most innovative agricultural practices since they have the advantage that they are eco-friendly, increasing agricultural sustainability. Microalgae extracts could very well replace micronutrient foliar fertilizers. Furthermore, these extracts are absorbed directly through the stomata and cuticular pores of the leaf, providing the greatest effectiveness. The response of plants to nutrients supplied through foliar sprays is considerably faster than when introduced as a soil amendment. In microalgae extracts these nutrients are available in a reasonable percentage, especially P, K, iron (Fe), magnesium (Mg), and Ca. In addition, microalgal extracts could indeed contain phytohormones, enzymes, and vitamins, which might also enhance nutrient assimilation and solute translocation, leading to significant crop yield improvements (Ronga et al., 2019).

3.19 EFFECT ON THE SOIL

Biostimulants used as bioremediants have been shown to increase levels of adenosine triphosphate (ATP) and phosphatase and urease activity, thus increasing the rate of soil xenobiotics degradation and improving beneficial microbial soil communities in semi-arid climates. Biostimulants can also lead to reducing the amount of hazardous agrochemicals, mainly by decreasing fertilizer and pesticide consumption (Yakhin et al., 2017), which allows sustainable crop productivity to be reached (Mahfoud, 2014).

3.20 EFFECT ON PLANT PROTECTION AGAINST STRESSES

Plants are subject to many environmental pressures over their entire life cycle, which adversely affect their productivity and survival (Chanda et al., 2020). The response of stress and defense for both terrestrial plants and algae contains phenolic compounds. These thus exhibit a significant biological impact–including antioxidant, radio-protective, antiviral, and antimicrobial activity, along with the ability to control enzyme-mediated responses (Chojnacka et al., 2018). Moreover, ROS is one of the most common indications of stress. Considering that ROS is a toxic molecule capable of destroying cells (Chanda et al., 2020).

Surprisingly, it has been shown that microalgae polysaccharides can reduce ROS toxicity in plants by boosting the ROS antioxidant enzyme activity output. This mechanism can be explained by the fact that polysaccharides contain neutral sugars which can interact as signal molecules with plasma membrane receptors and thus result in a series of biochemical reactions being activated. Most importantly, membrane receptors could perceive algal polysaccharides as microbial-associated molecular patterns (MAMPs) and thus induce MAMP-dependent signaling pathways involving the activation of Ca^{2+} influx, ROS production, lipoxygenase (LOX) and phenylalanine ammonia-lyase (PAL) pathways as key enzymes, and thus ROS scavenging enzymes (Chanda et al., 2020).

3.20.1 THEIR USES IN PLANT PROTECTION AGAINST BIOTIC STRESSES

Several studies have proved the positive impact of using microalgae extracts, notably polysaccharides/exopolysaccharides, in terms of the induction of the plant immune system. In general, the polysaccharide molecules perceived by membrane receptors of plants can activate a series of defense mechanisms. Moreover, it showed a considerable increase in the activity of β-1,3-glucanase, which is one of the pathogenesis-related (PR) proteins grouped into the PR-2 family that break down pathogens' cell wall components. The induction of PR proteins results from the activation of plant defense pathways that limit the entry or even further spread of pathogens (Chanda et al., 2020).

Some research revealed that crude polysaccharide application on plants affected the activity of PAL, showing an upregulation of the phenylpropanoid pathway. Moreover, phenylpropanoid pathway products can end up in the composition of phenolic substances such as phytoalexin, which are pathogens-toxic molecules (Chanda et al., 2020). Besides, polysaccharides had an impact on the accumulation of phytoalexin and on the expression of genes involved in the signaling pathways of salicylic acid (SA) and jasmonic acid (JA) that induce plant resistance to pathogens. Moreover, the stimulation of these signaling pathways is associated with an increased expression of genes that encode PR proteins, PAL, LOX, and ROS scavenging enzymes, and other enzymes required for the synthesis of antimicrobial terpenoids and/or alkaloids (Chanda et al., 2020). Thus, because of this eliciting function, algae extracts could become an essential ally in crop protection, in a context of growing concern for the environment.

3.20.2 THEIR USES IN PLANT PROTECTION AGAINST ABIOTIC STRESSES

Nowadays, various abiotic stresses damage crop production due to irrigation water deficiency, soil salinization, and low and high temperatures, which are among the major agricultural constraints, it is harder to fulfill the possible crop yields of cultivated plants (Ronga et al., 2019). In this context, the use of microalgae-based products could provide protection against various environmental stresses. Due to a significant decrease in superoxide radical production and low lipid peroxidation, extracts from some microalgae species mitigated salt stress during the seed germination process. Furthermore, applying biostimulants to stressed crops increased antioxidant capacity and improved root length growth, increased leaf number, and area, as well as improved salinity tolerance (Ronga et al., 2019).

Several results observed after using products based on phytohormones would function at the level of organ development. In contrast, polysaccharides would be implicated in stimulating the natural defensive reactions of plants (elicitors). Furthermore, betaines, derived from amino acids, are osmo-compatible molecules whose beneficial effects have been shown during their application on plants for adaptation to thermal, hydric, and saline stress (Mahfoud, 2014).

3.20.3 BIOSTIMULANT

Biostimulant is described as a formulated product of biological origin that further increases the productivity of plants as a result of the novel or emerging properties of the constituent complex—not as the ultimate consequence of the presence of known essential plant nutrients, plant growth regulators, or plant protection compounds (Yakhin et al., 2017). Agricultural biostimulants (ABs) include different formulations of compounds, substances as well as other products, including certain microorganisms, trace elements, enzymes, plant growth regulators (PGRs), and algae extracts (Sharma et al., 2014).

3.20.4 BIOCONTROL

The Rural and Maritime Fisheries Code (CRPM) defines biocontrol products as agents and products that use natural mechanisms as part of an integrated fight against crop enemies, including macro-organisms and phytopharmaceutical products containing microorganisms, chemical mediators such as pheromones and kairomones, and natural substances from plant, animal or mineral origin. However, biostimulants are distinct from biocontrol and induce disease resistance by concentrating on effects related to improved plant growth, stress tolerance, and quality (Farid et al., 2019).

3.21 MICROALGAE ECONOMIC INTEREST

3.21.1 A GLOBAL OVERVIEW

The 2030 Sustainable Development Agenda regularly mentions the bioeconomy as a way of accomplishing food security and nutrition through the use of natural resources in a manner that guarantees sustainable economic, social, and environmental development (Vigani, 2020). Bioeconomy, as mentioned at the Global Summit on Bioeconomy, is "the knowledge-based production and use of biological resources, innovative biological processes, and principles for the sustainable delivery of goods and services across all economic sectors." The bioresource-based economy is beneficial for the reasons mentioned: reduction of greenhouse gases, sustainable resources, improved environmental protection and lacking any sort of secondary pollution (Nagarajan et al., 2020).

The total blue economy (described as "ocean-based industries (such as marine biotechnology), as well as the natural resources and ecosystem services offered by the ocean.") was only 2.8% of the overall blue economy in 1950; it went up to 16.5% in 2016. The latter trends show a significant global growth in algae use as well as consumption. Increasingly recognizing the importance of marine resources for the security of food and nutrition suggests that algae perform a growing and considerable role in accomplishing the SDGs Agenda 2030 (Vigani, 2020).

About 50 countries across Africa, the Americas, Europe, Asia, and the Pacific have laid out initiatives and strategies to develop the bioeconomy at the national or regional level to achieve the objective of these policies. For example, in South Africa, most strategies focus on research and innovation for the bioeconomy. The US national strategy on the bioeconomy was primarily concerned with renewable energy production. The National Bioeconomy Blueprint outlines the core bioeconomy technologies, notably genetic engineering, DNA sequencing, and automated high-performance biomolecular manipulation (Vigani, 2020).

3.22 ORIENTATION OF FUTURE RESEARCH ON MICROALGAE

3.22.1 FUTURE PROSPECTS

This chapter has shown that microalgae are a potential candidate for a process of carbon capture, therefore a solution to climate change and, at the same time, a guarantee of sustainable agriculture through the substitute for synthetic products and fertilizers with microalgae-based products that are eco-friendly. Consequently, microalgae biotechnology opens up many research perspectives and many industry sectors: food, health, and nutrition; second and, in particular, environmental and industrial sectors. Numerous studies need to be carried out to carry out industrial development in a rigorous manner (Elmtili, 2014).

Until now, studies on microalgae in agriculture have highlighted their biostimulants properties in laboratory conditions on plants. Conversely, on-farm experiments and whole field studies are needed to provide support information for practical applications of microalgae and to evaluate the studies that have been carried out so far. The effect of microalgae products on a wider variety of plant species also needs to be investigated to understand their biostimulants effects on various types of plants (Chanda et al., 2019).

3.22.2 NEW MARKET OPPORTUNITIES

The interest of this study is thus to have paved the way for a possible new markets using microalgae potentialities at the industrial level:

- Microalgal biomass production from wastewater will transform waste nutrients into renewable and revolutionary biofertilizers with commercial opportunities in agricultural activities.
- The annual market for seaweed polysaccharides is not saturated and new biopolymers such as those provided by microalgae could

be added, taking into account their original structures and recent progress in their production.

3.22.3 ADVANTAGES AND CONSTRAINTS OF MICROALGAE

Compared to other sources of biomass the benefits of microalgae products are numerous:

- Ease of culture and harvesting of products, greater photosynthetic efficiency than terrestrial plants, higher biomass productivities due to their shorter reproductive cycle, faster growth rates than higher plants, and higher rates of CO_2 fixation and O_2 production.
- The possibilities to develop new biomass uses without creating market tensions.
- The possibility of cultivating algae in the sea or on nonarable soil without competing with food for land use (giving food crops priority for food safety purposes).
- Possibility to make environmental services parallel with their production (Thomsen, 2016).
- The development of technologies for the production and treatment of microalgae biomass using the biorefinery method has greatly reduced the cost of commercial microalgae products and paved the way for the cultivation of nonexploited microalgae.

The constraints of microalgae products are multiple:

- Production at uncompetitive costs compared to other alternative products produced either through chemical synthesis, simply as a result of metabolism by other microorganisms, or even as extracts from fossil raw resources.
- Tighter regulatory constraints in terms of quality specifications, safety assurance, and minimization of environmental impact (Camacho et al., 2019).

3.22.4 CHALLENGES

Regarding the encouraging use of algae and the sector's overall development, numerous challenges currently need to be identified to completely demonstrate its capacity for the bioeconomy and for the SDGs.

Such challenges include the advancement of new technologies to facilitate the commercial viability of algae products and biomass, and appropriate market and economic data to support industrial and institutional strategy. Furthermore, targeted policies should be implemented taking into consideration that algae and seaweed biomass should be managed sustainably in the context of climate change and anthropogenic pressures influencing marine ecosystems (Vigani, 2020).

3.23 CONCLUSION

There is an increasing necessity for eco-friendly strategies for sustainable cultivation of horticultural crops that could maximize crop productivity without inducing further degradation of the environment (e.g., eutrophication, soil infertility, and loss of biodiversity). This is due to a rapid increase in world population and the need to produce more food dramatically. Hence, microalgae may indeed be the new sustainable bioproducts for plant production, growth, and protection. Such photosynthetic microorganisms are not only sources of bioactive products that are renewable, but sustainable and economical (Chanda et al., 2019).

The use of algal extracts has been recommended as a solution to this crisis. In addition, supercritical algal extracts are healthy for plants as well as plant-derived products for humans and animals. Algal products are environmentally safe and nontoxic. There are no issues with residues, toxicity, ecotoxicology, or fate in the environment, such as synthetic plant protection products. As a result, it is expected that the simple destination for the certification of products from natural resources would be implemented in a mandatory role. Any use of algal extracts for plant growth is a reasonable alternative that guarantees the growth of food vegetables/fruits with no chemical contamination. With absolute respect for humans and the environment, such natural products ensure not only increased agricultural yield but quality as well (Chanda et al., 2019).

Since microalgae contain high levels of plant growth essential micronutrients and macronutrients, they have wide applications as biofertilizer, which enhances soil chemical and biological properties, promotes plant growth, and recovers soil it has a higher protein content and growth rate than macroalgae. Providing a multitude of options that allows strain selection with different compositions or adapted to live in different environments. Several microalgae species could be easily produced with wastewater, lowering the cost of production. In addition, microalgae, as somewhat of an organic fertilizer, have the ability to prevent the loss of nutrients through the slow release of N, P, and K in accordance with plant requirements. Moreover, microalgae produce active compounds such as free amino acids and organic acids, phytohormones, enzymes, bioactive secondary metabolites, vitamins, and vitamin precursors (Plaza et al., 2018).

In this regard, further attention is given to the synthesis of natural substances capable of promoting plant growth, considered "plant biostimulants." Microalgae are a possible sustainable option for agricultural crop enhancement and protection (Plaza et al., 2018). Microalgae's modes of action were being poorly studied. Comprehending their specific elicitor effects on plant growth and defense mechanisms may not only be a useful tool in plant biotechnology but also a key step towards overcoming the issues of modern agriculture confronting today's fast-increasing demand for food and changing climate (Chanda et al., 2019). When the microalgae products were added, plants showed different reactions. This would include vigorous growth, increased yield, increased nutrient uptake, increased resistance to biotic and abiotic stress (fungal diseases and insect attacks), improved quality, and longer fruit shelf life (Plaza et al., 2018).

The product's ultimate price would be the main factor deciding the use of microalgal formulations for agricultural uses. Actually, crop economy limits the use of microalgal based-products exclusively to high-value crops (e.g., fruit, protected crops). The future growth of the sector will be influenced by a range of factors, along with the availability of microalgae (Sharma et al., 2014). Morocco considers the green economy mostly as a national priority, as certainly outlined in the National Sustainable Development Strategy, which further leads to environmental conservation, the regulation of global warming, the restoration of resources, and job creation (Uneca, 2016).

KEYWORDS

- **biocontrol**
- **bioeconomy**
- **biostimulants**
- **microalgae**
- **sustainable agriculture**

REFERENCES

Abla, B. Modification Protéique et Enzymatique de la Souche *Acutodesmus* sp. strain SG-1 Durant le Processus de la Caroténogenèse. Diplôme de Master, Université des Frères Mentouri Constantine Faculté des Sciences de la Nature et de la Vie, 2015.

Arpita, R. A Review on Harvesting and Lipid Extraction Methods for Biodiesel Production from Microalgae. *Res. Rev. Biosci.* **2017,** *12* (3), 134.

Belair, V.; Barnabé, S.; Tremblay, R. Sélection de Souches de Microalgues. Biotendance (CQVB), 2012; p 20.

Bingtao, Z.; Yaxin, S. Macro Assessment of Microalgae-Based CO_2 Sequestration: Environmental and Energy Effects. *Algal Res.* **2020,** *51*, 102066. https://doi.org/10.1016/j.algal.2020.102066.

Camacho, F.; Macedo, A.; Malcata, F. Potential Industrial Applications and Commercialization of Microalgae in the Functional Food and Feed Industries: A Short Review. *Mar. Drugs* **2019,** *17* (6), 312.

Chanda, M. J.; Merghoub, N.; El Arroussi, H. Microalgae Polysaccharides: The New Sustainable Bioactive Products for the Development of Plant Bio-stimulants?. *World J. Microbiol. Biotechnol.* **2019,** *35* (11), 177.

Chevanton, M. L. Interactions Microalgues-bactéries en système expérimental bispécifique: Effets sur la croissance de *Dunaliella* sp. Thèse de doctorat, Université de nantes faculté des sciences et des techniques, 2013.

Chojnacka, K.; Wieczorek, P. P.; Schroeder, G.; Michalak, I. In *Algae Biomass: Characteristics, Towards Algae-based Products*; Springer International Publishing: Cham, Switzerland, 2018.

El Hifnawy, C. Identifier des microalgues pour le traitement des eaux usées et la production de biocarburant, 2016.

Elmtili, N. Les algues marines: nouvelle potentialité économique pour le Maroc. Quelle stratégie biotechnologique?, Cahiers UAE, 2014, pp 1–7.

Farid, R.; Mutale-Joan, C.; Redouane, B.; Mernissi Najib, E. L.; Abderahime, A.; Laila, S.; Arroussi Hicham, E. L. Effect of Microalgae Polysaccharides on Biochemical and Metabolomics Pathways Related to Plant Defense in *Solanum lycopersicum*. *Appl. Biochem. Biotechnol.* **2019,** *188* (1), 225–240.

Garcia-Gonzalez, J.; Sommerfeld, M. Biofertilizer and Biostimulant Properties of the Microalga *Acutodesmus dimorphus*. *J. Appl. Phycol.* **2016,** *28* (2), 1051–1061.

Kapoore, R. V.; Wood, E. E.; Llewellyn, C. A. Algae Biostimulants: A Critical Look at Microalgal Biostimulants for Sustainable Agricultural Practices. *Biotechnol. Adv.* **2021,** *49*, 107754. doi: 10.1016/j.biotechadv.2021.107.

Khan, M. I.; Shin, J. H.; Kim, J. D. The Promising Future of Microalgae: Current Status, Challenges, and Optimization of a Sustainable and Renewable Industry for Biofuels, Feed, and Other Products. *Microb. Cell Factories* **2018,** *17* (1), 36–36.

Leung, D. Y. C.; Caramanna, G.; Maroto-Valer, M. M. An Overview of Current Status of Carbon Dioxide Capture and Storage Technologies. *Renew. Sustain. Energy Rev.* **2018,** *39*, 426–443.

Lorion, J. B. T. e. M. V., 2019. e-monsite. http://algue-production-energie.e-monsite.com/pages/comment-produire-du-biocarburant/selection.html.

Mahfoud, B. Identification des algues du Sahara septentrional: L'effet des algues sur le stress salin (cas de la région de Ouargla), Universite kasdi merbah, ouargla faculte des sciences de la nature et de la vie departement des sciences agronomiques, 2014.

Mutale-Joan, C.; Redouane, B.; Najib, E.; Yassine, K.; Lyamlouli, K.; Laila, S.; Zeroual, Y.; Hicham, E. A. Screening of Microalgae Liquid Extracts for their Bio Stimulant Properties on Plant Growth, Nutrient Uptake and Metabolite Profile of *Solanum lycopersicum* L. *Sci. Rep.* **2020,** *10* (1), 2820.

Nagarajan, D.; Lee, D. J.; Chen C. Y.; Chang, J. S. Resource Recovery from Wastewaters using Microalgae-Based Approaches: A Circular Bioeconomy Perspective. *Bioresour. Technol.* **2020,** *302*, 122817.

Perez-Garcia, O.; Escalante, F.; de-Bashan, L.; Bashan, Y. Heterotrophic Cultures of Microalgae: Metabolism and Potential Products. *Water Res.* **2010,** *45*, 11–36.

Plaza, B. M. Effect of Microalgae Hydrolysate Foliar Application (*Arthrospira platensis* and *Scenedesmus* sp.) on *Petunia* x *hybrida* Growth. *J. Appl. Phycol.* **2018,** *30* (4), 2359–2365.

Procházková, G.; Brányiková, I.; Zachleder, V.; Brányik, T. Effect of Nutrient Supply Status on Biomass Composition of Eukaryotic Green Microalgae. *J. Appl. Phycol.* **2014,** *26*, 1359–1377. https://doi.org/10.1007/s10811-013-0154-9.

Rmiki, N. E. Reponses photoadaptatives de la chlorophycee *Haematococcus pluvialis* flotow et production d' ast axanthine, Universite des sciences et technologies lille i u.f.r. De sciences naturelles, 1999.

Romanenko, K.; Kosakovskaya, I. V.; Romanenko, P. A. Phytohormones of Microalgae: Biological Role and Involvement in the Regulation of Physiological Processes. Pt I. Auxins, Abscisic Acid, Ethylene. *Int. J. Algae.* **2015,** *17*, 275–289.

Ronga, D.; Biazzi, E.; Parati, K.; Carminati, D.; Carminati, E.; Tava, A. Microalgal Biostimulants and Biofertilisers in Crop Productions. *Agronomy* **2019,** *9* (4), 192.

Sharma, H. S. S.; Fleming, C.; Selby, C.; Rao, J. R.; Martin, T. Plant Biostimulants: A Review on the Processing of Macroalgae and Use of Extracts for Crop Management to Reduce Abiotic and Biotic Stresses. *J. Appl. Phycol.* **2014,** *26* (1), 465–490.

Shukla, P. S.; Borza, T.; Critchley A. T.; Prithiviraj, B. Carrageenans from Red Seaweeds as Promoters of Growth and Elicitors of Defense Response in Plants. *Front. Mar. Sci.* **2016,** *3* (81).

Shweta, T.; Krishna, M. P. Heavy Metal Detoxification Mechanisms by Microalgae: Insights from Transcriptomics Analysis. *Environ. Pollut.* **2021,** *285*, 117443. https://doi.org/10.1016/j.envpol.2021.117443.

Sialve, B.; Steyer, J. P. Les microalgues, promesses et défis. *Innovations Agronomiques*, **2013,** *26*, 25–39.

Tan, J. S.; Lee, S. Y.; Chew, K. W.; Lam, M. K.; Lim, J. W.; Ho S. H.; Show, P. L. A Review on Microalgae Cultivation and Harvesting, and their Biomass Extraction Processing Using Ionic Liquids. *Bioengineered* **2020,** *11* (1), 116–129.

Thomsen, L. Le secteur des micro-algues en Méditerranée perspectives et contribution au développement durable, institut de prospective économique du monde méditerranéen, 2016, p 94.

UNECA. The Green Economy in Morocco, A Strategic Goal Involving Partnership Dynamics and Intensified Coordination of Policies and Initiatives, in United Nations Economic Commission for North Africa, 2016.

Vigani, M. The Bioeconomy of Microalgae-Based Processes and Products. In *Handbook of Microalgae-Based Processes and Products*; Jacob-Lopes, E.; Maroneze, M. M.; Queiroz, M. I.; Zepka, L. Q., Academic Press, 2020; pp 799–821.
Wolff, A. L'utilisation des microalgues pour la fabrication de biocarburants: Analyse de la chaîne de valeur - contexte français et international, 2012.
Yakhin, O. I.; Lubyanov, A. A.; Yakhin, I. A.; Brown, P. H. Biostimulants in Plant Science: A Global Perspective. *Front. Plant Sci.* **2017,** *7* (2049).

CHAPTER 4

BIOFERTILIZERS AS AN EFFECTIVE TOOL FOR SUSTAINABLE AGRICULTURE

SAMAR A. GHANEM[1*], ALAA H. KHALLOUF[2,3], ABDEL RAHMAN MOHAMMAD AL-TAWAHA[4], JEYABALAN SANGEETHA[5], and DEVARAJAN THANGADURAI[6]

[1]*Soil and Water Department, Faculty of Agriculture, Tishreen University, Latakia, Syria*

[2]*General Commission for Scientific Agricultural Research (GCSAR), Damascus, Syria*

[3]*Soil and Land Reclamation Department, Faculty of Agriculture, Al-Baath University, Homs, Syria*

[4]*Department of Biological Sciences, Al Hussein bin Talal University, Maan, Jordan*

[5]*Department of Environmental Science, Central University of Kerala, Kasaragod, Kerala, India*

[6]*Department of Botany, Karnatak University, Dharwad, Karnataka, India*

**Corresponding author.*

ABSTRACT

Besides water and CO_2, soil nutrients are the essential elements for plant growth and land productivity, particularly in the 21st century, of the expanding population and urgent needs for food and water. Biofertilizer is a

Microbial Fertilizer Technology for Sustainable Crop Production. Jeyabalan Sangeetha, Abdel Rahman Mohammad Al-Tawaha, & Devarajan Thangadurai (Eds.)

DOI: 10.1201/9781003594826-4

promising practice of natural resources management, where it is an appropriate alternative to mineral fertilizers and the key for reducing soil and water pollution. Biofertilizers are the combination of one or more organisms that have the ability for promoting plant growth by increasing the amount of the available nutrients in soil solution, secreting some plant growth regulators, protecting the plant roots from soil-borne diseases, and assisting plants in resisting the environmental stresses as soil salinity, drought, and low soil fertility. There are many resources for biofertilizers such as N_2-fixing bacteria, phosphate, and potassium-solubilization bacteria, mycorrhiza. The efficiency of biofertilizers is based on the vitality of the organisms, its association with the host plant, and its ability for competitiveness in the rhizosphere. Biofertilizers are an effective solution for achieving sustainable development needs, conserving soil fertility, and getting high-quality crops yields at low costs in appraisal with mineral fertilizers.

4.1 INTRODUCTION

Farming production in many countries faces many problems related to several factors (Abebe et al., 2005; Abera et al., 2005; Al-Tawaha et al., 2005; Al-Kiyyam et al., 2008; Al-Tawaha et al., 2010a,b; Ananthi et al., 2017; Al-Tawaha et al., 2017; Al-Tawaha et al., 2018a,b; Khallouf et al., 2019). One of the most important issues is the availability of nutrients that are required for the plant in appropriate quantities and optimal times through proper management. Although there are huge quantities of nutrients found in the soil, most of them are unavailable forms for the plant intake. With the population increasing and increasing demand for food, lead to a change in land management systems and the increasingly usage of technologies which provide with greater crop production regardless of potential damage. The intense usage of chemical fertilizers has caused the apparently damage of soil, water contamination, in addition to heavy metals accumulation in plant tissues. Besides, soil organisms, which are highly active in the soil, are strongly damaged (Kumar and Yadav, 2001; Farnia and Hasanpoor, 2015). Despite the significant role of chemical fertilizers in improving crop production within short time, the unbalanced additions have negative effects on soil and the environment, moreover high costs (Khallouf et al., 2019; Qureshi et al., 2019). Recently, attention is directed toward sustainable agriculture to reach the optimum level of production, thus upgrading in quality and obtaining the maximum production that can be achieved based on clean production specifications, while preserving the fertility of the land and crop production (Shen, 2010). According to this framework, the technique of using biocontrol and biofertilizer is one of the most vital modern agricultural techniques used. Biofertilizer is substances that contain microorganisms applied to the soil or the surfaces of plants, roots, or seeds, to improve the uptake of nutrients. As it may inhabit the rhizosphere or pass in the plant, its function is to encourage plant growth by supplying micro and macronutrients (Vessey, 2003; Figueiredo et al., 2011; Suyal et al., 2016). Its success is influenced by the efficiency of the used organism, its corresponding with the host plant, its aggregation, and its viability (Swami, 2020). There are many different sources of biofertilizers, including nitrogen fixation bacteria (*Azotobacter*, *Rhizobium*) and soluble phosphorous organisms (*Pseudomonas* spp.), which stimulate and encourage the growth of plants (Malusa and Vassilev, 2014).

4.2 DIFFERENT TYPES OF BIOFERTILIZERS

Biofertilizers can be grouped into many types according to nature and function: Nitrogen (N_2) fixing bacteria, phosphorous solubilization bacteria, phosphate and nitrogen mobilizing, micro-nutrients solubilizing and plant growth promoting rhizobacteria (PGPR).

4.2.1 PHOSPHATE BIOFERTILIZERS

Phosphorus is the second essential plant nutrient following nitrogen, and its main role appears in plant growth. Gupta et al. (2012) mentioned that the fundamental metabolic reactions are photosynthesis, biological oxidation, and cell divisions. Phosphorous may be distinguished from other elements by its chemical peculiarity, in terms of its trend to fixed in acid soils (precipitated by iron and aluminum oxides), and calcareous soil (by its interaction with calcium, since calcium phosphate forms and adsorbs on the surface of calcium carbonate colloids), and it turns into insoluble compounds to the plant uptake (Awasthi et al., 2011). Generally, there are two groups of external factors that affect the availability of phosphorous: biological factors and host-related factors.

The first group relates to the activity of soil organisms and it includes some methods to facilitate phosphate nutrition in the soil in several ways such as:

- Plant root infection with mycorrhizal fungi and the best results can be reported in the poor and medium soil of available phosphorous (Marschner and Dell, 1994), where mycorrhizal hyphae can extend the soil that is free of plant roots and thus it plays the role of extra roots of the plant (Allouch and Sander, 1992).
- Root infection with rhizobium, where nitrogen-fixing leguminous can use phosphorous by its ability to release 0.4–0.8 mol of H^+ to the rhizosphere from each mole of N_2 that is in the root nodes (Mengel and Kirkby, 2001).
- Using phosphate-soluble bacteria and fungi, which can dissolve insoluble phosphorous (P) and convert it into a soluble form by excreting several acids (organic acids, mineral acids), humic substances, CO_2, and others that contribute to reducing pH (Illmer and Schinner, 1995; Quoreshi et al., 2019) (Table 4.1).
- Increasing the rhizosphere content of beneficial microorganisms, and increasing the activity of the microorganisms that decomposed the organic substances by enriching the rhizosphere with organic root materials (carbohydrates, amino acids, and organic acids).

Thus, the mycorrhizal hyphae can cause changes in the pH values by secreting organic acids or increasing the activity of the phosphatase enzyme that releases $H_2PO_4^-$ ions from mineral or organic forms (Smith and Reed, 1997).

4.2.2 RHIZOBIUM AS SYMBIOTIC NITROGEN FIXERS

Nitrogen is the first macro-nutrient for plant growth and its development. Its value shows in entering for protoplasm composition, also as a key component of all living cells and in the production of enzymes, amino nucleic acids, and other important compounds. Nitrogen differs from other mineral elements that exist in agricultural soils by its source which mostly is atmospheric air (as nitrogen makes up about 80% of the volume of the air), but it is not suitable for uptake by the plant. But, the parent materials and soil minerals do not contain N_2. Nitrogen is mainly uptaken by plants usually in the form of NH_4^+ and NO_3^- ions. The growing plants cannot directly benefit from N_2 gaseous nitrogen, and it must be converted into ammonia through a series of reactions by processes called nitrogen fixation (Azotication). It is usually accomplished by various microorganisms in the rhizosphere where they live either free or inside the root tissues.

TABLE 4.1 Biodiversity of Potential P Solubilizing Microorganisms.

Bacteria	*Alcaligenes* sp., *Aerobactor aerogenes*, *Achromobacter* sp., *Actinomadura oligospora*, *Agrobacterium* sp., *Azospirillum brasilense*, *Bacillus* sp., *Bacillus circulans*, *B. cereus*, *B. fusiformis*, *B. pumilus*, *B. megaterium*, *B. mycoides*, *B. polymyxa*, *B. coagulans*, *B. chitinolyticus*, *B. subtilis*, *Bradyrhizobium* sp., *Brevibacterium* sp., *Citrobacter* sp., *Pseudomonas* sp., *P. putida*, *P. striata*, *P. fluorescens*, *P. calcis*, *Flavobacterium* sp., *Nitrosomonas* sp., *Erwinia* sp., *Micrococcus* sp., *Escherichia intermedia*, *Enterobacter asburiae*, *Serratia phosphoticum*, *Nitrobacter* sp., *Thiobacillus ferroxidans*, *T. thioxidans*, *Rhizobium meliloti*, *Xanthomonas* sp.
Fungi	*Aspergillus awamori*, *A. niger*, *A. tereus*, *A. flavus*, *A. nidulans*, *A. foetidus*, *A. wentii*, *Fusarium oxysporum*, *Alternaria teneius*, *Achrothcium* sp., *Penicillium digitatum*, *P lilacinium*, *P. balaji*, *P. funicolosum*, *Cephalosporium* sp., *Cladosporium* sp., *Curvularia lunata*, *Cunnighamella*, *Candida* sp., *Chaetomium globosum*, *Humicola inslens*, *H. lanuginosa*, *Helminthosporium* sp., *Paecilomyces fusisporous*, *Pythium* sp., *Phoma* sp., *Populospora mytilina*, *Myrothecium roridum*, *Morteirella* sp., *Micromonospora* sp., *Oideodendron* sp., *Rhizoctonia solani*, *Rhizopus* sp., *Mucor* sp., *Trichoderma viridae*, *Torula thermophila*, *Schwanniomyces occidentalis*, *Sclerotium rolfsii*
Actinomycetes	*Actinomyces*, *Streptomyces*
Cyanobacteria	*Anabena* sp., *Calothrix braunii*, *Nostoc* sp., *Scytonema* sp.
VAM	*Glomus fasciculatum*

Source: Adapted from Sharma et al. (2013), Creative Commons Attribution 2.0 International.

Many problems result from the excessive use of mineral nitrogen fertilization, such as a reduction in the fixation of N_2 because of the presence of mineral nitrogen in sufficient quantity; contamination of groundwater through leaching and washing processes, and surface water pollution through the Eutrophication phenomenon. As a result, the leaching nitrogen in the water affects organisms and others (Bhattacharjee et al., 2008). Numerous microorganisms can fix N_2, but it differs in its mechanisms: some of the microorganisms can fix atmospheric nitrogen in a symbiotic nitrogen fixation method, including Rhizobia legumes, where the root nodes are the place where this relationship occurs and nitrogen conversion. This process is comparable in terms of its importance to the photosynthesis process. On the other hand, *Azotobacter* and *Azospirillum* can fix atmospheric nitrogen in a freeway; they have a process called Azotication because the main part of fixation is due to the group of *Azotobacter* bacteria. The *Azospirillum* inoculation is used for grassy seeds as a free nitrogen fixer in the soil, and experiments have shown that it can change the structure of the roots by the number of sidy roots and external rooty hairs increasing, that leads to expanding the root reaction area, which aids to uptaking sufficient nutrients (Bashan et al., 2004; Mahdipour et al., 2012). Some of symbiotic N_2-fixing organisms are as follows:

4.2.2.1 RHIZOBIUM

Rhizobiaceae family is the symbiotic N_2-fixing bacteria that usually exist in legume rhizosphere. *Rhizobium* inoculants have a great importance in sustainable agriculture, because of their ability of N_2 incorporation of specific legumes (Sheraz et al., 2010). It is reported that the amount of fixed N_2 by *Rhizobium* in legume's root nodules was estimated by 14 million tons at global level which is about 15% of industrial N. Choosing the suitable *Rhizobium* cultures is the key for the high yield of many legume crops, where the effective nodulation is related to selecting a certain species of *Rhizobium* in order to develop effective nodules. Many *Rhizobia* strains can infect various types of legumes, but only effective strains result in effective nodules and as a result improving in plant growth (Cooper, 2004). For each crop or tree type of legumes has particular *Rhizobium* inoculants. For instance, *R. trifolii* for berseem, *R. meliloti* for lucerne, *R. phaseoli* for green gram and black gram, *R. japonicum* for soya bean, *R. leguminosarum* for pea and lentil, *R. lupini* for chickpea, and *Rhizobium* spp. for cowpea.

4.2.2.2 MYCORRHIZA

In the late of the nineteenth century, Arbuscular Mycorrhizal Fungi (AMF) was first well defined by Albert Bernand. AMF refers to the relationship between plant roots and mycorrhiza fungi. AMF depends on providing plants root with uptake nutrients and receiving carbohydrates. This relation usually occurred between AMF and vascular flowering plants (Fisher and Long, 1992). AMF relationship involves three components: the roots of plant, the fungal elements between the root cells, and external hyphae in the rhizosphere (Gruhn et al., 2000). AMF have a major part of supplying plant with mineral nutrients (Jarecki et al., 2008). The significant role of AMF is increasing uptake of immobile phosphate in the soil, which usually interacts with cations including Ca^{2+}, Fe^{3+}, and Al^{3+} (Kloepper, 1978; Kacar et al., 2006) by rising exterior hyphae system, which may raise the surface area more than 40 times, and the amount of uptaking nutrients (Lewis et al., 1991) by forming enzymes and/or releasing organic compounds (Okon, 1994). The AMF mechanism to make phosphate available is by releasing phosphatases enzyme which leads to phosphate in organic P-compounds hydrolyzing (Ogunseitan, 2005; Ohno et al., 2005; Weller, 2007), consequently improving crop productivity under deficiency of phosphorus conditions (Jetiyanon et al., 2003). The external hyphae are important in the process of intake of NH^{4+}, insoluble microelements (Cu and Zn), and some soil cations (K^{+}, Ca^{2+}, Mg^{2+}, and Fe^{3+}) (Peix et al., 2001; Mantelin and Touraine, 2004). It is also reported that AMF, by improving plant nutrients, has a vital role in phytohormone balance, which in turn affects plant development and relieves the harsh impact of the environmental stresses. Consequently, biomass and crop yield increase (Stewart et al., 2005). AMF is the most widespread type of mycorrhizae, which infects about 90% plant species; in the contrast, the Brassicaceae (e.g., mustard, cabbage, and canola) and Chenopodiaceae (e.g., sugar beets and spinach) are not usually infected by AMF (Ikiz, 2003).

4.2.2.3 FREE-LIVING N_2 FIXATION BIOFERTILIZERS

4.2.2.3.1 *Azotobacter*

The *Azotobacter* bacteria, one of the genera of free-living bacteria, is used as a biofertilizer under the name (Azotobactrine) on various crops, due to its N_2-fixing capability; in addition to its secretion, it has some compounds that have the vital part in plant growth, like hormones, enzymes, vitamins, and growth regulators such as Uxinate (Sammarrai and Rahi, 2006).

4.2.2.3.2 *Azospirillum*

Azospirillum bacteria have long been known as one of the PGPR (Okon, 1994; Okon and Vanderleyden, 1997). It is described as microaerophilic organisms, but some of them are aerobic and gram-negative. Thus, they are fit for rice field conditions and growing. *Azospirillum* converts N_2 into NH^{4+} under micro-aerobic conditions at slight N_2 levels, where the nitrogenase complicated action (Burris and Roberts, 1993). The conversion of N_2 into NH_4^+ demands huge energy amount, in addition to the sensitivity of nitrogenase enzyme to oxygen. All lead to a limited biological N_2 fixation process (Steenhoudt and Vanderleyden, 2000). 20–40 kg N/ha is the amount of fixed N_2 by *Azospirillum* bacteria (Arun, 2007), but the most important role of *Azospirillum* apart from N_2 fixation is it releases growth-promoting substances, which aid to protect the plant root system from soil diseases, It also enhances the leaf area and crop productivity (Heinrich and Hess, 1985; Lopez-de-Victoria and Lovell, 1993; Lopez-de-Victoria et al., 1994). *Azospirillum* associates with the C_4-dicarboxylic pathway of photosynthesis plants; thus, plants such as maize, sorghum, sugarcane, pearl millet (Singh et al., 2016) are able to benefit from *Azospirillum*–root symbiosis association (Arun, 2007). The well-known varieties have been defined: *A. lipoferum*, *A. brasilense* (Tarrand et al., 1978), *A. amazonense* (Magalha-es et al., 1983), *A. halopraeferens* (Reinhold et al., 1987), and *A. irakense* (Khammas et al., 1989).

4.2.2.3.3 *Azolla*

Mahajan et al. (2003a) reported that nearly 100–150 kg N/ha/year can be fixed by *Azolla* in rice fields with *Anabaena*. *Azolla* is also introduced as a compost by applying it before rice cultivation (Yadav et al., 2014). It had been widely used in China as a green manure because of its features as a cheap source of nitrogen before the 1960s. Prior studies realized that at slight N levels, the ability of *Azolla* for raising the N reduction rates and rice productivity had been confirmed (Singh et al., 1988; Manna and Singh, 1989).

4.2.2.3.4 *Blue Green Algae (BGA)*

The combination of both BGA and *Azolla* is usually contributing together to increasing the rice yield by providing plant nutrients. It is found that

BGA can supply plants through biological fixation with 50–100 kg N/ha/ year. Also, it plays the key part in plant growth by providing plant growth regulating substances under paddled situations (Mahajan et al., 2003b).

4.2.2.4 PHOSPHORUS AND NITROGEN MOBILIZING BIOFERTILIZERS

Arbuscular Mycorrhiza (AM fungi), Ectomycorrhiza (ECM), Ericoid Mycorrhiza (EM), and Orchid Mycorrhiza (OM) are considered potential biofertilizer for many crops and plants. Fungal mycelia enter and associate with plant roots in several forms based on the involved mycorrhizal types. Therefore, it is increasing the surface area of roots, creating a symbiotic association, and increasing volume of soils accessible by the root system and improving the uptake of nutrients from the soil system.

4.3 USING BIOFERTILIZER IN ENVIRONMENTAL STRESSES AND SOIL FERTILITY PROBLEMS

4.3.1 ACIDIC SOILS

Fertility problems appear in acidic soils because of its low pH value. When the pH is less than 4, not only is the negative impact of an increase in hydrogen ions, but rather also the presence of other causes such as a lack or toxicity of some nutrients are necessary for plant growing. As the concentration of aluminum (Al^{+3}), both soluble and exchangeable, increases—along with elevated levels of manganese—deficiencies in phosphorus, calcium, and magnesium typically occur. Therefore, the deficiency of phosphorous, calcium, and magnesium usually occurs. Acidic soils are generally considered poor in nitrogen due to high rainfall which leads to leaching N_2.

Many studies have reported the significant role of the biofertilizer in facing the difficulties of phosphorous availability in acidic soils, as the phosphorous associated with iron and aluminum (Al-P) constitutes 1–25% of the total phosphorous in acid soils.

The phosphorous associated with calcium (Ca-P) constitutes 40% in calcareous soils in major soil orders (Dotaniya et al., 2013, 2014) such as Ultisols, Acrisols, and Ferralsols in the FAO classification (UNESCO and FAO, 1974).

Various studies carried out by Alloush et al. (2001) indicated that mycorrhizal hyphae can increase the utilization factor of plants by the phosphate rock which is introduced to acidic soils, and usually, the utilization factor is better in the incidence of organic substances (Alloush et al., 2000). The Vesicular arbuscular mycorrhiza (VAM), a symbiotic mycorrhiza, can infect the root and contribute to supply of mineral elements and water to plant roots, where the hyphae of this fungi extend over long distances, multiplying the root stretch, thereby reaching additional areas in the soil that plant roots do not reach, and it works to provide the necessary nutrients for the plant such as molybdenum, zinc, phosphorous, and copper. In soils under low acidic tropical conditions, it contains the necessary positive ions (calcium and magnesium) as well as in the case of coarse roots of limited growth (Swami, 2020) (Figure 4.1).

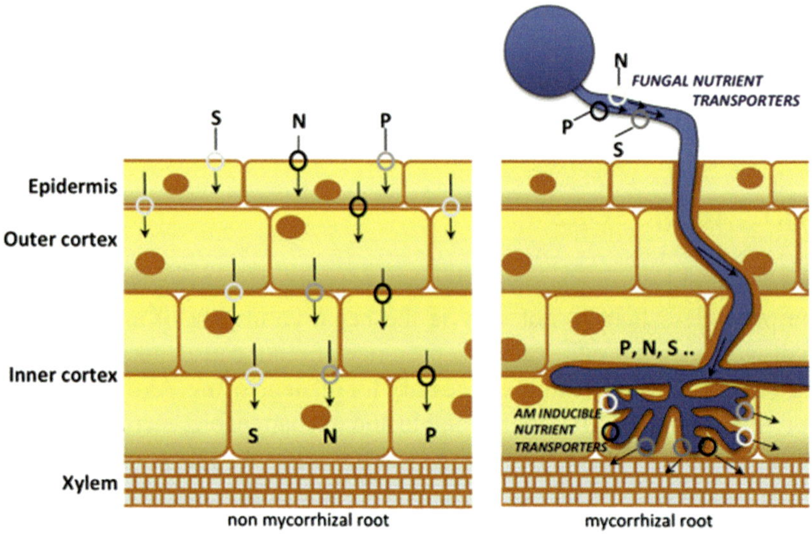

FIGURE 4.1 Direct and symbiotic plant nutrient uptake. Plants can absorb nutrients directly from the soil and through arbuscular mycorrhizal (AM) fungal mycelia. It is known that P is mainly absorbed throughout the fungus whereas the knowledge for the other macronutrient is still scattered. In the direct pathway, S, N, and P are absorbed by plant nutrient transporters, which accumulate in the epidermis and cortex (represented by white, grey, and black circles). In the mycorrhizal pathway, S, N, and P are taken up into AM fungal hyphae by fungal nutrient transporters (Tisserant et al., 2012) and are translocated to intracellular fungal structures in root cortical cells.

Source: Reproduced with permission from Giovannetti et al. (2017), Copyright © Elsevier (2017).

4.3.2 SOIL SALINITY—STRESS CONDITION

Soil salinity is a well-known phenomenon in arid and semi-arid lands all over the world, due to the deposition of salts in the soil profile. It causes an increase in the electrical conductivity of soils. The efficiency of nitrate and phosphate fertilizers added under these conditions also decreases. Experiments have reported that the use of the PGPR increased the resistance of plants to salinity by applying changes to the root structure where the level of indole-3-acetic acid (IAA) increased, which allowed plants to uptake more nutrients under soil salinity conditions (Vacheron et al., 2013; Goswami et al., 2014). Treatment of seeds with triple-enriched inoculation such as phosphate-dissolving bacteria and VAM (Vesicular arbuscular mycorrhiza) (VAM) 600 g/ ha gave the highest yield under conditions of salt stress (Kamaraj et al., 2018). Hussain et al. (2002) reported that *Rhizobium trifoli* inoculation with *Trifolium alexandrium* resulted from a huge biomass and great changes under the saline stress conditions (ACC) deaminase, since in drought-stress conditions it leads to a decrease in antioxidant activity, promoting proline production, amino acids, and carbohydrates in plants (Vardharajula et al., 2011).

4.3.3 DROUGHT-STRESS CONDITION

Many growth indicators are affected by drought-stress conditions, where the lack in the available water amount leads to a reduction in cell size and in the integrity of the cell membrane, a reduction in crop production where drought stimulates leaf senescence (Tiwari et al., 2015). The plants associated with microbes can overcome such drought conditions by many mechanisms including:

- Providing water content and supplying the plant with nutrients
- Supplying better conditions for growth through improving the drought-resistance system
- Producing some promoting growth substances for instance: IAA, cytokinins, and abscisic acid, since IAA increases the cell division, influences obviously in the plant tissues differentiation, including vascular and root tissues (Goswami et al., 2015)

- Producing bacterial substances which aid some plants to increase drought tolerance by increasing the amount of water in plant, protein, and carbohydrates concentrations (Naseem and Bano, 2014)
- Producing 1-aminocyclopropane-1-carboxylate (ACC) deaminase, since in drought-stress conditions it leads to a decrease in anti-oxidant activity, promoting proline production, amino acids, and carbohydrates in plants (Vardharajula et al., 2011)

4.3.4 CALCAREOUS SOIL CONDITIONS

More than 30% of surface earth land is described as calcareous soils that usually suffered from a shortage of plant nutrients, especially micronutrients (Fe, Zn, Cu, and Mn) due to high soil pH values and formation of immobile complexes of calcium-phosphate (Ca-P) (Marschner, 1995). The second issue observed in calcareous soils is chlorosis, caused by lime-induced iron deficiency. This condition results from impaired iron metabolism due to high calcium concentrations in the soil (Faust, 1989). Therefore, symptoms of iron and phosphorous deficiency appear in calcareous soils since high pH is the most serious explanation. As a conclusion, one-third of agricultural lands all over the world have a reduction of soil productivity (Wallace and Lunt, 1960). Phosphorus (P) is a key component of the energy transition in the plant, and it has significant part of nutrition for plants. The main functions of P are demonstrated in various physiological activities and biochemical reactions of all organisms (Rodriguez and Fraga, 1999; Vance et al., 2003). The mechanisms of insulating of P are usually precipitation and adsorption in the soil. P can be adsorbed on the soil particle surfaces and joined with soil cations, especially Fe, Al, and Ca, to develop unsolvable complexes (Wild, 1988). Consequently, majority of the used phosphorous become hard to uptaking by plant, but it is held in an unsolvable form. As a sequence, plant development and growth are mostly prohibited in calcareous soils, but the applied biofertilizers that contain PGPR can alleviate this condition. Biofertilizers can enhance plant development and its yield by direct and indirect methods. Biofertilizers might have opposed effects against phytopathogenic microorganisms in the rhizosphere (Bashan and de-Bashan, 2005). On the other hand, PGPR can excrete several growth-promoting substances like abscisic acids, cytokinins, gibberellins, and auxin, solubilizing inorganic phosphate and mineralization of organic phosphate and/or their nutrients and a symbiotic fixation of nitrogen (Glick, 1995; Zahir et al., 2004).

4.4 CONCLUSION

Nowadays, the effects of climate change and its associated conditions are impacting plant growth and productivity. Increasing usage of mineral fertilizers has caused great damage to both the soil and the environment, and therefore the urgent need for agricultural practices that achieve sustainability and solve environmental problems at the same time secure the population's food needs has emerged. Therefore, various types of biological fertilizers have proven their high ability to improve soil fertility and its quality, provide the necessary nutrients, and enhance plant growth. Thus, biofertilizers are the one of the most important low-cost, high-efficiency, and safest alternatives to chemical fertilizers in agriculture.

KEYWORDS

- **acidic soils**
- ***Azospirillum***
- ***Azotobacter***
- **biofertilizers**
- **blue green algae**
- **calcareous soils**
- **drought stress**

REFERENCES

Abebe, G.; Hattar, B.; Al-Tawah, A. M. Nutrient Availability as Affected by Manure Application to Cowpea (*Vigna unguiculata* L. Walp.) on Calacarious Soils. *J. Agric. Soc. Sci.* **2005,** *1*(1), 1–6.

Abera, T.; Feyisa, D.; Yusuf, H.; Nikus, O.; Al-Tawaha, A. R.; Grain Yield of Maize as Affected by Biogass Slurry and NP Fertilizer Rate at Bako, Western Oromiya, Ethopia. *Biosci. Res.* **2005,** *2*, 31–38.

Al-Kiyyam, M. A.; Turk, M.; Al-Mahmoud, M.; Al-Tawaha, A. R. Effect of Plant Density and Nitrogen Rate on Herbage Yields of Marjoram Under Mediterranean conditions. *J. Am. Eurasian J. Agric. Environ. Sci.* **2008,** *3*, 153–158.

Allouch, G.; Sanders, F. E. *The Role of Root Exudates in Iron and Phosphorus Acquisition by Fe-Stressed Chickpea Plants*. Science Week: Ddamascus, Syria, 1992; pp 39–61.

Alloush, G. A.; Clark, R. B. Maize Response to Phosphate Rock and Arbuscular Mycorrhizal Fungi in Acidic Soil. *Commun. Soil Sci. Plant Anal.* **2001,** *1-2*, 231–254.

Alloush, G. A.; Zeto, S. K.; Clark, R. B. Phosphorus Source, Organic Matter and Arbuscular Mycorriza Effects on Growth and Mineral Acquisition of Chickpea Grow in Acidic Soil. *J. Plant Nutr.* **2000,** *23*, 1351–1369.

Al-Tawaha, A. M.; Turk, M.; Hameed, K.; Assaf, T.; Tahhan, R.; Al-Jamali, A. F. Effects of Application of Olive Mill By-Products on Lentil Yield and their Symbiosis with Mycorrhizal Fungi Under Arid Conditions. *Crop Res.* **2010,** *40*(3), 66–75.

AL-Tawaha, A. R. M.; Nidal, O. Use of *Sorghum* and Maize Allelopathic Properties to Inhibit Germination and Growth of Wild Barley (*Hordeum spontaneum*). *Not. Bot. Horti Agrobo. Cluj-Napoca* **2010,** *38*(3), 124–127.

Al-Tawaha, A. R.; Al-Karaki, G.; Al-Tawaha, A. R.; Sirajuddin, S. N.; Makhadmeh, I.; Wahab, P. E. M.; Youssef, R. A.; Al Sultan, W.; Massadeh, A. Effect of Water Flow Rate on Quantity and Quality of Lettuce (*Lactuca sativa* L.) in Nutrient Film Technique (NFT) Under Hydroponics Conditions. *Bulg. J. Agric. Sci.* **2018a,** *245*, 791–798.

Al-Tawaha, A. R.; Al-Tawaha, A. R.; Alu'datt, M.; Al-Ghzawi, A. L.; Wedyan, M.; Al-Obaidy, S. D. A.; Al-Ramamneh, E. A. D. Effects of Soil Type and Rainwater Harvesting Treatments in the Growth, Productivity and Morphological Trains of Barley Plants Cultivated in Semi-Arid Environment. *Aust. J. Crop Sci.* **2018b,** *12*(6), 975–979.

Al-Tawaha, A. R.; McNeil, D. L.; Yadav, S. S.; Turk, M.; Ajlouni, M.; Abu-Darwish, M. S.; Al-Ghzawi, A. L. A.; Al-udatt, M.; Aladaileh, S. Integrated Legume Crops Production and Management Technology. In *Climate Change and Management of Cool Season Grain Legume Crops*; Yadav, S. S., McNeil, D. L., Redden, R., Sharanagouda. A, P., Eds.; Springer: Dordrecht, 2010; pp 325–349.

Al-Tawaha, A. R.; Turk, M. A.; Abu-Zaitoon, Y. M.; Aladaileh, S. H.; Al-Rawashdeh, I. M.; Alnaimat, S.; Al-Tawaha, A. R. M.; Alu'datt, M. H.; Wedyan, M. Plants Adaptation to Drought Environment. *Bulg. J. Agric. Sci.* **2017,** *23*(3), 381–388.

Ananthi, T.; Amanullah, M. M.; Al-Tawaha, A. R. M. S. A Review on Maize-Legume Intercropping for Enhancing the Productivity and Soil Fertility for Sustainable Agriculture in India. *Adv. Environ. Biol.* **2017,** *11*(5), 49–63.

Arun, K. S. *Bio-Fertilizers for Sustainable Agriculture: Mechanism of P-Solubilization*; Agrobios Publishers: Jodhpur, India, 2007; pp 196–197.

Awasthi, R.; Tewari, R.; Nayyar, H. Synergy Between Plants and P-Solubilizing Microbes in Soils: Effects on Growth and Physiology of Crops. *Int. Res. J. Microbiol.* **2011,** *12*, 484–503.

Bashan, Y.; de-Bashan, L. E.; Bacteria/Plant Growth-Promotion. In *Encyclopedia of Soils in the Environment*; Hillel, D., Ed., Elsevier: Oxford, 2005; pp 103–115.

Bashan, Y.; Holguin, G.; De-Bashan, L. E. *Azospirillum*-Plant Relationships: Physiological, Molecular, Agricultural, and Environmental Advances (1997–2003). *Can. J. Microbiol.* **2004,** *8*, 521–577.

Bhattacharjee, R. B.; Mukhopadhyay, S. N.; Singh, A. Use of Nitrogen-Fixing Bacteria as Biofertiliser for Non-Legumes: Prospects and Challenges. *Appl. Microbiol. Biotechnol.* **2008,** *80*, 199–209.

Burris, R. H.; Roberts, G. P. Biological Nitrogen Fixation. *Annu. Rev. Nutr.* **1993,** *13*, 317–335.

Cooper, J. Multiple Responses of *Rhizobia* to Flavonoids During Legume Root Infection. In *Advances in Botanical Research: Incorporating Advances in Plant Pathology*; Callow, J. A., Ed.; Academic Press: Cambridge, 2004; pp 1–62.

Dotaniya, M. L.; Kushwah, S. K.; Rajendiran, S.; Coumar, M. V.; Kundu, S.; Rao, A. S. Rhizosphere Effect of Kharif Crops on Phosphatases and Dehydrogenase Activities in a Typic Haplustert. *Natl. Acad. Sci. Lett.* **2014,** *2*, 103–106.

Dotaniya, M. L.; Pingoliya, K. K.; Meena, H. M.; Prasad, D. Status and Rational Use of Rock Phosphate in Agricultural Crop Production - A Review. *Agric. Sustain. Dev.* **2013,** *1*, 103–108.

Farnia, A.; Hasanpoor, K. Comparison Between Effect of Chemical and Biological Fertilizers on Yield and Yield Components in Wheat (*Triticum aestivum* L.). *Indian J. Nat. Sci.* **2015,** *30*, 7792–7800.

Faust, M. *Physiology of Temperate Zone Fruit Trees*; John Wiley & Sons, Inc.: New York, 1989.

Figueiredo, M. V. B.; Seldin, L.; de Araujo.; F. F.; Mariano, R. L. R. Plant Growth Promoting Rhizobacteria: Fundamentals and Applications. In *Plant Growth and Health Promoting Bacteria;* Maheshwari, D., Ed.; Springer: Berlin, Heidelberg, 2011; pp 21–43.

Fisher, R. F.; Long, S. R. Rhizobium-Plant Signal Exchange. *Nature* **1992,** *357*(6380), 655–660.

Giovannetti, M.; Volpe, V.; Salvioli, A.; Bonfante, P. Fungal and Plant Tools for the Uptake of Nutrients in Arbuscular Mycorrhizas. In *Mycorrhizal Mediation of Soil: Fertility, Structure, and Carbon Storage*; Nancy, C. J., Catherine, G., Jan, J., Eds.; Elsevier: Netherlands, 2017; pp 107–128. doi:10.1016/b978-0-12-804312-7.00007-3.

Glick, B. R. The Enhancement of Plant Growth by Free Living Bacteria. *Can. J. Microbiol.* **1995,** *41*, 109–117.

Goswami, D.; Dhandhukia, P.; Pate, I. P.; Thakker, J. N. Screening of PGPR from Saline Desert of Kutch: Growth Promotion in *Arachis hypogea* by *Bacillus licheniformis* A2. *Microbiol. Res.* **2014,** *1*, 66–75.

Goswami, D.; Thakker, J. N.; Dhandhukia, P. C. Simultaneous Detection and Quantification of Indole-3-Acetic Acid (IAA) and Indole-3-Butyric Acid (IBA) Produced by Rhizobacteria from l-Tryptophan (Trp) Using HPTLC. *J. Microbiol. Methods* **2015,** *110*, 7–14.

Gruhn, P.; Goletti, F.; Yudelman, M. Integrated Nutrient Management, Soil Fertility, and Sustainable Agriculture: Current Issues and Future Challenges. In *Food, Agriculture and the Environment Discussion Paper 32*; International Food Policy Research Institute: Washington DC., 2000.

Gupta, M.; Kiran, S.; Gulati, A.; Singh, B.; Tewari, R. Isolation and Identification of Phosphate Solubilizing Bacteria Able to Enhance the Growth and aloin-A Biosynthesis of *Aloe barbadensis* Miller. *Microbiol. Res.* **2012,** *6*, 358–363.

Heinrich, D.; Hess, D. Chemotactic Attraction of *Azospirillum lipoferum* by Wheat Roots and Characterization of Some Attractants. *Can. J. Microbiol.* **1985,** *31*, 26–31.

Hussain, N.; Mujeeb, F.; Tahir, M.; Khan, G. D.; Hassan, N. M.; Bari, A. Effectiveness of *Rhizobium* Under Salinity Stress. *Asian J Plant Sci.* **2002,** *1*, 12–14

Ikiz, O. The Effects of Mycorrhiza on Soilless Pepper Cultivation. PhD Thesis, Department of Horticulture, Institute of Natural and Applied Sciences, University of Cukurova, Adana, Turkey, 2003.
Illmer, P.; Schinner, F. Solubilization of Inorganic Calcium Phosphates - Solubilization Mechanisms. *Soil. Biol. Biochem.* **1995,** *3*, 257–263.
Jarecki, M. K.; Parkin, T. B.; Chan, A. S.; Hatfield, J. L.; Jones, R. Greenhouse Gas Emissions From Two Soils Receiving Nitrogen Fertilizer and Swine Manure Slurry. *J. Environ. Qual.* **2008,** *37*(4), 1432–1438.
Jetiyanon, K.; Fowler, W. D.; Kloepper, J. W. Broad-Spectrum Protection Against Several Pathogens by PGPR Mixtures Under Field Conditions. *Plant Dis.* **2003,** *87*, 1390–1394.
Kacar, B.; Katkat, B.; Ozturk, S. *Plant Physiology*; Nobel Publishing Distribution: Ankara, Turkey, 2006.
Kamaraj, A.; Padmavathi, S. Alleviation of Saline Salt Stress Through Pre-Sowing Biofertilizer Seed Treatment on Crop Growth and Seed Yield in Green Gram CV ADT3. *J. Pharmacogn. Phytochem.* **2018,** *1*, 2205–2209.
Khammas, K. M.; Ageron, E.; Grimont, P. A. D.; Kaiser, P. *Azospirillum irakense* sp. nov., A Nitrogen-Fixing Bacterium Associated with Rice Roots and Rhizosphere Soil. *Res. Microbiol.* **1989,** *140*, 679–693.
Kloepper, J. W.; Schroth, M. N. *Plant Growth-Promoting Rhizobacteria on Radishes.* Proceedings of the 4th International Conference on Plant Pathogenic Bacteria, Vol. 2., Station de Pathologie Vegetale et Phyto-Bacteriologie, Gilbert-Clarey, Tours, 1978; pp 879–882.
Kumar, A.; Yadav, D. S. Long-Term Effects of Fertilizers on the Soil Fertility and Productivity of a Rice–Wheat System. *J. Agron. Crop Sci.* **2001,** *1*, 47–54.
Lewis, J. A.; Papavizas, G. C. Biocontrol of Plant Diseases: The Approach for Tomorrow. *Crop Prot.* **1991,** *10*, 95–105.
Lopez-de-Victoria, G.; Lovell, C. R. Chemotaxis of *Azospirillum* Species to Aromatic Compounds. *Appl. Environ. Microbiol.* **1993,** *59*, 2951–2955.
Lopez-de-Victoria, G.; Fielder, D. R.; Zimmer-Faust, R. K.; Lovell, C. R. Motility Behavior of *Azospirillum* Species in Response to Aromatic Compounds. *Can. J. Microbiol.* **1994,** *40*, 705–711.
Magalhaes, F. M.; Baldani, J. I.; Souto, S. M.; Kuykendall, J. R.; & Do-bereiner, J. A New Acid-Tolerant *Azospirillum* Species. *An. Acad. Bras. Cienc.* **1983,** *55*, 417–429.
Mahajan, A.; Bhagat, R.M.; Trikha, A. Fertilizing Bio-Fertilizers. *Agric. Today* **2003a,** *6*, 36–37.
Mahajan, A.; Choudhary, A. K.; Jaggi, R. C.; Dogra, R. K. Importance of Bio-Fertilizers in Sustainable Agriculture. *Farm Forum.* **2003b,** *3*, 17–19.
Mahdipour, M. M.; Emtiazi, G.; Salehi, Z. Enhanced Auxin Production by *Azospirillum* Pure Cultures from Plant Root Exudates. *J. Agric. Sci. Technol.* **2012,** *14*, 985–994.
Malusá, E.; Vassilev, N. A. Contribution to Set a Legal Framework for Biofertilisers. *Appl. Microbiol. Biotechnol.* **2014,** *15*, 6599–6607.
Manna, A. B.; Singh, P. K. Rice Yields as Influenced by *Azolla* N_2 Fixation and Urea N Fertilization. *Plant Soil*, **1989,** *114*, 63–68.
Mantelin, S.; Touraine, B. Plant Growth-Promoting Bacteria and Nitrate Availability: Impacts on Root Development and Nitrate Uptake. *J. Exp. Bot.* **2004,** *55*, 27–34.
Marschner, H.; Dell, B. Nutrient Uptake in Mycorrhizal Symbiosis. *Plant Soil* **1994,** *159*, 89–102.

Marschner, H.; *Mineral Nutrition of Higher Plants*; Academic Press: London, UK, 1995.
Mengel, K.; Kirkby, E. A. *Principles of Plant Nutrition*; Kluwer Academic Publishers: Dordrecht, The Netherlands, 2001.
Naseem, H.; Bano, A. Role of plant Growth-Promoting Rhizobacteria and their Exopolysaccharide in Drought Tolerance of Maize. *J. Plant Interact.* **2014,** *9*, 689–701.
Ogunseitan, O. *Microbial Diversity: Form and Function in Prokaryotes*. Blackwell Science Ltd.: Massachusetts, USA, 2005, pp 142.
Ohno, T.; Griffin, T. S.; Liebman, M.; porter, G. A. Chemical Characterization of Soil Phosphorus and Organic Matter in Different Cropping Systems in Maine, USA. *Agric. Ecosyst. Environ.* **2005,** *105*, 625–634.
Okon, Y.; Vanderleyden, J. Root-Associated *Azospirillum* Species Can Stimulate Plants. *ASM News* **1997,** *63*, 366–370.
Okon, Y. *Azospirillum/Plant Associations*; CRC Press: Boca Raton, FL, 1994; p 175.
Peix, A.; Rivas-Boyero, A. A.; Mateos, P. F.; Rodriguez-Barrueco, C.; Martínez-Molina, E.; Velazquez, E. Growth Promotion of Chickpea and Barley by a Phosphate Solubilizing Strain of *Mesorhizobium mediterraneum* Under Growth Chamber Conditions. *Soil Biol. Biochem.* **2001,** *33*, 103–110.
Quoreshi, A. M.; Suleiman, M. K.; Manuvel, A. J.; Sivadasan, M. T.; Jacob, S.; Thomas, R. *Biofertilizers for Agriculture and Reclamation of Disturbed Lands: An Eco-Friendly Resource for Plant*; Environment and Life Sciences Research Center, Kuwait Institute for Scientific Research: Kuwait, 2019; pp. 231–243.
Reinhold, B.; Hurek, T.; Fendrik, I.; Pot, B.; Gillis, M.; Kersters, K.; Thielemans, S.; De Ley, J. *Azospirillum halopraeferens* sp. nov., a Nitrogen-Fixing Organism Associated with Roots of Kallar Grass (*Leptochloa fusca* (L.) Kunth). *Int. J. Syst. Bacteriol.* **1987,** *37*, 43–51.
Rodriguez, H.; Fraga, R. Phosphate Solubilizing Bacteria and their Role in Plant Growth Promotion. *Biotechnol. Adv.* **1999,** *17*, 319–339.
Sharma, S. B.; Sayyed, R. Z.; Trivedi, M. H.; Gobi, T. A. Phosphate Solubilizing Microbes: Sustainable Approach for Managing Phosphorus Deficiency in Agricultural Soils. *SpringerPlus* **2013,** *2*, 587. https://doi.org/10.1186/2193-1801-2-587.
Shen, J. P.; Zhang L. M.; Guo, J. F.; Ray, J. L.; He, J. Z. Impact of Long-Term Fertilization Practices on the Abundance and Composition of Soil Bacterial Communities in Northeast China. *Appl. Soil Ecol.* **2010,** *46*, 119–124.
Sheraz, M. S.; Hassan, G. I.; Samoon, S. A.; Rather, H. A.; Dar, S. A.; Zehra, B. Biofertilizers in Organic Agriculture. *J. Phytol.* **2010,** *2*, 42–54.
Singh, A. L.; Singh, P. K.; Lata, P. Effects of Different Levels of Chemical Nitrogen (urea) on *Azolla* and Blue-Green Algae Intercropping with Rice. *Fertil. Res.* **1988,** *17*, 47–59.
Singh, M.; Dotaniya, M. L.; Mishra, A.; Dotaniya, C. K.; Regar, K. L.; Lata, M. Role of Biofertilizers in Conservation Agriculture. In *Conservation Agriculture*, Bisht, J. K., Ed.; Springer: Singapore, 2016; pp 113–134.
Smith, S. E.; Reed, D. J. *Mycorrhizal Symbiosis*; Academic Press: London, 1997.
Steenhoudt, O.; Vanderleyden, J. *Azospirillum*, A Free-Living Nitrogen-Fixing Bacterium Closely Associated with Grasses: Genetic, Biochemical and Ecological Aspects. *FEMS Microbiol. Rev.* **2000,** *24*, 487–506.
Stewart, L. I.; Hamel, C.; Hogue, R.; Moutoglis, P. Response of Strawberry to Inoculation with Arbuscular Myccorrhizal Fungi Under Very High Soil Phosphorus Conditions. *Mycorrhiza* **2005,** *15*, 612–619.

Suyal, D. C.; Soni, R.; Sai, S.; Goel, R. Microbial Inoculants as Biofertilizers. In *Microbial Inoculants in Sustainable Agricultural Productivity*; Singh, D., Singh, H., Prabha, R., Eds.; Springer: New Delhi, 2016.

Swami, S. Soil Microbes for Securing the Future of Sustainable Farming. *Int. J. Curr. Microbiol. Appl. Sci.* **2020,** *4*, 2687–2706.

Tarrand, J. J.; Krieg, N. R.; Do-Bereiner, J. A Taxonomic Study of the *Spirillum lipoferum* Group, with Descriptions of a New Genus, *Azospirillum* gen. nov. and Two Species, *Azospirillum lipoferum* (Beijerinck) comb. nov. and *Azospirillum brasilense* sp. nov. *Can. J. Microbiol.* **1978,** *24*, 967–980.

Tiwari, S.; Lata, C.; Chauhan, P. S.; Nautiyal, C. S. *Pseudomonas putida* Attunes Morpho-physiological, Biochemical and Molecular Responses in *Cicer arietinum* L. During Drought Stress and Recovery. *Plant Physiol. Biochem.* **2015,** *99*, 108–117.

UNESCO and FAO *Soil Map of the World 1: 5000000*. UNESCO: Paris, 1974.

Vacheron, J.; Desbrosses, G.; Bouffaud, M. L.; Touraine, B.; Moënne-Loccoz, Y.; Muller, D. Plant Growth-Promoting Rhizobacteria and Root System Functioning. *Front. Plant Sci.* **2013,** *4*, 356.

Vance, C. P.; Uhde-Stone, C.; Allan, D. L. Phosphorus acquisition and use: Critical adaptations by plants for securing a nonrenewable resource. *New Phytol.* **2003,** *157*, 423–447.

Vardharajula, S.; Ali, S. Z.; Grover, M.; Reddy, G.; Bandi, V. Drought Tolerant Plant Growth Promoting *Bacillus* spp.: Effect on Growth Osmolytes, and Antioxidant Status of Maize Under Drought Stress. *J. Plant Interact.* **2011,** *6*, 1–14.

Vessey, J. K. Plant Growth Promoting Rhizobacteria as Bio-Fertilizers. *J. Plant Soil* **2003,** *43*, 571–586.

Wallace, A.; Lunt O. R. Iron Chlorosis in Horticultural Plants, A Review. *J. Am. Soc. Hortic. Sci.* **1960,** *75*, 819–841.

Weller, D. M. *Pseudomonas* Biocontrol Agents of Soilborne Pathogens: Looking Back Over 30 Years. *Phytopathology,* **2007,** *97*, 250–256.

Wild, A. Plant Nutrients in Soil: Phosphate. In *Soil Conditions and Plant Growth*; Longman Scientific and Technical: Burnt Mill, UK, 1988; pp 695–742.

Yadav, R. K.; Abraham, G.; Singh, Y. V.; Singh, P. K. Advancements in the Utilization of *Azolla-Anabaena* System in Relation to Sustainable Agricultural Practices. *Proc. Indian Natl. Sci. Acad.* **2014,** *80*, 301–316.

Zahir, A. Z.; Arshad, M.; Frankenberger, W. T. Plant Growth Promoting Rhizobacteria: Applications and Perspectives in Agriculture. *Adv. Agron.* **2004,** *81*, 97–168.

CHAPTER 5

RHIZOBIA AS PLANT GROWTH-PROMOTING BIOFERTILIZERS

HIBA ALATRASH[1*], ABDEL RAHMAN MOHAMMAD AL-TAWAHA[2], YAMAN JABBOUR[3], ROGHAYEH KARIMIRAD[4], ABHIJIT DEY[5], SIDDHARTHA PATI[6], ABDEL RAZZAQ AL-TAWAHA[7], PUTERI E. MEGAT WAHAB[8], KHALDOUN OTHMAN AL SANE[9], R. Z. SAYYED[10], DEVARAJAN THANGADURAI[11], JEYABALAN SANGEETHA[12], AMREENA SULTAN[13], KUMAR PANDEY[14], IMRAN[15], AMANULLAH[15], SHAH KHALID[15], SAMAR S. SHAWAQFEH[16], EKATERINA KOZUHAROVA[17], SABAA JAWAD ABDULKADHIM[18], SAMIA KHANUM[19], and IFTIKHAR ALI[20]

[1]*University of Helsinki, Helsinki, Finland*

[2]*Department of Biological Sciences, Al Hussein bin Talal University, Maan, Jordan*

[3]*General Commission for Scientific Agriculture Research, Syria*

[4]*Department of Horticultural Sciences, Faculty of Agriculture and Natural Resources, University of Mohaghegh Ardabili, Ardabīl, Iran*

[5]*Department of Life Sciences, Presidency University, Kolkata, India*

[6]*NatNov Bioscience Private Limited, Balasore, Odisha, India*

[7]*National Agricultural Research Center (NARC), Baq'a 19381, Jordan*

[8]*Department of Crop Science, Faculty of Agriculture, University Putra Malaysia, Serdang, Selangor, Malaysia*

[9]*Directorate of Plant Biodiversity and Plant Genetic Resources, National Agricultural Research Center (NARC), Jordan*

[10]*Department of Microbiology, PSGVP Mandal's Arts, Science and Commerce College, Shahada, Maharashtra, India*

Microbial Fertilizer Technology for Sustainable Crop Production. Jeyabalan Sangeetha, Abdel Rahman Mohammad Al-Tawaha, & Devarajan Thangadurai (Eds.)

DOI: 10.1201/9781003594826-5

[11]Department of Botany, Karnatak University, Dharwad, Karnataka, India

[12]Department of Environmental Science, Central University of Kerala, Kasaragod, Kerala, India

[13]Division of Vegetable Science, SKUAST-K, Shalimar, Srinagar, Jammu and Kashmir, India

[14]Department of Biotechnology, School of Bioengineering and Biosciences, Lovely Professional University, Phagwara, Punjab, India

[15]Department of Agronomy, The University of Agriculture, Peshawar, Pakistan

[16]Department of Plant Production and Protection, College of Agriculture, Jerash University, Jerash, Jordan

[17]Medical University of Sofia, Faculty of Pharmacy, Department of Pharmacognosy, Sofia, Bulgaria

[18]Al Furat Al-Awsat Technical University, Al-Mussaib Technical College, Iraq

[19]Department of Botany, University of the Punjab, Lahore, Pakistan

[20]Center for Plant Sciences and Biodiversity, University of Swat, Charbagh, Pakistan

*Corresponding author.

ABSTRACT

Nitrogen is an important plant nutrient element since it is a necessary component of all amino acids and nucleic acids. Farmers in resource-poor areas might use biological N_2 fixation (BNF) as a free supply of nitrogen for higher crop yields. There is a complex molecular communication between the legume host and the rhizobial microsymbiont that leads to root nodule production. The symbiotic connection between rhizobia and legumes that results in nodule development is a complex chemical conversation. Throughout this intricate process many phytohormones can be produced by rhizobia and their legume hosts. On the other hand, as a broad group of bacteria, plant growth-promoting bacteria (PGPR) provide plants with growth benefits through numerous processes. The rhizosphere soil of plants frequently contains PGPR that can release protons to solubilize soil-bound phosphorus for plant usage.

5.1 INTRODUCTION

When it comes to integrated soil fertility management, the goal is to improve crop output while also increasing the efficiency with which nutrients are used (Hayyawi et al., 2018; Amanullah et al., 2019; Hani et al., 2019). This can be accomplished by the use of grain legumes, which improve soil fertility by facilitating biological nitrogen fixation, as well as the application of synthetic fertilizers (Turk et al., 2003a,b; Aludatt et al., 2019). Increased soil fertility can be achieved through the use of cover crops, which add organic matter to the soil and promote a healthy, fertile soil (Turk et al., 2003c; Nikus et al., 2004a, 2004b; Tawaha and Turk 2004; Turk and Tawaha 2004). The use of green manure or the cultivation of legumes, which fix nitrogen from the air through the process of biological nitrogen fixation; the use of micro-dose fertilizer applications, which replace nitrogen losses due to plant uptake and other processes; and the prevention of losses through leaching below the crop rooting zone (Al-Tawaha et al., 2005; Lee et al., 2005; Sulpanjan et al., 2005; Tawaha et al., 2005; Assaf et al., 2006; Turk et al., 2006). In order for soil to be fertile, nutrients must be continuously cycled between organic and inorganic forms. Mineralization occurs when microorganisms break down organic matter, such as plant and animal manure, to release inorganic nutrients into the soil solution (Al-Kiyyam et al., 2008; Al-Ajlouni et al., 2009; Al-Tawaha et al., 2010, 2018a, 2018b). Additional changes to those nutrients might be made possible by soil microorganisms. A process known as immobilization occurs when microorganisms compete with plants for nutrients, locking up the nutrients in the microbial biomass and making them unavailable to plants. In agriculture, soil fertility refers to a soil's ability to support plant growth by delivering critical plant nutrients, as well as desirable chemical, physical, and biological qualities that serve as a conducive environment for plant growth. The macronutrients nitrogen, phosphorus, and potassium, as well as sulfur, calcium, and magnesium, are found in plant nutrition. Micronutrients are mostly composed of the elements boron, chlorine, copper, iron, manganese, molybdenum, zinc, and selenium (Turk and Tawaha 2001; Tawaha and Turk 2002a, 2002b; Turk and Tawaha 2002a, 2002b; Tawaha et al., 2003; Abebe et al., 2005; Abera et al., 2005; Al-Tawaha et al., 2005; Elser et al., 2007; Al-Kiyyam et al., 2008; Hameed et al., 2008; Al-Tawaha et al., 2010; Ananthi et al., 2017; Al-Tawaha et al., 2017; Abu Obaid et al., 2018). Rhizobia, or root nodule bacteria, refer to the root and stem nodule bacteria of legumes (Sprent, 2001). This name was derived from the genus name *Rhizobium* (Wang et al., 2019). These bacteria are a Gram-negative bacterium, soil habitat, aerobic in nature, and live symbiotically with the roots of leguminous plants (Datta et al., 2014) by fixing atmospheric nitrogen (N_2) and converting it into ammonia (NH_3) in root nodules of legumes (Wang et al., 2019), in order to be obtainable for plants, and in return leguminous plants supply nutrients to the bacteria (Datta et al., 2014). In general, Rhizobia can provide the soil with nitrogen at an average of 50–100 kg/ha (Mazid and Khan, 2014). This ecological and agronomic symbiotic association are considered a great environmental system because of their ability to fix a large amount of atmospheric nitrogen (Alemneh et al., 2020). From the first report by Beijerinck who isolated the root nodule bacteria which was responsible for the nitrogen fixation process (Willems, 2006) to this time, many Rhizobia have been identified. Hence, the taxonomy became needed to classify these bacteria into groups. Initial classification of these bacteria was based on their ability to nodulate specific legumes, a concept known as cross inoculation group which was described by Baldwin and Fred (1929) and Velázquez et al. (2017) but later abandoned because some rhizobial strains have the ability to nodulate a wide range of legumes (Giller, 2001). Generation, time, and growth rate classified the species into two groups: the fast-growing strains include genus (*Rhizobium*) which contained three species: *R. phaseoli*, *R. trifolii*, *R. meliloti*, nodulating *Phaseolus*, *Trifolium*, and *Melilotus*, respectively, as described by Dangeard (1926). The slow-growing strains include genus (*Bradyrhizobium*) as described by Jordan (1982), it comprised only one species, *B. japonicum*, nodulating all soybean strains (Velázquez et al., 2017). However, the results were not useful to classify species of rhizobia due to the possibility of natural transfer of symbiotic plasmids among bacterial strains in the soil (Nakatsukasa et al., 2008). After this, the phylogenetic analysis of 16 rRNA gene sequences (Woese and Fox, 1977) led to the description of a large number of rhizobia in a nonsystematic way and placed rhizobia within the alpha subdivision of Proteobacteria (Woese et al., 1984). The first species described on the basis of partial sequencing of the 16S rRNA gene was *Rhizobium tropici* (Martinez-Romero et al., 1991). Although the sequencing of 16S rRNA gene is still used widely to classify and identify rhizobia, it has some limitation and cannot be used alone to differentiate among the closest *Rhizobium* species (Ramirez-Bahena et al., 2008). Therefore, researchers suggest using the sequencing of several housekeeping genes such as recA and atpD to describe the closest-related species of rhizobia. Moreover, the analysis of multilocus sequences (MLSA) and multilocus sequence typing (MLST) helped with the description of new genera and species of rhizobia (Ribeiro et al., 2009). Recently, two new methods have been used for studying the taxonomy of legume-nodulating bacteria: comparative genomics using whole-genome sequence comparisons to define new taxonomy of rhizobia (Ormeno-Orrillo et al., 2015), and average nucleotide identity (ANI) of genome comparisons (Rashid et al., 2015).

In 1886, Hellriegel and Wilfarth discovered biological nitrogen fixation, and they reported that some legumes could fix nitrogen gas (N_2) from nodules on their roots (Franche et al., 2009). Two years later Beijerinck (1888) isolated a bacterium from plant root nodules, and named it *Bacillus radicicola*. One year later, Frank named this bacterium *Rhizobium leguminosarum* and identified new species belonging to the same group. However, the term "rhizobia" was first used to name bacteria belonging to the genus *Rhizobium* (Wang et al., 2019). Later many genera were identified, such as *Bradyrhizobium* (Jordan, 1982), *Sinorhizobium* (Chen et al., 1988), *Azorhizobium* (Dreyfus et al., 1988), *Devosia* (Nakagawa et al., 1996), and *Mesorhizobium* (Jarvis et al., 1997). Zakhia and de Lajudie (2001) briefed the classification of these bacteria into 6 rhizobia genera belonging to the α-Proteobacteria (*Rhizobium*, *Mesorhizobium*, *Sinorhizobium*, *Azorhizobium*, *Bradyrhizobium*, *Allorhizobium*) and 28 species. Followed by Sawada et al. (2003), announcing that 12 bacterial genera consist of 44 species belonging to α, and β-Proteobacteria can fix atmospheric nitrogen symbiosis with legume, which gathered in 9 groups based on the comparative analysis of 16S rDNA sequence: Group 1: includes *Rhizobium* and *Allorhizobium* which contain legume symbionts (intermixed with *Agrobacterium* and *Blastobacter* species, which are non-symbionts); Group 2: *Sinorhizobium*, *Ensifer* species (unclassified nonsymbionts); Group 3: *Mesorhizobium* species (with nonsymbiotic *Aminobacter*, *Pseudaminobacter* species); Group 4: *Bradyrhizobium* species, *Blastobacter denitrificans* (with non-symbiotic *Agromonas*, *Nitrobacter*, *Afipia* and *Rhodopsedudomonas* species); Group 5: *Methylobacter nodulans* (with non-symbiotic *Methylobacter* species); Group 6: *Azorhizobium* species (with non-symbiotic *Xanthobacter* and *Aquabacter* species); Group 7: *Devosia neptuniae* (with nonsymbiotic *Devosia* species and unclassified nonsymbionts); Group 8: symbiotic

Burkholderia strain (with nonsymbiotic *Burkholderia* species); Group 9: *Ralstonia taiwanensis* (with nonsymbiotic *Ralstonia* species). The first seven groups belong to α–Proteobacteria, while the last two belong to β-Proteobacteria. After that, Willems (2006) reported that rhizobia included 53 bacterial species. 16 species from them belong to the genus *Rhizobium*, 11, 11, 7, 5, 2, and 1 species belong to the genera, *Sinorhizobium*, *Mesorhizobium*, *Bradyrhizobium*, *Agrobacterium*, *Azorhizobium*, and *Allorhizobium* respectively. Till 2013, all the descriptive rhizobial species belong to eleven genera of α-proteobacteria and three genera of β-proteobacteria (Laranjo et al., 2014). Berrada and Fikri-Benbrahim (2014) added 45 new species and 2 genera (*Shinella*, *Microvirga*) to the previous classification. They reported that there are 98 species of legume-nodulating bacteria belonging to 14 genera, followed by Shamseldin et al. (2017), who described about 205 species which are symbiotic, and 33 species which are nonsymbiotic bacteria. They are distributed among 18 genera, divided into two classes, the class α-proteobacteria include 16 genera (*Rhizobium*, *Ensifer* (formerly *Sinorhizobium*), *Allorhizobium*, *Shinella*, *Pararhizobium*, *Mesorhizobium*, *Phyllobacterium*, *Aminobacter*, *Bradyrhizobium*, *Blastobacter*, *Photorhizobium*, *Devosia*, *Azorhizobium*, *Methylobacterium*, *Microvirga* and *Ochrobactrum*), whereas the class β-proteobacteria include *Burkholderia*, *Ralstonia* (former *Cupriavidus*). He also noted that there is a *Pseudomonas* sp. belonging to the γ-Proteobacteria that can nodulate *Robinia pseudoacacia*. At present, Wang et al. (2019) reported that all the symbiotic nitrogen-fixing bacteria belong to the classes α, β-proteobacteria and maybe also γ-proteobacteria, with about 180 nodulating species in 21 genera.

5.2 THE ROLE OF RHIZOBIA IN PROMOTING PLANT GROWTH

The concept of plant growth promoting rhizobacteria (PGPR), introduced in the late 1970s (Kloepper, 1978; Kloepper et al., 1980), refers to a group of bacteria that can be found in the rhizosphere (Ahmad et al., 2008) and shows an important role in enhancing plant growth (Burdman et al., 2000). Both symbiotic and nonsymbiotic PGPR have been applied in agricultural practices to promote growth and yields of legume crops (Ahmad et al., 2013; Khaitov et al., 2016) and non-legume crops (García-Fraile et al., 2012; Ziaf et al., 2016). Generally, PGPR can be defined as any root-colonizing bacteria that can improve plant growth either directly, by facilitating nutrients to the plant, or indirectly, by helping plants against the pathogens (Glick, 2012). In many cases, it is a combination of both (Vargas et al., 2017).

Rhizobia, the famous mutualistic symbiotic bacteria, establishes symbiotic associations with leguminous plants by producing root structures known as nodules and transforming atmospheric nitrogen (N_2) into usable N (ammonia) so the plants can use it easily (Lindström and Martinez-Romero, 2005). These bacteria were not involved in the PGPR group at first, though it was later included in the intracellular category of PGPR (iPGPR) "bacteria living in specialized nodular structures" (Gray and Smith, 2005). This includes several endophytic and symbiotic rhizobia (*Allorhizobium*, *Azorhizobium*, *Bradyrhizobium*, *Mesorhizobium*, etc.) and *Frankia* species (Wang and Martinez-Romero, 2000; Laranjo et al., 2014). On the other hand, some rhizobia have also been grouped as extracellular PGPR (ePGPR), which forms associations with non-legumes (Vargas et al., 2017) and includes bacteria from genera such as *Azotobacter*, *Bacillus*, *Chromobacterium*, *Agrobacterium*, *Caulobacter*, *Pseudomonas*, and *Serratia* (Gray and Smith, 2005). The most widely and practically investigated organisms are found among the PGPR rhizobia due largely to their ability to form effective symbiosis with legumes crops. In addition to their beneficial effect on pulses, rhizobia have also been found useful in promoting the quantity and quality of non-legume crops (Vargas et al., 2017). As they can be safely used in agricultural practices, reducing the dependence on chemicals used in agricultural practices (Vargas et al., 2017), it has become widely used as a microbial inoculant to enhance plant growth promotion in different production systems and is considered the best option for microbial technology in agricultural practices (Vargas et al., 2017). For example, in Brazil, inoculation of soybean (*Glycine max*) fields with rhizobia has been reported to supply up to 300 kg/ha of N, resulting in savings of N fertilizers estimated to be around US$ 3 billion.

The successful utilization of PGPR is reliant on many factors: its existence in soil, the compatibility with the crop on which it is inoculated, the interaction ability with native microflora in soil, and environmental conditions (Martinez-Viveros et al., 2010) (Table 5.1).

TABLE 5.1 Plant Growth-Promoting Substances Released by some Commonly Employed Nitrogen-Fixing Plant Growth-Promoting Rhizobacteria.

Rhizobia	Source	Plant growth regulators	References
Rhizobium undicola, *Rhizobium* spp.	Nodules of aquatic legume	ACC deaminase, indole acetic acid	Bhagat et al. (2014); Ghosh et al. (2015)
Mesorhizobium, *R. leguminosarum*, *Bradyrhizobium*, *Sinorhizobium meliloti*	*Neptunia oleracea*, *Pisum sativum*, *Trifolium alexandrinum* L., *Cicer arietinum* L., *Trigonella foenum-graecum* L., *Medicago sativa* L., *Indigofera* spp., birdsfoot trefoil (*Lotus corniculatus*)	Exopolysaccharides, N_2 fixation, P solubilization, siderophores, ammonia, hydrogen cyanide, antifungals, volatile antifungal compounds, protease	Ma et al. (2004); Sahasrabudhe (2011); Bhattacharjee et al. (2012); Machado et al. (2013)
Azotobacter	Rhizosphere soil	P solubilization, siderophores, ammonia, hydrogen cyanide, IAA	Prasad et al. (2014)
Sinorhizobium sp. strain MRR101-KC428651, *Rhizobium* sp. strain 103-JX576499, *Sinorhizobium kostiense* strain MRR104-KC428653	Root nodules of *Vigna trilobata* plants	P solubilization, antifungal activity	Kumar et al. (2014)
Azotobacter	Rhizosphere soil	IAA	Kumar et al. (2014)
Azotobacter	Rhizosphere soil	Siderophores	Muthuselvan and Balagurunathan (2013)
Rhizobium psm6	Agricultural soil	P solubilization	Karpagam and Nagalakshmi (2014)
Mesorhizobium	Tunisian soils	P solubilization	Imen et al. (2015)
Rhizobium BICC 651	Root nodule of chickpea	Siderophores	Datta and Chakrabartty (2014)

TABLE 5.1 *(Continued)*

Rhizobia	Source	Plant growth regulators	References
Mesorhizobium spp.	Native isolates	HCN, siderphores, protease, cellulose, volatile antifungal compounds	Bhagat et al. 2014
Azospirillum brasilense	-	Siderophores, IAA antifungal activity	Zakharova et al. (1999); Tortora et al. (2011)

Source: Reproduced with permission from Vargas et al. (2017); Copyright © 2017, Springer International Publishing.

5.3 MECHANISMS OF PLANT GROWTH PROMOTION BY RHIZOBIA

The working mechanisms of PGPR can be separated into direct and indirect mechanisms (Yu et al., 2012).

5.3.1 *DIRECT MECHANISMS*

5.3.1.1 FIXATION OF NITROGEN

Rhizobia are an excellent example of bacteria that can fix atmospheric nitrogen in symbiotic systems with legume crops (Wang et al., 2019). Some rhizobia that belong to nonsymbiotic N_2 fixation, however, can fix N in association with even non-legume plants; for example, nitrogen-fixing activity was detected in wild rice (*Oryza breviligulata*) plants inoculated with a photosynthetic *Bradyrhizobium* (Chaintreuil et al., 2000; Machado et al., 2013).

5.3.1.2 SECRETION OF PLANT GROWTH-PROMOTING SUBSTANCES

Plant growth regulators, also named plant exogenous hormones, are organic molecules similar to natural plant hormones, which, at low concentrations, cause a physiological response and regulate the growth of plants. These materials have been grouped into six categories: (1) auxins; (2) cytokinins; (3) gibberellins; (4) ethylene; (5) a group called inhibitors, which includes abscisic acid, phenolics, and alkaloids (Mishra et al., 2006); and (6) brassinosteroids (Bajguz and Tretyn, 2003; Ghosh et al., 2015).

5.3.1.3 SOLUBILIZATION OF INORGANIC INSOLUBLE PHOSPHATES AND MINERALIZATION OF ORGANIC PHOSPHATES

Some PGPR have the ability to solubilize insoluble phosphate (Wani et al., 2007), so the plants can absorb it easily resulting in phosphate-solubilizing bacteria (PSB) including rhizobia such us *Mesorhizobium*,

R. leguminosarum, *Bradyrhizobium*, and *Sinorhizobium meliloti* (Kumar and Raghu Ram, 2014; Prasad et al., 2014).

5.3.2 INDIRECT MECHANISMS

Based on PGPR release of bioactive molecules that reduce or stop the harmful effects of phytopathogens (Datta and Chakrabartty, 2014), these mechanisms include:

5.3.2.1 PRODUCTION OF SIDEROPHORES

Siderophores are a low molecular weight (400–1000 Da) iron-chelating compound (Lukkani and Reddy, 2014). This kind of mechanism (production of siderophores) can enhance plant growth either by providing plants with Fe (Solano et al., 2008) that exists in nature in the form of Fe^{3+}, which is highly insoluble. When Fe is limited, microbial siderophores enhance plant growth either by providing plants with Fe (Vejan et al., 2016) or by inhibiting growth of pathogens in the rhizosphere by limiting the availability of iron to the pathogens (Solano et al., 2008). Another possible positive effect of siderophore production is the complexation of toxic aluminum (Roy and Chakrabartty, 2000).

5.3.2.2 RELEASE OF ANTIBIOTIC SUBSTANCES

Antibiotic substance release especially bacteriocins (antibiotic-like proteinaceous toxins) which can be produced by many rhizobia (Anthony, 2012; Prabhavati and Chen et al., 2012) and hydrocyanic acid (Prasad et al., 2014).

5.3.2.3 PRODUCTION OF LYTIC ENZYMES

Synthesis of lytic enzymes (Nabti et al., 2014) such as β-1,3-glucanase, proteases (Compant et al., 2005), and chitinases (Kacem et al., 2009) which are considered important mechanisms of biological control for plant pathogenic microorganisms adopted by PGPR.

5.3.2.4 INDUCED SYSTEMIC RESISTANCE (ISR)

Rhizobia are able to stimulate reactions of plant defense against pathogens, as demonstrated by Elbadry et al. (2006) study on the occurrence of induced systemic resistance (ISR) and its important role against bean yellow mosaic potyvirus (BYMV) in faba beans (*Vicia faba*) inoculated with *P. fluorescens* FB11 and *R. leguminosarum* bv. *viciae* FBG05. Plants that were inoculated showed a significant reduction in percent disease incidence and a significant reduction in virus concentration (Vargas et al., 2017).

Rhizobia are the most effective bacteria among PGPR because of their importance in controlling plant pathogens. Chao (1990) mentioned that rhizobia belonging to genera *Rhizobium*, *Bradyrhizobium*, and *Sinorhizobium* showed variable growth inhibitory activities against fungal isolates. In another study, for example, some rhizobial strains dissolved the fungal mycelium at the initial stage and used the lyzed contents as nutrients for their growth and development (Hossain and Mårtensson, 2008).

Hemissi et al. (2011) mentioned that among the 42 rhizobial isolates used to control *Rhizoctonia solani,* 24 bacterial isolates could effectively inhibit the growth of fungi *in vitro*, while two rhizobial isolates reduced the disease by more than 80% under glasshouse conditions. Furthermore, ten rhizobial isolates solubilized insoluble P (Bhagat et al., 2014) (Figure 5.1).

5.4 THE ROLE OF *RHIZOBIA* IN LEGUME CROPS

The main purpose of legume seed inoculation with *Rhizobia* is to provide nitrogenous compounds in order to enhance growth and increase crop yields (Laranjo et al., 2014) by improving soil health and increasing plant growth-promoting enzymes, hormones, and auxins (Fact sheet, 2016), while simultaneously decreasing the use of chemical fertilizers and consequently reducing environmental pollution (Laranjo et al., 2014). The use of *Rhizobium* inoculants in legume seeds could be the oldest known agro biotechnological application of this process (Lindström et al., 2010). Many studies have shown that several *Rhizobium* strains have the ability to improve productivity of leguminous crops and soil fertility in both dry lands (Abdelgani et al., 2003) and saline soils (Elsheikh, 1998a, 2001). Using *R. tropici* co-inoculated with *Paenibacillus polymyxa*, plant height

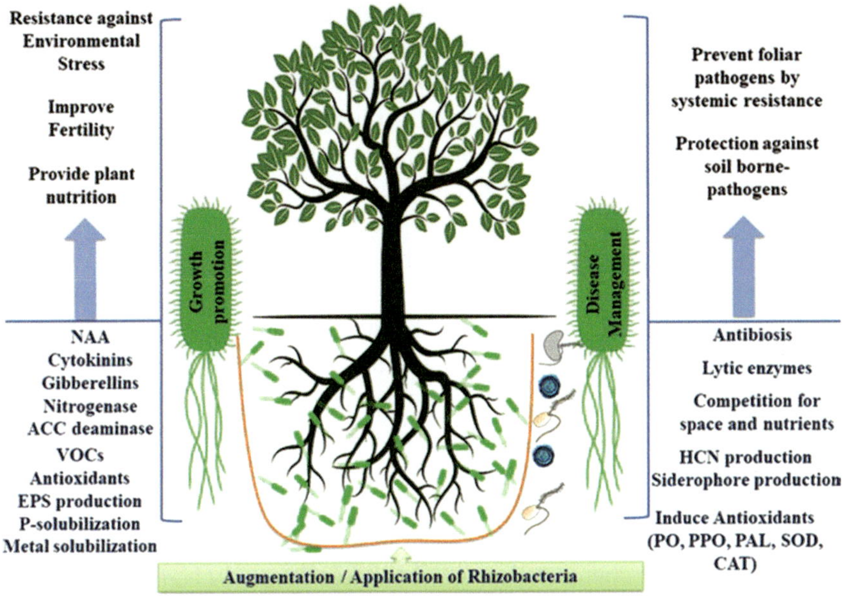

FIGURE 5.1 PGPR-mediated plant growth promotion and disease management.

Source: Reproduced with Permission from Harish et al. (2019), Copyright © 2019, Springer Nature, Singapore.

can also be enhanced, and shoot dry weight and nodule number increased under drought stress conditions (Figueiredo et al., 2008). Moreover, inoculating Peanut plants with *Bradyrhizobium* strains SEMIA6144 and ESA 123 reduced the water stress effects on some genotypes, indicating that *Bradyrhizobium* strains could be used as inoculants under water stress conditions (Barbosa et al., 2018), also, as a biological control agent, some plant diseases and viruses, which decrease faba bean infection by Bean Yellow Mosaic Virus (Babiker et al., 1995; Elbadry et al., 2006) and Broad Bean Mottle Bromo virus (Elsheikh and Osman, 1995). Several additional studies show that inoculating faba bean (*Vicia faba* L.) with *Rhizobium* enhanced nodulation, yield, yield attributes, and seed quality (Osman et al., 2010; Rugheim and Abdelgani, 2012), and increased hundred-seed weight by 11.44%, number of pods per plant from 7 to 13.08 nodules, and dry weight by 34.04% when compared with non-inoculated plants (Bezabih, 2018). In the same instance, Adissie et al. (2020) revealed in their study that *Rhizobium leguminosarum* (var. *viciae*) strains and micronutrients significantly affected yield and yield components of faba beans

(*Vicia faba* L.), except for the number of seed pods; the main effect of micronutrients significantly affects only the weight of one hundred seeds. On the other hand, inoculation with strains alone results in statistically significant differences in plant height, plant dry weight, and number of pods per plant as compared to non-inoculated treatments. Additionally, Allito et al. (2020) mentioned an enhanced nitrogen fixation, soil nitrogen balance, and nutrient uptake in *Vicia faba* L. inoculated with *Rhizobium* strains. Sistani et al. (2017) reported that *Rhizobium leguminosarum* positively affected the yield of pea (*Pisum sativum* L.) through the increase of dry weights based on better seed filling. It also reduces seed infection severity with the pathogen *Didymella pinodes* by inducing several proteins and metabolites involved in pathogen response. In Canada, rhizobia inoculants of lentil crop positively affect seed yields, which increased by 15–45% compared to non-inoculated crops (Gan et al., 2002). Khatkar et al. (2007) found in their experiment on the application of biofertilizers that the dual inoculation of *Rhizobium* and PSB along with sulfur significantly improved growth characters and yield of black gram (*Vigna mungo* L.). Additionally, inoculation of *Rhizobium* and PSB positively affected plant height, number of branches per plant, dry matter per plant, and grain yield of pigeon pea (Singh and Yadav, 2008). Sajid et al. (2011) reported that *Rhizobium* inoculation increased plant height of mung beans and ground nuts. Bahadur and Tiwari (2014) highlighted the positive effects of inoculating mung beans with *Rhizobium* and PSB, which showed significant response to all growth and yield parameters. Dar et al. (2020) also conducted an experiment to study the effect of different microorganisms (*Rhizobium*, *Azotobacter*, Arbuscular mycorrhizae) as biofertilizers under different levels of phosphorus (20 and 40 kg/ha) application on yield and nutrient content in seeds of the common bean (*Phaseolus vulgaris*). Their findings highlighted that the number of pods, pod length and weight, and the nutrient content in seeds of all the inoculated common bean plants were much higher than with the control plants. Htwe et al. (2019) indicated that the application of an effective biofertilizer (SAY3-7, BLY3-8, and P4) on efficient soybean variety is a promising method for promoting plant growth, nodulation, nitrogen fixation, and seed yield. Furthermore, Zeffa et al. (2020) observed a significant increase in nodule number (11.40%), nodule biomass (6.47%), root biomass (12.84%), and shoot biomass (6.53%) in soybeans when inoculated with co-inoculation of *Bradyrhizobium* and other PGPR. According to Crusciol et al. (2019), supplying *Bradyrhizobium* as an inoculant combined with molybdenum (Mo) fertilizer on the peanut (*Arachis hypogea*) crop enhanced nodulation and increased the yield of pods and kernels. Ben Romdhane et al. (2007) reported an increased number of nodules and shoot dry weight induced by rhizobia strains in different chickpea cultivars. Furthermore, it seems to activate chickpea genes involved with the production of phenolic compounds/phytoalexins and higher levels of these compounds that help the plant by limiting disease development (Arfaoui et al., 2007). Likewise, Verma et al. (2013) confirmed that the chickpea co-inoculated with *Mesorhizobium* sp. and *Pseudomonas aeruginosa* showed an increase of 32% in grain yield, compared to the non-inoculated control. Khaitov et al. (2020) concluded from their experiment that inoculating chickpea with different symbiotic *Rhizobium* strains (viz. R4, R6, R9, and IC53) in saline soils enhanced the average biomass by 3.3–33.9%, and also increased chickpea plant growth and yield, as well as improving seed protein and oil content.

5.5 THE ROLE OF *RHIZOBIA* IN CEREAL CROPS

The role and importance of rhizobia as biofertilizer in cereal crops have been studied by several researchers. *Rhizobia* produces phytohormones such as auxins, cytokinin, gibberellins, and abscisic acid. It is likely that their release into cropping systems promotes plant growth and possibly increases yield, even though no N_2 fixation by rhizobia has been detected in non-legumes (Matiru and Dakora, 2004) (Figure 5.2). Moreover, it releases lumichrome and lipo-chitooligosaccharides (LCOs) which stimulate growth of crop plants (Phillips et al., 1999; Dakora, 2003). Dakora et al. (2002) stated that applying 5 nm concentrations of lumichrome increased plant growth in sorghum, soybean, and cowpea genotypes. Judge and Hickey (2019) demonstrated that *Rhizobium* bacteria can stimulate faster germination in the *Poaceae* species as a result of the natural release of LCOs, which motivate faster germination by imitating the effect of gibberellins and increasing release of hydrolytic enzymes such as α-amylase from the aleurone layer into the endosperm, thereby ceasing seed dormancy.

Many researchers suggest that rhizobia such as *B. japonicum*, *R. leguminosarum*, *S. meliloti*, and arctic Rhizobia can colonize the roots of non-leguminous plants and are able to survive in the internal tissue system.

These species of bacteria have the ability to participate with legumes and non-legume plants (Patil et al., 2017).

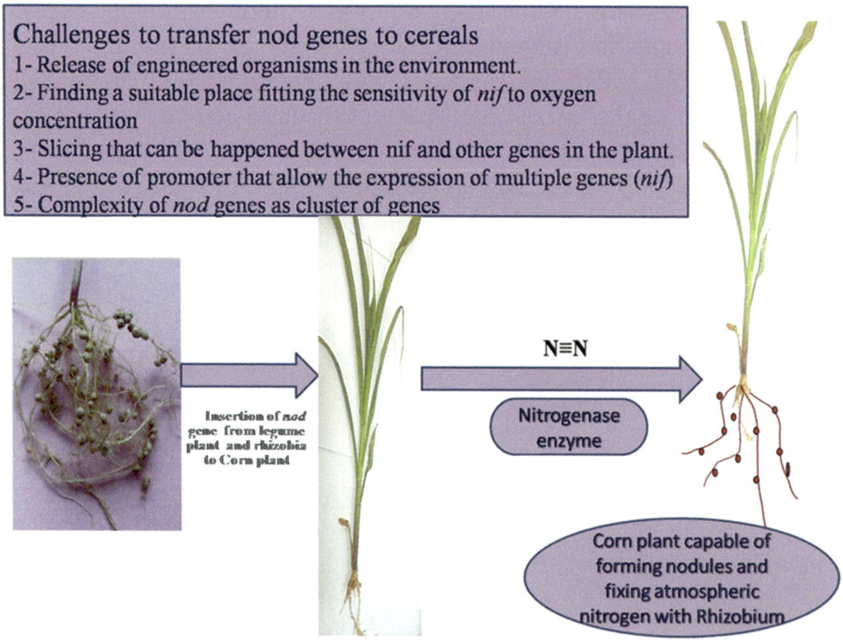

FIGURE 5.2 Challenges in engineering of cereals with *nod* machinery system.

Source: Reproduced with permission from Shamseldin (2022); Copyright © 2022, Springer Nature.

Matiru and Dakora (2004) reported from their study, which aimed to investigate the potential use of diverse rhizobial strains (*Bradyrhizobium japonicum*, *Azorhizobium caulinodans*, *Rhizobium* NGR234, *Sinorhizobium meliloti*, *Rhizobium leguminosarum* bv. *viciae* Cn6, and *R. leguminosarum* bv. *viciae* strain 30) as promoters of plant growth for increased yield of sorghum and millet landraces, that sorghum roots easily infected by rhizobial, growth were increased in response to inoculation. However, where it was difficult to establish rhizobial presence in millet roots, plant growth was unaffected by rhizobial inoculation. Moreover, because P uptake by sorghum increased with inoculation, it would seem that the bacteria made some changes in phosphate transporter activity of sorghum root plasma membrane, leading to improved P nutrition.

In addition, it has been found from various experimental results that rice inoculated with *Rhizobium leguminosarum* has increased root weight and yield production (Biswas et al., 2000a, 2000b), increased biomass, grain size, and yield production (Hussain et al., 2009), and enhanced yield production by supplying *R. vietnamiensis* inoculation (Tran Van et al., 2000). Furthermore, applying *R. trifolii* or *Sinorhizobium* spp. inoculation improves the yield production of maize (Riggs et al., 2001), and applying *R. etli* bv. *phaseoli* improves the total biomass (Zamora and Matinez-Romero, 2001). Afzal and Asghari (2008) reported an increase of 30–40% in grain yield as a result of the application of single or dual inoculums + P fertilizer as compared to a single application of P fertilizer on wheat. Furthermore, Mehboob et al. (2011) stated that the application of different *Rhizobium* spp. on wheat increased growth and yield attributes compared with non-inoculated control. Saber et al. (2020) revealed that inoculation of wheat with a combination of different species of rhizobia inoculation (*Rhizobium leguminosarum* bv. *viciae*, *Rhizobium leguminosarum* bv. *phaseoli*, *Bradyrhizobium*, and *Mesorhizobium ciceri*) with NPK fertilizer had the potential to increase grain number, grain yield, harvest index, weight per 1000 grain, and biological yield, as well as improve soil fertility.

5.6 CONCLUSIONS

As a required component of all amino acids and nucleic acids, nitrogen is a critical plant nutrition element in the production of plants. Plants that fix nitrogen from the atmosphere (biological N_2 fixation, or BNF) can be used by farmers in resource-limited settings to increase agricultural yields. There is a complex molecular communication pathway between the legume host and the rhizobial microsymbiont that results in the development of root nodules in the legume. The symbiotic relationship between rhizobia and legumes that results in nodule growth is a complex chemical conversation that takes place between both organisms. Rhizobia and their legume hosts have the ability to create a wide range of phytohormones during this complex process. As a group of bacteria, plant growth-promoting bacteria (PGPR) are beneficial to plants in a variety of ways, and they are classified as such by the World Health Organization (WHO). It is common for plants to have PGPR in their rhizosphere soil, which can release protons and thereby solubilize soil-bound phosphorus for use by the plants.

KEYWORDS

- **biofertilizers**
- **biological nitrogen fixation**
- **legume**
- **rhizobia**

REFERENCES

Abdelgani, M. E.; Osman, A. G.; Mohamed, S. S. Restoring Soil Fertility of Desertified Lands through Biological Nitrogen Fixation. In *Desertification in the Third Millennium*; Dubai, UAE, 2003.

Abebe, G.; Hattar, B.; Al-Tawah, A. M. Nutrient Availability as Affected by Manure Application to Cowpea (*Vigna unguiculata* L. Walp.) on Calacarious Soils. *J. Agric. Soc. Sci.* **2005,** *1* (1), 1–6.

Abera, T.; Feyisa, D.; Yusuf, H.; Nikus, O.; Al-Tawaha, A. R. Grain Yield of Maize as Affected by Biogass Slurry and NP Fertilizer Rate at Bako, Western Oromiya, Ethopia. *Biosci. Res.* **2005,** *2*, 31–38.

Adissie, S.; Adgo, E.; Feyisa, T. Effect of Rhizobial Inoculants and Micronutrients on Yield and Yield Components of Faba Bean (*Vicia faba* L.) on Vertisol of Wereillu District, South Wollo, Ethiopia. *Cogent Food Agric.* **2020,** *6*, 1747854.

Afzal, A.; Asghari, B. *Rhizobium* and Phosphate Solubilizing Bacteria Improve the Yield and Phosphorus Uptake in Wheat (*Triticum aestivum* L.). *Int. J. Agric. Biol.* **2008,** *10*, 85–88.

Ahmad, F.; Ahmad, I.; Khan, M. S. Screening of Free-Living Rhizospheric Bacteria for their Multiple Plant Growth Promoting Activities. *Microbiol. Res.* **2008,** *163*, 173–181.

Ahmad, E.; Khan, M. S.; Zaidi, A. ACC Deaminase Producing *Pseudomonas putida* Strain PSE3 and *Rhizobium leguminosarum* Strain RP2 in Synergism Improves Growth, Nodulation and Yield of Pea Grown in Alluvial Soils. *Symbiosis* **2013,** *61*, 93–104.

Al-Ajlouni, M. M.; Al-Ghzawi, A. L. A.; Al-Tawaha, A. R. Crop Rotation and Fertilization Effect on Barley Yield Grown in Arid Conditions. *J. Food Agric. Environ.* **2010,** *88* (3), 869–872.

Alemneh, A. A.; Zhou, Y.; Ryder, M. H.; Denton, M. D. Mechanisms in Plant Growth-Promoting Rhizobacteria that Enhance Legume–Rhizobial Symbioses. *J. Appl. Microbiol.* **2020,** *129* (5), 1133–1156.

Al-Juthery, H. W.; Habeeb, K. H.; Jawad, F.; Altaee, K.; AL-Taey, D. K. A.; Al-Tawaha, A. R. M. Effect of Foliar Application of Different Sources of Nano-Fertilizers on Growth and Yield of Wheat. *Biosci. Res.* **2010,** *15* (4), 3988–3997.

Al-Kiyyam, M. A.; Turk, M.; Al-Mahmoud, M.; Al-Tawaha, A. R. Effect of Plant Density and Nitrogen Rate on Herbage Yields of Marjoram under Mediterranean Conditions. *J. Am.-Eurasian J. Agric. Environ. Sci.* **2008,** *3*, 153–158.

Allito, B. B.; Ewusi-Mensah, N.; Logah, V. Legume-*Rhizobium* Strain Specificity Enhances Nutrition and Nitrogen Fixation in Faba Bean (*Vicia faba* L.). *Agronomy* **2020,** *10*, 826.

Al-Tawaha, A. R.; Al-Karaki, G.; Al-Tawaha, A. R.; Sirajuddin, S. N.; Makhadmeh, I.; Wahab, P. E. M.; Youssef, R. A.; Al Sultan, W.; Massadeh, A. Effect of Water Flow Rate on Quantity and Quality of Lettuce (*Lactuca sativa* L.) in Nutrient Film Technique (NFT) under Hydroponics Conditions. *Bulg. J. Agric. Sci.* **2018a,** *245*, 791–798.

Al-Tawaha, A. R.; Al-Tawaha, A. R.; Alu'datt, M.; Al-Ghzawi, A. L.; Wedyan, M.; Al-Obaidy, S. D. A.; Al-Ramamneh, E. A. D. Effects of Soil Type and Rainwater Harvesting Treatments in the Growth, Productivity and Morphological Trains of Barley Plants Cultivated in Semi-Arid Environment. *Aust. J. Crop Sci.* **2018b,** *12* (6), 975–979.

Al-Tawaha, A. R.; McNeil, D. L.; Yadav, S. S.; Turk, M.; Ajlouni, M.; Abu-Darwish, M. S.; Al-Ghzawi, A. L. A.; Al-udatt, M.; Aladaileh, S. Integrated Legume Crops Production and Management Technology. In *Climate Change and Management of Cool Season Grain Legume Crops*; Yadav, S. S., David, L., McNeil, Robert, R., Sharanagouda, A. P., Eds.; Springer: Dordrecht, 2010; pp 325–349.

Al-Tawaha, A. R. M.; Nidal, O. Use of Sorghum and Maize Allelopathic Properties to Inhibit Germination and Growth of Wild Barley (*Hordeum spontaneum*). *Not. Bot. Horti Agrob. Cluj-Napoca* **2010,** *38* (3), 124–127.

Al-Tawaha, A. R.; Turk, M. A.; Abu-Zaitoon, Y. M.; Aladaileh, S. H.; Al-Rawashdeh, I. M.; Alnaimat, S.; Al-Tawaha, A. R. M.; Alu'datt, M. H.; Wedyan, M. Plants Adaptation to Drought Environment. Plants Adaptation to Drought Environment. *Bulg. J. Agric. Sci.* **2017,** *23* (3), 381–388.

Al-Tawaha, A. M.; Turk, M.; Hameed, K.; Assaf, T.; Tahhan, R.; Al-Jamali, A. F. Effects of Application of Olive Mill by-Products on Lentil Yield and their Symbiosis with Mycorrhizal Fungi under Arid Conditions. *Crop Res.* **2010,** *40* (3), 66–75.

Alu'datt, M. H.; Rababah, T.; Alhamad, M. N.; Al-Tawaha, A.; Al-Tawaha, A. R.; Gammoh, S.; Ereifej, K. I.; Al-Karaki, G.; Hamasha, H. R.; Tranchant, C. C.; Kubow, S. Herbal Yield, Nutritive Composition, Phenolic Contents and Antioxidant Activity of Purslane (*Portulaca oleracea* L.) grown in Different Soilless Media in a Closed System. *Ind. Crops Prod.* **2019,** *141*, 111746.

Ananthi, T.; Amanullah, M. M.; Al-Tawaha, A. R. M. S. A Review on Maize-Legume Intercropping for Enhancing the Productivity and Soil Fertility for Sustainable Agriculture in India. *Adv. Environ. Biol.* **2017,** *11* (5), 49–63.

Arfaoui, A.; El Hadrami, A.; Mabrouk, Y.; Sifi, B.; Boudabous, A.; El Hadrami, I. Treatment of Chickpea with *Rhizobium* Isolates Enhances the Expression of Phenylpropanoid Defense-Related Genes in Response to Infection by *Fusarium oxysporum* f. sp. *ciceris*. *Plant Physiol. Biochem.* **2007,** *45*, 470–479.

Babiker, E. E.; Elsheikh, E. A. E.; Osman, A. G.; Eltinay, A. H. Effect of Nitrogen Fixation, Nitrogen Fertilization and Viral Infection on Yield, Tannin and Protein Contents and in vitro Protein Digestibility of Faba Bean. *Plant Foods Hum. Nutr.* **1995,** *47*, 257–263.

Bahadur, L.; Tiwari, D. D. Nutrient Management in Mung Bean (*Vigna radiata* L.) through Sulfur and Biofertilizers. *Legume Res.* **2014,** *37*, 180–187.

Bajguz, A.; Tretyn, A. The Chemical Characteristic and Distribution of Brassinosteroids in Plants. *Phytochemistry* **2003,** *62*, 1027–1046.

Baldwin, I. L.; Fred, E. B. Nomenclature of the Root Nodule Bacteria of the Leguminosae. *J. Bacteriol.* **1929,** *17* (2), 141–150.

Barbosa, D. D.; Brito, S. L.; Fernandes, P. D.; Fernandes-Júnior, P. I.; Lima, L. M. Can *Bradyrhizobium* Strains Inoculation Reduce Water Deficit Effects on Peanuts? *World J. Microbiol. Biotechnol.* **2018,** *34* (7), 87–98.

Ben, R. S.; Tajini, F.; Trabelsi, M.; Aouani, M.; Mhamdi, R. Competition for Nodule Formation between Introduced Strains of *Mesorhizobium ciceri* and the Native Populations of Rhizobia Nodulating Chickpea (*Cicer arietinum*) in Tunisia. *World J. Microbiol. Biotechnol.* **2007,** *23*, 1195–1201.

Berrada, H.; Fikri-Benbrahim, K. Taxonomy of the Rhizobia: Current Perspectives. *Brit. Microbiol. Res. J.* **2014,** *4*, 616–639.

Bezabih, W. Response of Faba Bean (*Vicia faba* L.) to *Rhizobium* Inoculation and Potassium Fertilizer Rate at Alicho Wuriro Highland, Southern Ethiopia. *Int. J. Res. Agric. For.* **2018,** *5* (5), 43–47.

Bhagat, D.; Sharma, P.; Sirari, A.; Kumawat, K. C. Screening of *Mesorhizobium* spp. for Control of *Fusarium* Wilt in Chickpea in vitro Conditions. *Int. J. Curr. Microbiol. Appl. Sci.* **2014,** *3*, 923–930.

Bhattacharjee, R. B.; Philippe, J.; Clémence, C.; Bernard, D.; Aqbal, S.; Satya, N. M. Indole Acetic Acid and ACC Deaminase-Producing *Rhizobium leguminosarum* bv. *trifolii* SN10 Promote Rice Growth, and in the Process Undergo Colonization and Chemotaxis. *Biol. Fertil. Soils* **2012,** *48*, 173–182.

Biswas, J. C.; Ladha, J. K.; Dazzo, F. B. Rhizobia Inoculation Improves Nutrient Uptake and Growth of Lowland Rice. *Soil Sci. Soc. Amer. J.* **2000a,** *64*, 1644–1650.

Biswas, J. C.; Ladha, J. K.; Dazzo, F. B.; Yanni, Y. G.; Rolfe, B. G. Rhizobial Inoculation Influences Seedling Vigor and Yield of Rice. *Agronomy J.* **2000b,** *92*, 880–886.

Burdman, S.; Jurkevitch, E.; Okon, Y. Recent Advances in the use of Plant Growth Promoting Rhizobacteria (PGPR) in Agriculture. In *Microbial Interactions in Agriculture and Forestry*; Science Publishers: Enfield, NH, USA, 2000; pp 229–250.

Chaintreuil, C.; Giraud, E.; Prin, Y.; Lorquin, J.; Bâ, A.; Gillis, M.; de Lajudie, P.; Dreyfus, B. Photosynthetic *Bradyrhizobia* are Natural Endophytes of the African Wild Rice *Oryza breviligulata. Appl. Environ. Microbiol.* **2000,** *66*, 5437–5447.

Chao, W. L. Antagonistic Activity of *Rhizobium* spp. Against Beneficial and Plant Pathogenic Fungi. *Lett. Appl. Microbiol.* **1990,** *10*, 213–215.

Chen, N.; Jin, M.; Qu, H. M.; Chen, Z. Q.; Chen, Z. L.; Qiu, Z. G.; Wang, X. W.; Li, J. W. Isolation and Characterization of *Bacillus* sp. Producing Broad-Spectrum Antibiotics Against Human and Plant Pathogenic Fungi. *J. Microbiol. Biotechnol.* **2012,** *22*, 256–563.

Chen, W. X.; Yan, G. H.; Li, J. L. Numerical Taxonomic Study of Fast-Growing Soybean Rhizobia and a Proposal that *Rhizobium fredii* be Assigned to *Sinorhizobium* gen. nov. *Int. J. Syst. Bacteriol.* **1988,** *38* (4), 392–397.

Compant, S.; Duffy, B.; Nowak, J.; Clément, C.; Barka, E. A. Use of Plant Growth-Promoting Bacteria for Biocontrol of Plant Diseases: Principles, Mechanisms of Action, and Future Prospects. *Appl. Environ. Microbiol.* **2005,** *71*, 4951–4959.

Crusciol, C. A. C.; Neto, J. F.; Mui, T. S.; Franzluebbers, A. J. D.; Costa, C. H. M.; Castro, G. S. A.; Ribeiro, L. C.; Costa, N. R. Rhizobial Inoculation and Molybdenum Fertilization in Peanut Crops grown in a no Tillage System after 20 Years of Pasture. *Rev. Bras. Cienc. Solo.* **2019,** *43*, e0170399.

Dakora, F. D. Defining New Roles for Plant and Rhizobial Molecules in Sole and Mixed Plant Cultures Involving Symbiotic Legumes. *New Phytol.* **2003,** *158*, 39–49.

Dakora, F. D.; Matiru, V.; King, M.; Phillips, D. A. Plant Growth Promotion in Legumes and Cereals by Lumichrome, a Rhizobial Signal Metabolite. In *Nitrogen Fixation: Global Perspectives*; Finan, T. M., O'Brian, M. R., Layzell, D. B., Vessey, K., Newton, W. E., Eds.; CABI Publishing: Wallingford, UK, 2002; pp 321–322.

Dar, M. H.; Singh, N.; Khan, F. A. Impact of Different Microorganisms as a Biofertilizers on Yield and Nutrient Content of Common Bean (*Phaseolus vulgaris* L.) under Intercropping System. In *Microbes for Humankind and Application*; Deepak, K. M., Meenu, R., Divya, B., Shikha, J., Eds.; Daya Publishing House: New Delhi, 2020; pp 49–57.

Datta, B.; Chakrabartty, P. K. Siderophore Biosynthesis Genes of *Rhizobium* sp. Isolated from *Cicer arietinum* L. *3Biotech* **2014,** *4*, 391–401.

Datta, A.; Singh, R. K.; Kumar, Sh.; Kumar, S. An Effective and Beneficial Plant Growth Promoting Soil Bacterium "*Rhizobium*": A Review. *Ann. Plant Sci.* **2014,** *4* (1), 933–942.

Dreyfus, B.; Garcia, L.; Gillis, M. Characterization of *Azorhizobium caulinodans* gen. nov. sp. nov., A Stem-Nodulating Nitrogen-Fixing Bacterium Isolated from *Sesbania rostrata*. *Int. J. Syst. Bacteriol.* **1988,** *38* (1), 89–98.

Elbadry, M.; Taha, R.; Eldougdoug, K.; Gamal-Eldin, H. Induction of Systemic Resistance in Faba Bean (*Vicia faba* L.) to Bean Yellow Mosaic Poty Virus (BYMV) via Seed Bacterization with Plant Growth Promoting Rhizobacteria. *J. Plant Dis. Protect.* **2006,** *113*, 247–251.

Elsheikh, E. A. E. Response of Legume-*Rhizobium* Symbiosis to Salinity in the Sudan: A Review. *Univ. Khartoum J. Agric. Sci.* **1998a,** *6* (2), 142–156.

Elsheikh, E. A. E. Effect of Inoculation with *Rhizobium* on the Seed Chemical and Physical Properties of Legumes. *Asp. Appl. Biol.* **2001,** *63*, 151–163.

Elsheikh, E. A. E.; Osman, A. G. *Rhizobium leguminosarum* Inoculation Decreases Damage to Faba Bean (*Vicia faba* L.) Caused by Broad Bean Mottle Bromovirus and Bean Yellow Mosaic Potyvirus. *World J. Microbiol. Biotechnol.* **1995,** *11*, 223–227.

Figueiredo, M.; Martinez, C.; Burity, H.; Chanway, C. Plant Growth-Promoting Rhizobacteria for Improving Nodulation and Nitrogen Fixation in the Common Bean (*Phaseolus vulgaris* L.). *World J. Microb. Biotechnol.* **2008,** *24* (7), 1187–1193.

Franche, C.; Lindström, K.; Elmerich, C. Nitrogen-Fixing Bacteria Associated with Leguminous and Non-leguminous Plants. *Plant Soil* **2009,** *321*, 35–59.

Gan, Y.; Selles, F.; Hanson, K. G.; Zentner, R. P.; McDonald, C. L. Response of Lentil to *Rhizobium* in a Semiarid Environment; *Agriculture and Agri-Food Canada*, 2002; pp 1–3.

García-Fraile, P.; Carro, L.; Robledo, M.; Ramírez-Bahena, M. H.; Flores-Félix, J. D.; Fernández, M. T.; Mateos, P. F.; Rivas, R.; Igual, J. M.; Martínez-Molina, E.; Peix, A.; Velázquez, E. *Rhizobium* Promotes Non-Legumes Growth and Quality in Several Production Steps: Towards a Biofertilization of Edible Raw Vegetables Healthy for Humans. *PLoS One* **2012,** *7*, 38122.

Ghosh, P. K.; De, T. K.; Maiti, T. K. Production and Metabolism of Indole Acetic Acid in Root Nodules and Symbiont (*Rhizobium undicola*) Isolated from Root Nodule of Aquatic Medicinal Legume *Neptunia oleracea* Lour. *J. Bot.* **2015,** *2015*, 1–11.

Giller, K. E. In *Nitrogen Fixation in Tropical Cropping Systems*; CABI Publishing, 2001.

Glick, B. R. Plant Growth-Promoting Bacteria: Mechanisms and Applications. *Scientifica* **2012,** *963401*, 1–15.

Gray, E.; Smith, D. Intracellular and Extracellular PGPR: Commonalities and Distinctions in the Plant–Bacterium Signaling Processes. *Soil Biol. Biochem.* **2005,** *37*, 395–412.

Harish, S.; Parthasarathy, S.; Durgadevi, D.; Anandhi, K.; Raguchander, T. Plant Growth-Promoting Rhizobacteria: Harnessing its Potential for Sustainable Plant Disease Management. In *Plant Growth Promoting Rhizobacteria for Agricultural Sustainability*; Springer: Singapore, 2019. https://doi.org/10.1007/978-981-13-7553-8_8.

Hemissi, I.; Mabrouk, Y.; Abdi, N.; Bouraoui, M.; Saidi, M.; Sifi, B. Effects of some *Rhizobium* Strains on Chickpea Growth and Biological Control of *Rhizoctonia solani*. *Afr. J. Microbiol. Res.* **2011,** *5*, 4080–4090.

Hossain, M. S.; Mårtensson, A. Potential Use of *Rhizobium* spp. to Improve Fitness of Non-Nitrogen- Fixing Plants. *Acta Agric. Scand.* **2008,** *58*, 352–358.

Htwe, A. Z.; Moh, S. M.; Moe, K.; Yamakawa, T. Biofertilizer Production for Agronomic Application and Evaluation of its Symbiotic Effectiveness in Soybeans. *Agronomy* **2019,** *9*, 162.

Hussain, M. B.; Mehboob, I.; Zahir, Z. A.; Naveed, M.; Asghar, H. N. Potential of *Rhizobium* spp. for Improving Growth and Yield of Rice (*Oryza sativa* L.). *Soil Environ.* **2009,** *28* (1), 49–55.

Imen, H.; Neila, A.; Adnane, B.; et al. Inoculation with Phosphate Solubilizing *Mesorhizobium* Strains Improves the Performance of Chickpea (*Cicer arietinum* L.) under Phosphorus Deficiency. *J. Plant Nutr.* **2015,** *38*, 1656–1671.

Jarvis, B. D. W.; Van, B. P.; Chen, W. X.; Nour, S. M.; Fernandez, M. P.; Cleyet-Marel, J. C.; Gillis. M. Transfer of *Rhizobium loti*, *Rhizobium huakuii*, *Rhizobium ciceri*, *Rhizobium mediterraneum*, and *Rhizobium tianshanense* to *Mesorhizobium* gen. nov. *Int. J. Syst. Bacteriol.* **1997,** *47*, 895–898.

Jordan, D. C. Transfer of *Rhizobium japonicum* Buchanan 1980 to *Bradyrhizobium* gen. nov., a Genus of Slow-Growing, Root Nodule Bacteria from Leguminous Plants. *Int. J. Syst. Evol. Microbiol.* **1982,** *32*, 136–139.

Judge, C.; Hickey, E. Microbial Signalling of *Rhizobium* Bacteria as a Method for Cereal Crop Germination Enhancement. *BioRxiv.* **2019**. doi: https://doi.org/10.1101/788125.

Kacem, M.; Kazouz, F.; Merabet, C.; Rezki, M.; de Lajudie, P.; Bekki, A. Antimicrobial Activity of *Rhizobium* sp. Strains Against *Pseudomonas savastanoi*, the Agent Responsible for the Olive Knot Disease in Algeria. *Grasas Aceites* **2009,** *60*, 139–146.

Karpagam, T.; Nagalakshmi, P. K. Isolation and Characterization of Phosphate Solubilizing Microbes from Agricultural Soil. *Int. J. Curr. Microbiol. Appl. Sci.* **2014,** *3*, 601–614.

Khaitov, B.; Karimov, A.; Abdiev, A.; Farrukh, J.; Park, K. W. Beneficial Effect of *Rhizobium* Inoculation on Growth and Yield of Chickpea (*Cicer arietinum* L.) in Saline Soils. *Bulg. J. Agric. Sci.* **2020,** *26* (1), 96–104.

Khaitov, B.; Kurbonov, A.; Abdiev, A.; Adilov, M. Effect of Chickpea in Association with *Rhizobium* to Crop Productivity and Soil Fertility. *Eurasian J. Soil Sci.* **2016,** *5*, 105–112.

Khatkar, R.; Abraham, T.; Joseph, S. A. Effect of Biofertilizers and Sulfur Level on Growth and Yield of Black Gram (*Vigna mungo* L.). *Legume Res.* **2007,** *30* (3), 233–234.

Kloepper, J. W. In *Plant Growth-Promoting Rhizobacteria on Radishes*, Proceedings of the Fourth International Conference on Plant Pathogenic Bacteria; Schroth, M. N., Ed.; 1978, pp 879–882.

Kloepper, J. W.; Leong, J.; Teintze, M.; Schroth, M. N. Enhanced Plant Growth by Siderophores Produced by Plant Growth-Promoting Rhizobacteria. *Nature* **1980,** *286*, 885–886.

Kumar, A.; Kumar, K.; Kumar, P.; et al. Production of Indole Acetic Acid by *Azotobacter* Strains Associated with Mungbean. *Plant Arch.* **2014,** *14*, 41–42.

Kumar, G.; Raghu Ram, M. Phosphate Solubilizing Rhizobia Isolated from *Vigna trilobata*. *Am. J. Microbiol. Res.* **2014,** *2*, 105–109.

Laranjoa, M.; Alexandre, A.; Oliveiraa, S. Legume Growth-Promoting *Rhizobium*: An Overview on the *Mesorhizobium* Genus. *Microbiol. Res.* **2014,** *169*, 2–17.

Lindström, K.; Martinez-Romero, M. International Committee on Systematics of Prokaryotes; Subcommittee on the Taxonomy of Agrobacterium and Rhizobium Minutes of the Meeting, 2004, Toulouse, France. *Int. J. Syst. Evol. Microbiol.* **2005,** *55*, 1383–1383.

Lindström, K.; Murwira, M.; Willems, A.; Altier, N. The Biodiversity of Beneficial Microbe-Host Mutualism: The Case of Rhizobia. *Res. Microbiol.* **2010,** *161*, 453–463.

Lukkani, N. J.; Reddy, E. C. S. Evaluation of Plant Growth Promoting Attributes and Biocontrol Potential of Native Fluorescent *Pseudomonas* spp. Against *Aspergillus niger* Causing Collar Rot of Ground Nut. *Int. J. Plant Anim. Environ. Sci.* **2014,** *4*, 267–262.

Ma, W.; Charles, T. C.; Glick, B. R. Expression of an Exogenous 1-Aminocyclopropane-1-Carboxylate Deaminase Gene in *Sinorhizobium meliloti* Increases its Ability to Nodulate Alfalfa. *Appl. Environ. Microbiol.* **2004,** *70*, 5891–5897.

Machado, R. G.; Sá, E. L. S.; Bruxel, M.; Giongo, A.; Santos, N. S.; Nunes, A. S. Indoleacetic Acid Producing Rhizobia Promote Growth of Tanzania Grass (*Panicum maximum*) and Pensacola Grass (*Paspalum saurae*). *Int. J. Agric. Biol.* **2013,** *15*, 827–834.

Martinez-Romero, E.; Segovia, L.; Mercante, F. M.; Franco, A. A.; Graham, P.; Pardo, M. A. *Rhizobium tropici*, A Novel Species Nodulating *Phaseolus vulgaris* L. Beans and *Leucaena* sp. Trees. *Int. J. Syst. Bacteriol.* **1991,** *41* (3), 417–26.

Martinez-Viveros, O.; Jorquera, M. A.; Crowley, D. E.; Gajardo, G.; Mora, M. L. Mechanisms and Practical Considerations Involved in Plant Growth Promotion by Rhizobacteria. *J. Soil Sci. Plant Nutr.* **2010,** *10*, 293–319.

Matiru, V. N.; Dakora, F. D. Potential use of Rhizobial Bacteria as Promoters of Plant Growth for Increased Yield in Landraces of African Cereal Crops. *Afr. J. Biotechnol.* **2004,** *3* (1), 1–7.

Mazid, M.; Khan, T. A. Future of Bio-fertilizers in Indian Agriculture: An Overview. *Int. J. Agric. Food Res.* **2014,** *3*, 10–23.

Mehboob, I.; Zahir, Z.; Arshad, M.; Tanveer, A.; Azam, F. Growth Promoting Activities of different *Rhizobium* spp. in Wheat. *Pak. J. Bot.* **2011,** *43*, 1643–1650.

Mishra, R. P.; Singh, R. K.; Jaiswal, H. K.; Kumar, V.; Maurya, S. Rhizobium-Mediated Induction of Phenolics and Plant Growth Promotion in Rice (*Oryza sativa* L.) *Curr. Microbiol.* **2006,** *52*, 383–389.

Muthuselvan, I.; Balagurunathan, R. Siderophore production from *Azotobacter* sp. and its application as biocontrol agent. *Int. J. Curr. Res. Rev.* **2013,** *5* (11), 23–35.

Nabti, E.; Bensidhoum, L.; Tabli, N.; Dahel, D.; Weiss, A.; Rothballer, M.; Schmid, M.; Hartman, A. Growth Inhibition of Barley and Biocontrol Effect on Plant Pathogenic Fungi by a *Cellulosimicrobium* Isolated from Salt Affected Rhizosphere Soil in Northwestern Algeria. *Eur. J. Soil Biol.* **2014,** *61*, 20–26.

Nakatsukasa, H.; Uchiumi, T.; Kucho, K.; Suzuki, A.; Higashi, S.; Abe, M. Transposon Mediation Allows a Symbiotic Plasmid of *Rhizobium leguminosarum* bv. *Trifolii* to become a Symbiosis Island in *Agrobacterium* and *Rhizobium*. *J. Gen. Appl. Microbiol.* **2008,** *54*, 107–118.

Ormeno-Orrillo, E.; Servín-Garciduenas, L.; Rogel, M. A.; González, V.; Peralta, H.; Mora, J.; Martinez-Romero, J.; Martínez-Romero, E. Taxonomy of Rhizobia and Agrobacteria from the Rhizobiaceae Family in Light of Genomics. *Syst. Appl. Microbiol.* **2015,** *38*, 287–291.

Osman, A. G.; Abd-Elaziz, F. I.; Elhassan, G. A. Effects of Biological and Mineral Fertilization on Yield, Chemical Composition and Physical Characteristics of Faba Bean (*Vicia faba* L.) Cultivar Seleim. *Pak. J. Nutr.* **2010,** *9* (7), 703–708.

Patil, A.; Kale, A.; Ajane, G.; Sheikh, R.; Patil, S. Plant Growth-Promoting Rhizobium: Mechanisms and Biotechnological Prospective; Springer International Publishing, 2017. doi: 10.1007/978-3-319-64982-5_7.

Phillips, D. A.; Joseph, C. M.; Yang, G. P.; Martinez-Romero, E.; Sanborn, J. R.; Volpin, H. Identification of Lumichrome as a *Sinorhizobium* Enhancer of Alfalfa Root Respiration and Shoot Growth. *Proc. Natl. Acad. Sci. USA.* **1999,** *96*, 12275–12280.

Prabhavati, E.; Anthony, J. Bacteriocin Production by Rhizobia Isolated from Root Nodules of Horse Gram. *Bangladesh J. Med. Sci.* **2012,** *11*, 28–32.

Prasad, J. S.; Reddy, R. S.; Reddy, P. N.; Rajashekar, A. U. Isolation, Screening and Characterization of *Azotobacter* from Rhizospheric Soils for Different Plant Growth Promotion (PGP) and Antagonistic Activities and Compatibility with Agrochemicals: An in vitro Study. *Ecol. Environ. Conserv.* **2014,** *20*, 959–966.

Ramirez-Bahena, M. H.; Garcia-Fraile, P.; Peix, A.; Valverde, A.; Rivas, R.; Igual, J. M.; Mateos, P. F.; Martinez-Molina, E.; Velazquez, E. Revision of the Taxonomic Status of the Species *Rhizobium leguminosarum* (Frank 1879) Frank 1889AL, *Rhizobium phaseoli* Dangeard1926AL and *Rhizobium trifolii* Dangeard 1926AL. *R. trifolii* is a Later Synonym of *R. leguminosarum*. Reclassification of the Strain *R. leguminosarum* DSM30132 (5NCIMB11478) as *Rhizobium pisi* sp. nov. *Int. J. Syst. Evol. Microbiol.* **2008,** *58*, 2484–2490.

Rashid, M. H.; Young, J. P.; Everall, I.; Clercx, P.; Willems, A.; Braun, M. S.; Wink, M. Average Nucleotide Identity of Genome Sequences Supports the Description of *Rhizobium lentis* sp. nov., *Rhizobium bangladeshense* sp. nov. and *Rhizobium binae* sp. nov. from lentil (*Lens culinaris*) nodules. *Int. J. Syst. Evol. Microbiol.* **2015,** *65*, 3037–3045.

Ribeiro, R. A.; Barcellos, F. G.; Thompson, F. L.; Hungria, M. Multilocus Sequence Analysis of Brazilian *Rhizobium* Microsymbionts of Common Bean (*Phaseolus vulgaris* L.) Reveals Unexpected Taxonomic Diversity. *Res. Microbiol.* **2009,** *160*, 297–306.

Riggs, P. J.; Chelius, M. K.; Iniguez, A. L.; Kaeppler, S. M.; Triplett, E. W. Enhanced Maize Productivity by Inoculation with Diazotrophic Bacteria. *Aust. J. Plant Physiol.* **2001,** *28*, 829–836.

Roy, N.; Chakrabartty, P. K. Effect of Aluminum on the Production of Siderophore by *Rhizobium* sp. (*Cicer arietinum*). *Curr. Microbiol.* **2000,** *41*, 5–10.

Rugheim, A. M. E.; Abdelgani, M. E. Effects of Microbial and Chemical Fertilization on Yield and Seed Quality of Faba Bean (*Vicia faba* L). *Int. Food Res. J.* **2012,** *19*, 417–422.

Saber, T. D.; Khursheed, M. Q. Improvement of Wheat Quality and Soil Fertility by Integrates Chemical Fertilizer with Rhizobial Bacteria. *ZJPAS* **2020,** *32* (2), 178–191.

Sahasrabudhe, M. M. Screening of Rhizobia for Indole Acetic Acid Production. *Ann. Biol. Res.* **2011,** *2*, 460–468.

Sajid, M.; Rab, A.; Wahid, F. I.; Shah, S. N. M.; Jan, I.; Khan, M. A.; Hussain, S. A.; Khan, M. A.; Iqbal, Z. Influence of *Rhizobium* Inoculation on Growth and Yield of Groundnut Cultivars. *Sarhad J. Agric.* **2011,** *27*, 573–576.

Sangeetha, G.; Thangavelu, R.; Usha, R. S.; Muthukumar, A.; Udayakumar, R. Induction of Systemic Resistance by Mixtures of Antagonist Bacteria for the Management of Crown Rot Complex on Banana. *Acta Physiol. Plant* **2010,** *32*, 1177–1187.

Sawada, H.; Kuykendal, L. D.; Young, J. M. Changing Concepts in the Systematics of Bacterial Nitrogen-Fixing Legume Symbionts. *J. Gen. Appl. Microbiol.* **2003,** *49*, 155–179.

Shamseldin, A. Future Outlook of Transferring Biological Nitrogen Fixation (BNF) to Cereals and Challenges to Retard Achieving this Dream. *Curr. Microbiol.* **2022,** *79*, 171. https://doi.org/10.1007/s00284-022-02852-2.

Shamseldin, A.; Abdelkhalek, A.; Sadowsky, M. J. Recent Changes to the Classification of Symbiotic, Nitrogen-Fixing, Legume-Associating Bacteria: A Review. *Symbiosis* **2017,** *71*, 91–109.

Singh, R. S.; Yadav, M. K. Effect of Phosphorous and Biofertilizers on Growth, Yield and Nutrient Uptake of Long Duration Pigeon Pea under Rainfed Condition. *J. Food Legumes* **2008,** *21* (1), 46–48.

Sistani, N. R.; Kaul, H. P.; Desalegn, G.; Wienkoop, S. Rhizobium Impacts on Seed Productivity, Quality, and Protection of *Pisum sativum* upon Disease Stress Caused by *Didymella pinodes*: Phenotypic, Proteomic, and Metabolomic Traits. *Front. Plant Sci.* **2017,** *8*, 1961.

Solano, B. R.; Maicas, J. B.; Fjg, M. Physiological and Molecular Mechanisms of Plant Growth Promoting Rhizobacteria (PGPR). In *Plant-Bacteria Interactions: Strategies and Techniques to Promote Plant Growth*; Ahmad, I., Pichtel, J., Hayat, S., Eds.; Wiley, Weinheim, Germany, 2008; pp 41–52.

Sprent, J. I. Nodulation in Legumes. *Ann. Bot.* **2001,** *89* (6), 797–798.

Squartini, A.; Struffi, P.; Doring, H.; Selenska-Pobell, S.; Tola, E.; Giacomini, A.; Vendramin, E.; Velazquez, E.; Mateos, P. F.; Martinez-Molina, E.; Dazzo, F. B.; Casella, S.; Nuti, M. P. *Rhizobium sullae* sp. nov. (formerly *Rhizobium hedysari*), The Root-Nodule Microsymbiont of *Hedysarum coronarium* L. *Int. J. Syst. Evol. Microbiol.* **2002,** *52*,1267–1276.

Tawaha, A. M.; Singh, V. P.; Turk, M. A. A Review on Growth, Yield Components and Yield of Barley as Influenced by Genotypes, Herbicides and Fertilizer Application. *Res. Crop* **2003,** *4* (1), 1–9.

Tawaha, A. M.; Turk, M. A. Lentil (*Lens culinaris* Medic.) Productivity as Influenced by Rate and Method of Phosphate Placement in a Mediterranean Environment. *Acta Agron. Hung.* **2002a,** *50* (2), 197–201.

Tawaha, A. M.; Turk, M. A. Field Pea Seeding Management for Semi-Arid Mediterranean Conditions. *J. Agron. Crop Sci.* **2004,** *190*, 86–92.

Tawaha, A. R. M.; Turk, M. A.; Lee, K. D.; Zheng, W. J.; Ababneh, M.; Abebe, G.; Musallam, I. W. Impact of Fertilizer and Herbicide Application on Performance of Ten Barley Genotypes grown in Northeastern Part of Jordan. *Int. J. Agric. Biol.* **2005,** *7* (2), 162–166.

Tortora, M. L.; Díaz-Ricci, J. C.; Pedraza, R. O. *Azospirillum brasilense* Siderophores with Antifungal Activity against *Colletotrichum acutatum*. *Arch. Microbiol.* **2011,** *193*, 275–286.

Tran, V. V.; Bergo, O.; Ke, S. N.; Balandreau, J.; Heulin, T. Repeated Beneficial Effects of Rice Inoculation with a Strain of *Burkholderia vietnamiensis* on Early and Late Yield Components in Low Fertility Sulphate Acid Soils of Vietnam. *Plant Soil* **2000,** *218*, 273–284.

Turk, M. A.; Hmaeed, K. M.; Aqeel, A. M.; Tawaha, A. M. Nutritional Status of Durum Wheat Grown in Soil Supplemented with Olive Mill by-Products. *Agrochimica* **2003c,** *47*(5-6), 209–219.

Turk, M. A.; Lee, K. D.; Tawaha, A. M. Inhibitory Effects of Aqueous Extracts of Black Mustard on Germination and Growth of Radish. *Res. J. Agric. Biol. Sci.* **2005,** *1* (3), 227–231.

Turk, M. A.; Tawaha, A. M. Onion (*Allium cepa* L.) as Influenced by Rate and Method of Phosphorus Placement. *Crop Res.* **2002a,** *23* (1), 105–107.

Turk, M. A.; Tawaha, A. M. Allelopathic Effect of Black Mustard (*Brassica nigra* L.) on Germination and Growth of Wild Oat (*Avena fatua* L.). *Crop Prot.* **2003a,** *22* (4), 673.

Turk, M. A.; Tawaha, A. M. Allelopathic Effect of Black Mustard (*Brassica nigra* L.) on Germination and Growth of Wild Barley (*Hordeum spontaneum*). *J. Agron. Crop Sci.* **2003b,** *189* (5), 298–303.

Turk, M. A.; Tawaha, A. M.; Samara, N. Effects of Seedling Rate and Date, and Phosphorous Application on Growth and Yield of Carbon Vetch (*Vicia narbonensis*). *Agronomie* **2003b,** *23*, 1–4.

Turk, M. A.; Tawaha, A. M.; Shatnawi, M. Lentil (*Lens culinaris* Medik) Response to Plant Density, Sowing Date, Phosporous Fertilization and Ethephon Application in the Absence of Moisture Stress. *J. Agron. Crop Sci.* **2003a,** *189* (1), 1–6.

Vargas, L. K.; Volpiano, C. G.; Lisboa, B. B.; Giongo, A.; Beneduzi, A.; Passaglia, L. M. P. Potential of Rhizobia as Plant Growth Promoting Rhizobacteria. In *Microbes for Legume Improvement*; Almas, Z., Mohammad, S. K., Javed, M., Eds.; Springer International Publishing, 2017; pp 154–174.

Vejan, P.; Abdullah, R.; Khadiran, T.; Ismail, S.; Boyce, A. N. Role of Plant Growth Promoting Rhizobacteria in Agricultural Sustainability - A Review. *Molecules* **2016,** *21*, 573.

Velázquez, E.; García-Fraile, P.; Ramírez-Bahena, M.; Rivas, R.; Martínez-Molina, E. Current Status of the Taxonomy of Bacteria Able to Establish Nitrogen-Fixing Legume Symbiosis. In *Microbes for Legume Improvement*; Springer International Publishing, 2017; pp 1–43.

Verma, J. P.; Yadav, J.; Tiwari, K. N.; Kumar, A. Effect of Indigenous *Mesorhizobium* spp. and Plant Growth Promoting Rhizobacteria on Yields and Nutrients Uptake of Chickpea (*Cicer arietinum* L.) under Sustainable Agriculture. *Ecol Eng.* **2013,** *51*, 282–286.

Wang, E. T.; Cheng, F, T.; Chen, W. F.; Young, J. P. W.; Chen, W. X. Ecology and Evolution of *Rhizobia*: Principles and Applications; Springer Nature Singapore Pte Ltd., 2019.

Wang, E. T.; Martinez-Romero, E. *Sesbania herbacea–Rhizobium huautlense* Nodulation in Flooded Soils and Comparative Characterization of *S. herbacea*-Nodulating Rhizobia in Different Environments. *Microbial Ecol.* **2000,** *40* (1), 25–32.

Wani, P. A.; Khan, M. S.; Zaidi, A. Synergistic Effect of the Inoculation with Nitrogen-Fixing and Phosphate-Solubilizing Rhizobacteria on Performance of Field-Grown Chickpea. *J. Plant Nutr. Soil Sci.* **2007,** *170*, 283–287.

Willems, A. The Taxonomy of Rhizobia: An Overview. *Plant Soil* **2006,** *287*, 3–14.

Woese, C. R.; Fox, G. E. Phylogenetic Structure of the Prokaryotic Domain: The Primary Kingdoms. *Proc. Natl. Acad. Sci. USA* **1977,** *74*, 5088–5090.

Woese, C. R.; Stackebrandt, E.; Weisburg, W. G.; Paster, B. J.; Madigan, M. T.; Fowler, V. J.; Hahn, C. M.; Blanz, P.; Gupta, R.; Nealson, K. H.; Fox, G. E. The Phylogeny of Purple Bacteria: The Alpha Subdivision. *Syst. Appl. Microbiol.* **1984,** *5*, 315–326.

Yu, X.; Liu, X.; Zhu, T.; Liu, G.; Mao, C. Co-inoculation with Phosphate-Solubilizing and Nitrogen-Fixing Bacteria on Solubilization of Rock Phosphate and their Effect on Growth Promotion and Nutrient Uptake by Walnut. *Euro. J. Soil Biol.* **2012,** *50*, 112–117.

Zaidi, A.; Khan, M. S.; Saif, S.; Rizvi, A.; Ahmed, B.; Shahid, M. Role of Nitrogen-Fixing Plant Growth-Promoting Rhizobacteria in Sustainable Production of Vegetables: Current Perspective. In *Microbial Strategies for Vegetable Production*; Zaidi, A., Khan, M., Eds.; Springer: Cham, 2017; pp 50–79. https://doi.org/10.1007/978-3-319-54401-4_3.

Zakharova, E.; Shcherbakov, A.; Brudnik, V.; Skripko, N.; Bulkhin, N.; Ignatov, V. Biosynthesis of Indole-3-Acetic Acid in *Azospirillum brasilense*, Insights from Quantum Chemistry. *Eur. J. Biochem.* **1999,** *259*, 572–576.

Zakhia, F.; de, L. P. Taxonomy of Rhizobia. *Agronomie* **2001,** *21*, 569–576.

Zamora, G. M. L.; Martinez-Romero, E. Natural Endophytic Association between *Rhizobium etli* and Maize (*Zea mays* L.). *J. Biotechnol.* **2001,** *91*, 117–126.

Zeffa, D. M.; Fantin, L. H.; Koltun, A.; de Oliveira, A. L. M.; Nunes, M. P. B. A.; Canteri, M. G.; Goncalves, L. S. A. Effects of Plant Growth-Promoting Rhizobacteria on Co-inoculation with *Bradyrhizobium* in Soybean Crop: A Meta-Analysis of Studies from 1987 to 2018. *Peer J.* **2020**. doi: 10.7717/peerj.7905.

Ziaf, K.; Latif, U.; Amjad, M.; Shabir, M. Z.; Asghar, W.; Ahmed, S.; Ahmad, I.; Jahangir, M. M.; Anwar, W. Combined Use of Microbial and Synthetic Amendments can Improve Radish (*Raphanus sativus*) Yield. *J. Environ. Agric. Sci.* **2016,** *6*, 10–15.

CHAPTER 6

THE ROLE OF AZOLLA IN BIOFERTILIZATION AND ANIMAL NUTRITION

RAZZAGH KARIMIRAD[1*], ROGHAYEH KARIMIRAD[2], ABDEL RAHMAN MOHAMMAD AL-TAWAHA[3], DEVARAJAN THANGADURAI[4], JEYABALAN SANGEETHA[5], and VISHNU D. RAJPUT[6]

[1]*Department of Animal Science, Lorestan University, Khorramabad, Iran*

[2]*Department of Horticultural Sciences, Faculty of Agriculture and Natural Resources, University of Mohaghegh, Ardabili, Ardabil, Iran*

[3]*Department of Biological Sciences, Al Hussein bin Talal University, Maan, Jordan*

[4]*Department of Botany, Karnatak University, Dharwad, Karnataka, India*

[5]*Department of Environmental Science, Central University of Kerala, Kasaragod, Kerala, India*

[6]*Academy of Biology and Biotechnology, Southern Federal University, Rostov-on-Don, Russia*

**Corresponding author.*

ABSTRACT

Today, the best novel tool for agriculture science is biofertilizers. Biofertilizers are used in agricultural science as a substitution for conventional fertilizers.

Microbial Fertilizer Technology for Sustainable Crop Production. Jeyabalan Sangeetha, Abdel Rahman Mohammad Al-Tawaha, & Devarajan Thangadurai (Eds.)

DOI: 10.1201/9781003594826-6

Biofertilizer being essential ingredients of organic agriculture plays a fundamental function in preserving soil sustainability. *Azolla* is a small aquatic fern that commonly grows and reproduces on the surface of the ponds, and numerous kinds of streams. *Azolla* is a versatile aquatic fern to protect ecosystem and cyanobacteria. Nitrogen-fixing possible of cyanobacterial symbionts differs (30 and 60 kg N/ha), which introduces *Azolla* as a significant biological nitrogen source for livestock and agriculture industry. This fern grows enormously, under optimum conditions, and its biomass every 3 days doubles. The application of the *Azolla* plant is critical in agricultural science, mainly as organic compost and biofertilizer, aids the agronomist to decrease the expense of livestock feed complement, and it is beneficial as a supplement for the livestock industry and in industrial fields as wastewater purification, and biofuel.

6.1 INTRODUCTION

We live in a world in which the continually rising human population, together with a reduction in all-natural resources, is undoubtedly the main reason for the earth's damage around us. The requirement for enhancing plants production has become a quick demand to keep our everyday life. Previous evidence demonstrated that chemical fertilizers have resulted in contaminating nature along with groundwater; additionally, detection trapped in harvest products and moved to our systems (Chan and Griffiths, 1988). Alexandratos and Bruinsma (2012) predicated that an increase from 2005 to 2050 in the universal request for livestock products has been evaluated between 48% and 57%, while animal protein requests will grow even more. Proteins from beef, poultry, swine, and dairy crops must duplicate, whereas fish generation should nearly triple until 2050 in feed formulation. Such growth in animal production will generate a rise in supplement provided, evaluated to be more than 1.3 billion tons. In 2018, the universal livestock supplements production was assessed at 1.103 billion tons, related to a value of more than US$400 billion. Europe 28 contributed 277 million tons, a lower amount (394.9 million tons) in contrast to the values recorded for the Asia-Pacific region but higher (198.9 million tons) than values recorded for North America; the United States, China, Mexico, Spain, India, Brazil, Russia, and Turkey with each other produced 55% of the global supplement production (Alltech, 2019).

The globe population in 2025 is supposed to increase from 7 billion to 8.3 billion. The world by 2050 will need (70–100) percent more food; request area upon the agronomy industry to provide future food will be one of the multiple questions facing the agricultural group's in order to meet this challenge; concentrating on the soil system is allowing well realize the complex activities and relations governing the adjustment of crop generation together with soil health (Khan et al., 2009). Various agricultural management structures may influence the stability of food systems, as they may impact human and animal health, food safety and environmental condition sustainability. The food production approach is not always simple to categorize. This problem stems from not only the number and different forms of conventional and organic agronomic systems but also resulting from the overlap of these systems (WHO, 2012). Organic agriculture has appeared as the primary preference worldwide given the growing request for certain products and longtime sustainability, and concerns about environmental contamination related to the indiscriminate application of chemical fertilizers.

Even though the application of chemical material in cultivation is inescapable to meet the developing request for food worldwide, there are chances in chosen products and place areas where natural generation can be encouraged to tape the domestic export market. Biofertilizers, being a vital ingredient of organic agriculture, are the preparations comprising latent or live cells of effective strains of nitrogen (N) fixing, phosphate (PO_4^{3-}) solubilizing with the purpose of increasing the number of microbes and expedite those microbial mechanisms which increase the access of nutrients that can be quickly assimilated by many plants (Venkatashwarlu, 2008). Biofertilizers are environment-friendly fertilizers, used to raise quality and soil fertility. Previous research has highlighted the beneficial effects of biofertilizers on various crops, including vegetables, beans, and cereals. Many of these studies have confirmed the positive impacts of biofertilizers on plant growth. Because of the increasing request for natural plants, scientists researched new options for conventional fertilizers that resulted in some fitness and health concerns. Therefore, biofertilizers can dissolve the problem of nutrition and an increasing universal population at a period when agronomy is facing different environmental conditions and changes. Here, we discuss floating freshwater ferns from the genus *Azolla* as another crop for the production of protein food (Chan and Griffiths, 1988; Atiyeh et al., 2000; Van Hove and Lejeune, 2002; Suhag, 2016).

Azolla, a genus of aquatic ferns, includes seven species and fixes atmospheric nitrogen in connection with nitrogen-fixing *Anabaena azollae*. The broad utilization of *Azolla* as a biofertilizer for rice growing and other purposes such as animal feed and bioremediation supply a solid inducement to research many features of its biology (Vaishampayan, 2001; Qiu and Yu, 2003).

6.2 BIOFERTILIZERS

One of the main challenges in the 21st century is increasing production while maintaining the ecosystem's security (Berg, 2009). According to the previous document, fertilizers have been applied considerably to improve production from arable land. Increasing utilization of a chemical substance as fertilizers in agronomy may make a state self-sufficient in crop production, but this material has an unfavorable influence on the habitat of microorganisms. Further, these materials are costly; they impact the soil texture, decrease soil productivity, cause a disparity in the soil material, and affect unsuitable water contamination levels (Sprent and Sperent, 1990).

At this time, there is a growing concern about environmental damages and warnings about sustainable agronomy. Because of the above mention, the long-term utilization of biofertilizers demonstrated to be economical, environmental-friendly, more effective, cost-effective, and available to small farmers over chemical fertilizers. The requirement for the use of biofertilizer thus arises primarily for two causes. First, an increase in the use of fertilizers causes increased crop fertility, secondary because the increased use of chemical fertilizer causes injury in soil texture and raises other environmental concerns (Kumar et al., 2017). For optimal plant extension, nutrients must be accessed inadequate and moderate quantities (Chen, 2006). Soil infertility is the excellent main restriction limiting product yield in developing countries, and mostly among resource-poor farmers. Vlek and Vielhauer (1994) reported that soil productivity could be reformed successfully via adopting the integrated soil fertility management (ISFM), encompassing a procedure for nutrient management based on physical resource protection, potential biological nitrogen fixation (BNF), and increased performance of the inputs. There are seventy essential elements necessary for high yield of many plants.

Among the elements, nitrogen (N), phosphorous (P), and potassium (K) are required in high amounts (Mishra and Dash, 2014). Some microbes are often used as biofertilizers containing P-solubilizing bacteria, N-fixing soil bacteria, and applied with fungi and molds (Umesha et al., 2018). Additionally, biofertilizers, defined as microbial inoculants, are natural crops including particular microbes, which are obtained from root areas of the plants. These bio-inoculants colonize the soil zone and the inside of the plant, stimulating plant extension by 10–40% (Kawalekar, 2013) when used to plant surface, seed, and soil texture (Raghuwanshi, 2012). These colonize not only increase soil productivity and crop yield, growth of the rhizosphere, develop its life, degrade the harmful materials, and decrease the time to flowering (Youssef and Eissa, 2014), but also preserve them from many diseases. Another advantageous feature is that after continual use of biofertilizers (for 3–4 years), there is no requirement for their help, as parental inoculation is adequate for proliferation (Bumandalai and Tserennadmid, 2019).

In addition, plant hormone-synthesizing microbes are also applied in biofertilizer combinations. They supply growth-stimulating material such as amino acids, auxin, and vitamins to the plant and make better the productivity of the soil texture while preserving the crop efficiency (Parikh and James, 2012).

6.3 CLASSIFICATION OF BIOFERTILIZER

Various microbes and their interaction with plants are being utilized in the efficiency of biofertilizers. Biofertilizers can be classified into multiple ways based on their nature and action.

6.3.1 RHIZOBIUM

This belongs to the family Rhizobiaceae, is symbiotic in nature, colonizing with legumes fix atmospheric N 50–100 kg/ha. It is beneficial for pulse legumes (red-gram, pea, and black gram) forages legumes (berseem and Lucerne), and oil-seed legumes (groundnut and soybean). Better nodulation of leguminous products by *Rhizobium* mostly depends on the accessibility of suitable strain for a specific legume (Venkatashwarlu, 2008).

6.3.2 AZOTOBACTER

This belongs to the family Azotobacteriaceae, an obligate aerobe, heterotrophic. They are existence in PH = 7 or PH>7 soil. *A. chroococcum* is the main generally occurring species in arable soils. *A. beijerinckii*, *A. macrocytogenes*, *A. vinelandii*, and, *A. insignis* are other famous species. The occurrence of *Azotobacter* has been presented from the root zone of many crop plants such as rice, bajra, maize, and sugarcane vegetables. These crops play a principal role in the N cycle in nature, are unavailable to plants, and free it in the form of NH_4 in the soil. *Azotobacter* has a great range of enzymes that require N fixation: ferredoxin (FD), hydrogenase, and nitrogenase (Wani et al., 2015).

6.3.3 AZOSPIRILLUM

A. brasilense and *A. lipoferum* population are early inhabitants of soil and rhizosphere of plants and the intercellular place of the root layer of graminaceous. *Azospirillum* expands associative symbiotic connection with graminaceous plants. In addition, N fixation, growth-stimulating substance production as auxin, drought tolerance, and disease resistance are some of the extra benefits of inoculation with *Azospirillum* (Kumar et al., 2017).

6.3.4 CYANOBACTERIA

N_2 fixing cyanobacteria are the broad extension N_2 fixers. Cyanobacteria (blue-green algae) are the various classes of prokaryotes. The activities of N-fixing microorganisms supply a main origin of nitrogen to the marine ecosystem (Gonzalez et al., 2005). Peter et al. (2002) reported that cyanobacteria grow and fix N in terrestrial environment conditions, from rainfall forests to desert. They can remain in extreme environmental conditions and have the capacity to fix N due to their ability to fix N; cyanobacteria are used as biofertilizer. Apart from fixing N, cyanobacteria add organic matter, secrete growth-stimulating substances such as IAA, vitamins, mobilize insoluble P, and better physiochemical condition of the soil.

6.3.5 PHOSPHATE SOLUBILIZING MICROORGANISMS (PSM)

Some fungi and soil bacteria, mainly *Penicillium*, *Aspergillus*, *Pseudomonas*, and *Bacillus*, emit organic acids and decrease the pH in their area to bring about the dissolution of bound $(PO_4)^{3-}$ in texture of the soil. Improved efficiency of potato and wheat was reported to cause inoculation of the organic layer of soil-based cultures of *Pseudomonas striata* and *Bacillus polymyxa* (Kumar et al., 2017).

6.3.6 AM FUNGI

The shift of elements, mostly P, Zn, and S, from the soil texture to the root layer is controlled by the genera *Glomus*, *Sclerocystis*, *Acaulospora*, *Gigaspora*, and *Endogone*, which have vesicles for storage of element and arbuscles for funneling these substances into the root area. The usual genus seems to be glomus, which has various species diffuse in soil texture (Kumar et al., 2017).

6.3.7 SILICATE SOLUBILIZING BACTERIA (SSB)

Microbes can decompose silicates and aluminum silicates (Al_2SiO_5). During the chemical reaction of microorganisms, several organic acids such as oxalic acid and keto acids are made, and these have a double role in silicate weathering. They provide hydrogen ions to the medium and stimulate hydrolysis. Organic acids, citric and hydroxy carbolic acids, which form composite with cations, promote their elimination and preservation in the medium in a soluble state (Kumar et al., 2017).

6.3.8 PLANT GROWTH-PROMOTING RHIZOBACTERIA (PGPR)

PGPR is the class of bacteria that colonize roots and are beneficial to products. The PGPR inoculants raise growth through subduing of plant disease (bioprotectants), plant hormone production (biostimulants), and enhanced nutrient acquisition (biofertilizers) (Kumar et al., 2017).

6.3.9 *AZOLLA*

Azolla genus belongs to the family of Azollaceae and is generally found in the tropics and sub-tropics region of the world (Lumpkin, 1984). It grows originally in stagnant water in canals, ditches, drains, rivers, and ponds (Ansari and Sharma, 1991). In the world, the most species are known of *Azolla*, namely *A. filiculoides*, *A. rubra*, *A. microphylla*, *A. pinnata*, *A. caroliniana*, and *A. imbricate* (Raja et al., 2012). Kathirvelan et al. (2015) reported that *Azolla* comprises several valuable phytochemical substances such as flavonoids, phenols, vitamins (Vitamin A, Vitamin B, beta carotene, vitamin B_{12}), A, β-carotenes, alkaloids, hormones, amino acid, triterpenoid derivatives, and fatty acid types (Shaaban et al., 2020). These bioactive compounds contribute to various functional and medicinal properties, such as antioxidant, anti-inflammatory, anticarcinogenic, hepato and gastro-protective antiviral, antidiabetic, antihypertensive, neuro-protective, and cardio-protective (Maswada et al., 2020). *Azolla* contains higher critical amino acid composition 0.53% (leucine and lysine), 0.17% (methionine), 0.55% (threonine), 0.15% (tryptophan) (Leterme et al., 2009). *Azolla* contains 0.8–6.7% crude fat (Ali and Lesson, 1995) 14–16% crude fiber (Cherryl et al., 2014) 46–62% neutral detergent fiber (NDF), 31%, nitrogen free (Basak et al., 2002) 15–16% Ash (Buckingham et al., 1978). The concentrations of Ca, P, K, and Mg were 1.16, 1.29, 1.25, and 0.25%, respectively, while those of Na, Mn, Fe, Cu, and Zn were 23.79, 174.42, 755.73, 16.74, and 87.59 ppm, respectively. *Azolla* also contains 6.5% starch, 3.4–3.5 % sugar, 0.34–0.55% chlorophyll and Fe content is 0.26% on dry matter basis (Kathirvelan et al., 2015).

6.4 FUNCTIONS OF BIOFERTILIZERS

Some of the main functions of biofertilizers in agriculture are summarized as follows:

1. Application of biofertilizers causes improved mineral and water uptake, root expansion, vegetative growth, and N fixation.
2. Some biofertilizers (*Azotobacter* sp., Rhizobium BGA) promote the production of growth-stimulating materials like Auxin (IAA), Gibberellic acids (GA), and vitamin-B complex.

3. Vesicular-arbuscular-mycorrhiza (VAM fungi), when applied as biofertilizers, raise uptake of P, S, Zn, and water, cause constant crop development and improved yield, and also increase resistance to many root diseases.
4. They release growth-stimulating material and vitamins and help to preserve soil productivity.
5. P mobilizing, N-fixing, and cellulolytic microbes in biofertilizer raise the accessibility of plant nutrients in the soil texture and thus maintain agricultural production.
6. They are inexpensive, contamination-free, and renewable energy sources and improved physical features of soil, soil tilth, and soil fertility.
7. Bio-inoculants, including ligninolytic and cellulolytic microbes, increase the decomposition of organic matter in soil and also raise the rate of degradation in compost.
8. BGA plays a vital role in the N, recycling of plant nutrients, and economy of rice crops in tropical areas.
9. Azotobacter inoculants, when used most non-leguminous products, stimulate seed germination and primary potency of plants by generating growth-promoting material.
10. Blue-green algae, such as *Anabaena*, *Scytonema*, and *Nostoc*, are primarily employed in the resurrection of alkaline soils (Rachael et al., 2020).

6.5 APPLICATION OF *AZOLLA* IN CROPS

Azolla has been recognized for its applications in the field of agronomy because of its features that enable it to raise crops production. A symbiotic interaction with cyanobacteria is the fixation of N to be ingested by the *Azolla*. The symbiosis is unique among any plant-cyanobacteria communication as it is continuous. Its development and symbiotic relationship can be controlled by plant hormones (Abd et al., 2020) (Figure 6.1). Several reports indicating the use of *Azolla* in agriculture are summarized below.

The physiological function of *Azolla filiculoides* extracts in improving physiology, efficiency, and N uptake capability of maize plant and its growth and yield, mainly under irrigation deficiency. The results indicated *Azolla*

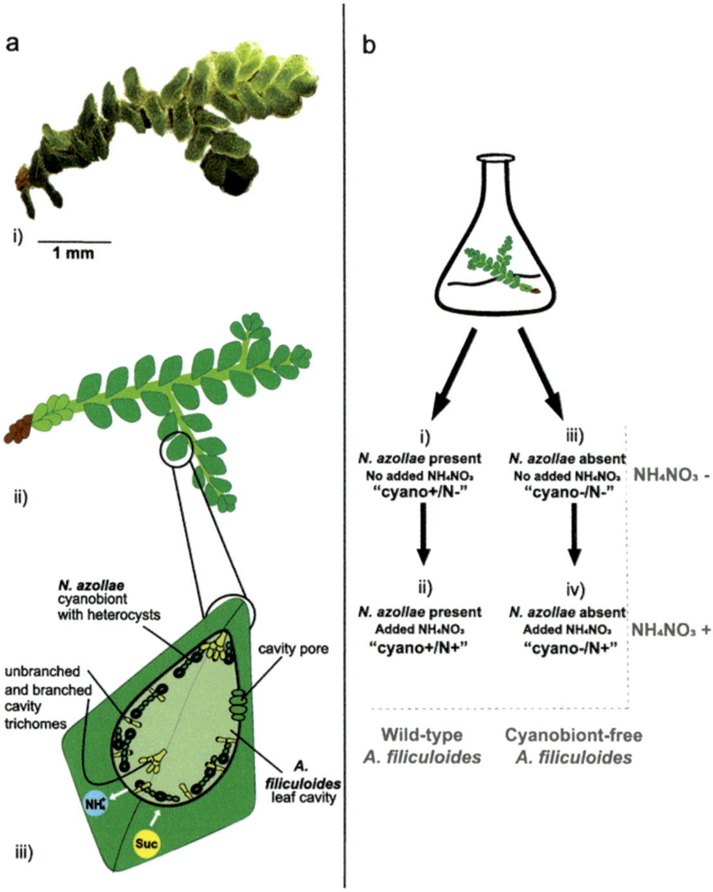

FIGURE 6.1 *Azolla filiculoides*, plant details (a) and growth culture conditions (b).

(a) (i) Close-up of plant, consisting of a primary stem with alternate leaves at regular intervals along the stem; secondary (branching) stems develop in the axil of certain leaves (image courtesy of Macroscopic Solutions, www.macroscopicsolutions.com), (ii) Cartoon representation of A. filiculoides, (iii) Cartoon representation of an A. filiculoides leaf showing leaf cavity and contents; direction of sucrose and ammonium nutrient exchange between N. azollae and A. filiculoides is indicated.

(b) Growth-culture conditions for A. filiculoides prior to RNA sequencing: (i) Wild-type *A. filiculoides*, without added NH_4NO_3, (ii) Wild type *A. filiculoides*, with added NH_4NO_3, (iii) Cyanobiont free *A. filiculoides*, without added NH_4NO_3, (iv) Cyanobiont free *A. filiculoides* with added NH_4NO_3. To obtain the putative symbiosis gene set, focused on comparing gene expression differences between the optimal growth condition (1) and the three other growth conditions (2–4).

Source: Reproduced with Permission from Eily et al. (2019), Copyright © 2019, Springer Nature.

extract is a strong biofertilizer, and its application in connection with deficit irrigation restricted any substantial reduction in the yields of N-deficient maize plant grain. It should be noted that *Azolla* is an economically sustainable and cost-effective organic substance that replaces more than 30% of urea (CH_4N_2O) fertilizers without affecting grain efficiency (Maswada et al., 2020). Ismail et al. (2017) reported the production of cowpea (*Vigna unguiculata*) under greenhouse conditions increased by the application of *Azolla caroliniana* and *Azolla pinnata* as biofertilizers. *Azolla* biofertilizers treatments improved the biochemical features of Alisa (commercial tomato) when the seeds were saturated in an equal concentration 24 h before sowing (Hanafy and El-Emary, 2018). Rabie et al. (2020) indicated *Azolla* treatments considerably enhanced the number of branches, plant height, and fresh and dry weights of chamomile plant, and also, *Azolla* extract as foliar had a favorable impact on chemical compounds and enhancement of the yield production by raising the active materials of chamomile plants. The grain improvement for foliar spraying with *Azolla* extract relative to control treatments was 57.37% and 51.71% (grain yield) and 37.66% and 21.57% (biological yield) for wheat in the first and second seasons, respectively.

The spraying of Egyptian bread wheat with *Azolla* extract was recorded, in comparison to mineral fertilization, to preserve high yield, simultaneously reduce production costs, and decrease production costs (Altai et al., 2019). *Azolla* is used as green manure in cultivating many crops such as arrowhead, taro, water bamboo, wheat, and rice (Teckle-Haimanot and Doku, 1995).

Chemical N fertilizer plays a dominant role in supplying the nutrient requirement of rice in Asia (Safriyani et al., 2020). It was also noted that the nutrient management system in the agricultural field is one of the most significant parameters that influence the yield and fertility of rice crops because the soil texture is one of the leading natural components. The extent of its quality plays a significant role in rice cultivation. Additionally, agricultural intensification connected with the increase of high amounts of mineral fertilizers negatively influences the biological and physiochemical features of soil (Allen and Pilbeam, 2006). Roy et al. (2016) indicated basal *Azolla* application rate for rice crop ranged from 10 to 12 tones/ha, enhanced soil N amount by 50–60 kg/ha, and decreased 30–35 kg of nitrogenous fertilizer requirement of this crop. It is found that either the application of 500 kg/ha through green *Azolla* increases the soil N content (50 kg/ha) and decreases the N fertilizer by 20–30 kg/ha. Yao et al. (2018) demonstrated that the application of *Azolla* to flooded water reduces the ammonia evaporation by 12–42%. Because of its high yield rate, *Azolla*, by providing a high level of organic matter on its incorporation into the soil, improves the physicochemical structure of its texture (Subedi and Shrestha, 2015). Additionally, Azolla aids in dissolving manganese (Mn), iron (Fe), and zinc (Zn), making these nutrients accessible to rice crops. It also releases phytohormones and vitamins, reduces evaporation rates from rice fields, decreases disease occurrence, and ultimately enhances rice crop yields (Biswa et al., 2005; Monajjem and Hajipour, 2010). Samal et al. (2020) found an increase in rice yield by 15–20% due to *Azolla* being used as a mono-cropping, whereas 14–40% grain yield by growing *Azolla* as a dual crop with rice. Razavipour et al. (2018) indicated that the application of 5.0% through *Azolla* compost produced a rice yield, 13.8% higher than that of the control treatment.

The main factor in using *Azolla* as a biofertilizer for many crops such as rice is its rapid degradation in soil and efficient accessibility of its nitrogen to rice plants. The quick proliferation rate and quick degradation ability of *Azolla* have become only a prime vital factor to use as green manure cum biofertilizer in rice fields. Incubation of *Azolla* as green manure in water-logged soil causes quick mineralization with the diffusion of 60–80% of the N within 14 days (Ito and Watanabe, 1985). Molotoks et al. (2018) reported that yield per land increase of cultivated land for rice crops could increase rice production but cause to the lack of desirable area. It is not possible to develop cultivated land. Therefore, the main focus is raising yield per land, contributing highly nutrient accessibility. Macroelements (N) are vital yield-restricting nutrients of this crop (Saito et al., 2019) (Figure 6.2).

6.6 *AZOLLA* AS NUTRITIONAL SUPPLEMENT FOR LIVESTOCK

The request for livestock products has been increasing and animal husbandry as a helpful occupation is expanding. Because of the decreasing area under forest and grasslands, there is a remarkable reduction in forage production. Additionally, the fodder accessibility from different crops has also reduced, mainly causing the presentation of high-yielding dwarf varieties. The deficiency of forage is, so, being recompensed with commercial feed, resulting in increased cost of livestock production. In addition, as commercial feed is blended with carbamide and other artificial milk boosters, it has a disadvantageous impact on the longevity of the livestock and the quality of milk produced, which in turn causes many diseases in human beings (Pillai et al., 2002).

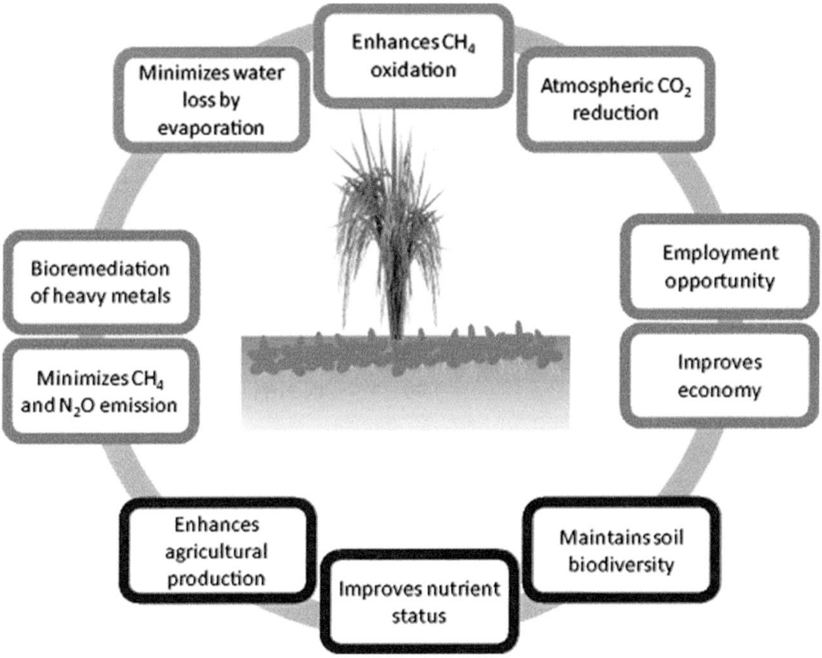

FIGURE 6.2 Socioeconomic and ecosystem services rendered by *Azolla* when applied to agriculture.

Source: Reproduced with Permission from Kollah et al. (2016), Copyright © 2015, Springer-Verlag, Berlin.

It is reported that 1.5–2 kg *Azolla*, if mixed with concentrate in 1:1 with sorghum, produces extra 16.9% milk in buffalo and 20.96% in cows (Biswas and Sarkar, 2013). *Azolla*, when fed directly to milk or mixed with concentrate, a 1:1 ratio along with concentrate, has increased milk production by 20.15–20% and decreased the requirement of 20–25% commercial feed (Pillai et al., 2002), while Rawat et al. (2015) revealed that *Azolla*, if mixed with concentrate in 1:1 in crossbred cows, produce 11.85% milk. The biopolymers and other growth stimulate that exist in *Azolla* are primarily responsible for improved milk production (Chatterjee et al., 2013). They revealed that on feeding *Azolla* (60 g/animal/day), there is improved growing calves rate. Increased milk production has been reported in growing calves by providing *Azolla* 700 g/animal/day and 600 g/animal/day mustard cake (Khare et al., 2014). It was reported that (Murthy et al., 2013) feeding of *Azolla* (2 Kg) instead of concentrate decreased feed and labor costs and (16.6%) milk production cost (18.5%) in crossbred calves. In various researches, an increase of 10% in milk production has been indicated in lactating crossbred cows and buffaloes on feeding *Azolla* (Gowda et al., 2015). It is to be noted that *Azolla* can be effectively used as a feed for non-ruminant (Seth et al., 2013). *Azolla* amount in the ratio of broilers indicated an improvement in yield at 5–15%, while other researchers found this improvement at 5% level (Naghshi et al., 2014). The inclusion level of *Azolla* at 5% in broiler chicken ration as a feed ingredient is most economical (Ara et al., 2015) due to reasonable cost and accessibility of quality fodder is important to the prosperous poultry industry. Some researchers found (Basak et al., 2002) that the use of *Azolla* up to 5% as a feed component to replace sesame meal in the ratio of 2–6 week-old broilers improved performance and energy efficiency without disadvantageous impact on palatability and mortalities.

Azolla can be used as a bit of substitution of protein source for growing-fattening pigs (Bacerra et al., 1995). Azolla (20%) was reported as a replacement for commercial feed in chicken as Azolla supplemented could improve the body weight (Subudhi and Singh, 1978). Saikia et al. (2014) detected improved body weight gain was in the group fed with 5% *Azolla* while the lower one was in the group supplemented with 15%, and they suggested that increasing the fiber content in a high quantity of *Azolla* negatively influences the appetite of the birds and therefore decreases the growth rate. Seth et al. (2013) demonstrated that the highest weight gain was in the treatment supplement on 5–10% in Vanaraja chicken fed compared to control. Prabina and Kumar (2010) suggested the consumed supplement was lower in broilers fed on *Azolla* (7.5%). Seth et al. (2013) also detected increased weight gain in chicken (Vanaraja) supplement on 5–10% *Azolla* compared with control. Alalade et al. (2007) indicated that *Azolla pinnata* feed in growing pullet ration can be safely included up to 15% with no health problem. In pullet chick diet, 10% inclusion amount has given the best yield. Enhancement in FCR and as well as body weight was found after the addition of *Azolla* 7.5% level (Kumar et al., 2018) and up to 10 % level in feed (Acharya et al., 2015). Furthermore, the feed cost of production by adding *Azolla* at either level was remarkably decreased. The addition of *Azolla* at the 10% level indicated a high economic yield. It was noted that fresh *Azolla* feed (30 g/bird/day) in 7-week-old turkeys decreased feed consumption with no influence on growth yield features caused by the mineral content and high protein amount of *Azolla* (Henry et al., 2017). A study carried out by Rana et al. (2017) showed the supplement of dried *Azolla* up to 5% to Vencobb broilers ration can be used safely without any harmful effects, but 2.5% feed level of *Azolla* also is effective in increasing biochemical factors. It was suggested a supplement of *Azolla* (5% or 7%) is helpful for safe and effective production of broilers, because to decrease mortalities and production costs, as well as enhancement benefits (Islam and Nishibor, 2017). Shukla et al. (2018) and Samad et al. (2020) found that the addition of *Azolla* resulted in growth yield without adverse effects on immune features, nutrient digestibility, and blood biochemistry of broiler chickens. Kannaiyan and Kumar (2005) showed that the inclusion of *Azolla* up to (100 g/bird/day) increased egg production. Similar high egg production and increasing nutrient value with the feed of *Azolla* in layers diet was detected by Lakshmanan et al. (2017). It was indicated that supplementation of *Azolla* up to 20% in the ratio of layer chicken diets had no significant effect on aspartate aminotransferase (AST) and alanine aminotransferase (ALT) levels, which shows no negative impact on *Azolla cristata* feed (Ara et al., 2018). Shamna et al. (2013) investigated that the addition of quails on *Azolla* at 5% substitution level of the basal ratio improved FCR and the growth rate. Rathod et al. (2013) and Varadharajan et al. (2019) concluded that supplementation of *Azolla* (6%) can include in quails' diet without a negative effect on carcass features and feed consumption. In the study of Acharya et al. (2015) *Azolla* as a feed source at level 10% could be supplemented in white Pekin broiler ducks. In an experiment with ducks, Bacerra et al. (1995) indicated that the addition of *Azolla* powder (15–60%) in ducks' diet provides 15.2–30.3% of the total protein; the FCR is reduced with raising *Azolla* consumption.

6.7 CONCLUSION

These days, there is a significant concern about meeting the expanding requests of the crops without corrupting the environment and soil well-being within the long term. There is a gap of ten million tons of plant supplements between the expulsion of crops and supply through chemical fertilizers. Moreover, biofertilizers can replace a portion of chemical fertilizer usage, reducing both the quantity and cost of chemical fertilizers, thereby helping to prevent environmental contamination associated with their extensive application. With the use of organic and natural fertilizers, a low-input system can be performed, and it can be helped to achieve farm sustainability. Considering these viewpoints, use of *Azolla* as biofertilizers can be a reasonable choice for the rice producer because it increases rice efficiency and additionally progresses soil wellbeing reasonably. *Azolla* also incorporates potential as an excellent and economical feed for diverse livestock species. It is trusted that within the coming days, *Azolla* innovation will be taken up enormously, by the dairy farmer, mainly, by those who involve land-scarce conditions for feed generation.

KEYWORDS

- ***Azolla***
- **biofertilizers**
- **livestock**
- **organic farming**
- **plant**
- **yield**

REFERENCES

Abd, E. A. S.; Mousa, A. A.; Orabi, S. H.; Gad-Allah, S. M.; Eldaim, M. A. A. Ameliorative Effect of *Azolla pinnata* Ethanolic Extract on Ranitidine-Induced Hepatotoxicity in Rats. *J. Curr. Vet. Res.* **2020,** *2*, 21–29.

Acharya, P.; Mohanty, G. P.; Pradhan, C. R.; Mishra, S. K.; Beura, N. C.; Moharana, B. Exploring the Effects of Inclusion of Dietary Fresh *Azolla* on the Performance of White Pekin Broiler Ducks. *Vet. World.* **2015,** *8*, 1293–1299.

Alalade, O. A.; Iyayi, E. A.; Alalade, T. O. The Nutritive Value of *Azolla* (*Azolla pinnata*) Meal in Diets for Growing Pullets and Subsequent Effect on Laying Performance. *J Poult Sci.* **2007,** *44*, 273–277.

Alexandratos, N.; Bruinsma, J. World Agriculture Towards 2030/2050: The 2012 Revision. *ESA Working Paper*; FAO: Rome (Italy), 2012, pp 12–03.

Ali, M. A.; Leeson, S. The Nutritive Value of Some Indigenous Asian Poultry Feed Ingredients. *Anim. Feed Sci. Technol.* **1995,** *55*, 227–237.

Allen, V. B.; Pilbeam, D. J. Plant Nutrition. Department of Plant Science, University of Massachusetts, USA, 2006, pp 293–328.

Alltech. Global Feed Survey. *Alltech*: Nicholasville (KY), 2019.

Altai, D.; Alhasany, A.; Noaema, A.; Idan, W.; Al-Farhan, I. Effectiveness of *Azolla* Extract in Reducing use of Mineral Fertilizers (NPK) and Increasing Productivity of Wheat under New Reclaimed Soils. *J. Plant Prod.* **2019,** *10*, 859–866.

Ansari, M. A.; Sharma, V. P. Role of *Azolla* in Controlling Mosquito Breeding in Ghaziabad District Village (U.P.). *Indian J. Malariol.* **1991,** *28*, 51–54.

Ara, S.; Adil, S.; Banday, M. T.; Khan, M. Feeding Potential of Aquatic Fern-*Azolla* in Broiler Chicken Ration. *J. Poult. Sci. Tech.* **2015,** *3*, 15–19.

Ara, S.; Adil, S.; Khan, M. A. Effect of Aquatic Fern, *Azolla cristata* in Diet on Growth, Serum Biochemistry and Laying Performance of Chicken. *Pak. J. Zool.* **2018,** *50*, 2325–2329.

Atiyeh, R. M.; Dominguez, J.; Subler, S.; Edwards, C. A. Biochemical Changes in Cow Manure Processed by Earthworms (*Eisenia andrei*) and their Effects on Plant-Growth. *Pedobiologia* **2000,** *44*, 709–724.

Bacerra, M.; Preston, T. R.; Ogle, B. Effect of Replacing Whole Boiled Soya Beans with *Azolla* in the Diets of Growing Ducks. *Livest. Res. Rural. Dev.* **1995,** *7*, 1–11.

Basak, B.; Pramanik, M. A. H.; Rahman, M. S.; Tarafdar, S. U.; Roy, B. C. *Azolla* (*Azolla pinnata*) as a Feed Ingredient in Broiler Ration. *Int. J. Poult. Sci.* **2002,** *1*, 29–34.

Berg, G. Plant–Microbe Interactions Promoting Plant Growth and Health: Perspectives for Controlled use of Microorganisms in Agriculture. *Appl. Microbiol. Biotechnol.* **2009,** *84*, 11–18.

Biswas, M.; Parveen, S.; Shimozawa, H.; Nakagoshi, N. Effects of *Azolla* Species on Weed Emergence in a Rice Paddy Ecosystem. *Weed Biol. Manag*. **2005,** *5*, 176–183.

Biswas, S.; Sarkar, S. *Azolla* Cultivation: A Supplementary Cattle Feed Production through Natural Resource Management. *Agric. Update* **2013,** *8* (4), 670–672.

Buckingham, K. E.; Stepher, W. E.; James, G. M.; Charles, R. G. Nutritive Value of Nitrogen Fixing Aquatic Fern *Azolla filiculoides*. *J. Agric. Food Chem.* **1978,** *26*, 1230–1234.

Bumandalai, O.; Tserennadmid, R. Effect of *Chlorella vulgaris* as a Biofertilizer on Germination of Tomato and Cucumber Seeds. *Int. J. Aquat. Biol.* **2019,** *7*, 95–99.

Chan, P. L. S.; Griffiths, D. A. The Vermicomposting of Pre-treated Pig Manure. *Biol. Wastes* **1988,** *24*, 57–69.

Chatterjee, A.; Sharma, P.; Ghosh, M. K.; Mandal, M.; Roy, P. K. Utilization of *Azolla microphylla* as Feed Supplement for Crossbred Cattle. *Int. J. Food Sci. Technol.* **2013,** *4*, 207–214.

Chen, J. In *The Combined use of Chemical and Organic Fertilizer and or Biofertilizer for Crop Growth and Soil Fertility*, International Workshop on Sustained Management of the Soil Rhizosphere System for Efficient Crop Production and Fertilizer Use, Thailand, 2006; pp 16–20.

Cherryl, D. M.; Prasad, R. M. V.; Jagadeeswara, R. S.; Jayalaxmi, P.; Srinivas, K. D. A Study on the Nutritive Value of *Azolla pinnata*. *Int. J. Livest. Res.* **2014,** *2*, 13–15.

Eily, A. N.; Pryer, K. M.; Li, F. W. A First Glimpse at Genes Important to the *Azolla–Nostoc* Symbiosis. *Symbiosis* **2019,** *78*, 149–162. https://doi.org/10.1007/s13199-019-00599-2.

Gonzalez, L. J.; Rodelas, B.; Pozo, C.; Salmeron, V.; Martinez, M. V.; Salmeron, V. Liberation of Amino Acids by Heterotrophic Nitrogen Fixing Bacteria. *Amino Acids* **2005,** *28*, 363–367.

Gowda, N. K. S.; Manegar, A.; Verma, S.; Valleesha, N. C. *Azolla* (*Azolla pinnata*) as a Green Feed Supplement for Dairy Cattle - An on Farm Study. *Anim. Nutr. Feed Technol.* **2015,** *15*, 283–287.

Hanafy, A.; El-Emary, G. A. E. Role of *Azolla pinnata* Biofertilizer Extract in Producing Healthy Tomatoes. *Asian J. Res. Biochem.* **2018,** *3*, 1–8.

Henry, A. C. E.; Reetha, T. L.; Paramasivam, A.; Mehala, C. Effect of *Azolla* Supplementation on Production Performance of Nandanam–II Turkey Growers. *Indian Vet. J.* **2017,** *94*, 28–30.

Islam, M. A.; Nishibori, M. Use of Multivitamin, Acidifier and *Azolla* in the Diet of Broiler Chickens. *Asian-Australas. J. Anim. Sci.* **2017,** *30*, 683–689.

Ismail, A. E. Impact of *Azolla caroliniana* and *A. pinnata* as Soil Amendments on *Rotylenchulus reniformis* and Plant Growth of Cowpea in Egypt. *J. Adv. Agric.* **2017,** *5*, 10–14.

Ito, O.; Watanabe, I. Availability to Rice Plants of Nitrogen Fixed by *Azolla*. *Soil Sci. Pl. Nutr.* **1985,** 34, 91–104.

Kannaiyan, S.; Kumar, K. *Azolla* in Animal Feeding. In *Azolla Biofertiliser For Sustainable Rice Production*; Daya Publishing House: Delhi, India, 2005.

Kathirvelan, C.; Banupriya, S.; Purushothaman, M. R. Azolla - An Alternate and Sustainable Feed for Livestock. *Int. J. Sci. Environ.* **2015,** *4*, 1153–1157.

Kawalekar, J. S. Role of Biofertilizers and Biopesticides for Sustainable Agriculture. *J. Bio. Innov.* **2013,** *2*, 73–78.

Khan, A. A.; Jilani, G.; Akhtar, M. S.; Naqvi, S. M. S.; Rasheed, M. Phosphorus Solubilizing Bacteria: Occurrence, Mechanisms and their Role in Crop Production. *J. Agric. Biol. Sci.* **2009,** *1*, 48–58.

Khare, A.; Baghel, R. P. S.; Gupta, R. S.; Nayak, S.; Khare, V.; Patil, A. K.; Sharma, R.; Tomar, R.; Singh, V. P. Milk Production of Indigenous Cattle for Supplement of Mustard Oil Cake or a *Azolla* Meal (*Azolla filliculoides*). *Livest. Res. Rural. Dev.* **2014,** *26*, 32–36.

Kollah, B.; Patra, A. K.; Mohanty, S. R. Aquatic Microphylla *Azolla*: A Perspective Paradigm for Sustainable Agriculture, Environment and Global Climate Change. *Environ. Sci. Pollut. Res.* **2016,** *23*, 4358–4369. https://doi.org/10.1007/s11356-015-5857-9.

Kumar, M.; Dhuria, R. K.; Jain, D.; Sharma, T.; Nehra, R.; Prajapat, U. K. Effect of Feeding *Azolla pinnata* on the Growth and Performance of Broiler Chicks. *Int. J. Chem. Stud.* **2018,** *6*, 3284–3290.

Kumar, R.; Kumawat, N.; Kumar, S. Y. Role of Biofertilizers in Agriculture. *Popular Kheti.* **2017,** *5*, 63–66.

Lakshmanan, A.; Kumar, K.; Latha, P. Azolla - A Low Cost and Effective Feed Supplement to Poultry Birds. *Int. J. Curr. Microbiol. Appl. Sci.* **2017,** *6*, 3622–3627.

Leterme, P.; Londono, A. M.; Munoz, J. E.; Suarez, J.; Bedoya, C. A.; Souffrant, W. B.; Buldgen, A. Nutritional Value of Aquatic Ferns (*Azolla filiculoides* Lam. and *Salvinia molesta* Mitchell) in Pigs. *Anim. Feed Sci. Technol.* **2009,** *149*, 135–148.

Lumpkin, T. A. Assessing the Potential for *Azolla* use in Humid Tropics. *Int. Rice Commission News.* **1984,** *33*, 30–33.

Maswada, H. F.; Abd El-Razek, U. A.; El-Sheshtawy, A. N. A.; Mazrou, Y. S. A. Effect of *Azolla filiculoides* on Growth, Physiological and Yield Attributes of Maize Grown under Water and Nitrogen Deficiencies. *J. Plant Growth Regul.* **2020,** *40*, 558–573.

Mishra, P.; Dash, D. Rejuvenation of Biofertilizer for Sustainable Agriculture and Economic Development. *Consilience* **2014,** 41–61.

Molotoks, A.; Stehfest, E.; Doelman, J.; Albanito, F.; Fitton, N.; Dawson, T. P.; Smith, P. Global Projections of Future Cropland Expansion to 2050 and Direct Impacts on Biodiversity and Carbon Storage. *Glob. Change Biol.* **2018,** *24*, 5895–5908.

Monajjem, S.; Hajipour, A. In *The Role of Azolla in Improving of Rice Fields Stability*; Proceedings of 5th National Conference on Sustainable Agriculture and Healthy Products. Isfahan Research Center for Agriculture and Natural Resources, 2010; pp 303–307.

Murthy, T. N. K.; Ashok, M.; Thrrumalesh, T.; Umesh, B. U.; Nataraju, O. R. Effect of Partial Supplement of Azolla for Concentrate Supplementation on Lactating Cross Bred Cow. *Ecol. Environ.* **2013,** *31*, 415–417.

Naghshi, H.; Khojasteh, S.; Jafari, M. Investigation of the Effect of Different Levels of *Azolla* (*Azolla pinnata*) on Performance and Characteristics of Cobb Broiler Chicks. *IJFAS* **2014,** *3*, 45–49.

Parikh, S. J.; James, B. R. Soil: The Foundation of Agriculture. *Nat. Educ. Knowl.* **2012,** *3*, 2.

Peter, V. M.; Cassman, K.; Cleveland, C.; Crews, T.; Christopher, B. F.; Grimm, B. N.; Howarth, W. R.; Marinov, R.; Martinelli, L.; Rastetter, B.; Sprent, I. J. Towards an Ecological Understanding of Biological Nitrogen Fixation. *Biogeochemistry* **2002,** *57*, 1–45.

Pillai, P. K.; Premalatha, S.; Rajamony, S. *Azolla* - A Sustainable Feed Substitute for Livestock. *LEISA India* **2002,** *4*, 15–17.

Prabina, B. J.; Kumar, K. Dried *Azolla* as a Nutritionally Rich Cost Effective and Immuno-Modulatory Feed Supplement for Broilers. *Asian J. Anim. Sci.* **2010,** *5*, 20–22.

Qiu, Y. L.; Yu, J. *Azolla* - A Model Organism for Plant Genomic Studies. *Geno. Prot. Bioinfo.* **2003,** *1*, 15–25.

Rabie, K. A. E.; El-Sherif, M. H.; El-Shahat, R. M.; Ali, F. S. I. *Azolla* as an Alternative of Mineral Nitrogen for Chamomile Plant *(Matricaria chamomilla* L.) Fertilization in Sandy Soil. *Asian J. Soil Sci. Plant Nutr.* **2020,** *5*, 1–13.

Rachael A. C.; Gospel, A. C.; Chukwudi, T. N. K.; Gabriel, O. C. A Review on the Role of Biofertilizers in Reducing Soil Pollution and Increasing Soil Nutrients. *Hmlyn. J. Agr.* **2020,** *1*, 34–38.

Raghuwanshi, R. Opportunities and Challenges to Sustainable Agriculture in India. *Nebio* **2012,** *3*, 78–86.

Raja, W.; Rathaur, P.; John, S. A.; Ramteke, P. W. *Azolla-Anabaena* Association and its Significance in Supportable Agriculture. *J. Biol. Chem.* **2012,** *40*, 1–6.

Rana, D.; Katoch, S.; Mane, B. G.; Rani, D.; Sankhyan, V. Biological Evaluation of *Azolla* in Ration of Commercial Chicken Broiler. *J. Anim. Res.* **2017,** *7*, 601–607.

Rathod, G. R.; Pramod, T.; Praveen, T.; Mandal, A. B.; Shinde, A. S. Feeding Value of *Azolla* (*Azolla pinnata*) Meal in Growing Japanese Quail. *Indian J. Poult. Sci.* **2013,** 48, 154–158.

Rawat, N.; Kumari, K.; Singh, F.; Gilhare, V. R. Effect of *Azolla* Supplemented Feeding on Milk Production of Cattle and Production of Broiler. *Appl. Biol. Res.* **2015,** *17*, 214–218.

Razavipour, T.; Moghaddam, S. S.; Doaei, S.; Noorhosseini, S. A.; Damalas, C. A. *Azolla* (*Azolla filiculoides*) Compost Improves Grain Yield of Rice (*Oryza sativa* L.) under Different Irrigation Regimes. *Agric. Water Manag.* **2018,** *209*, 1–10.

Roy, D. C.; Pakhira, M. C.; Bera, S. A Review on Biology, Cultivation and Utilization of *Azolla*. *Adv Life Sci.* **2016,** *5*, 11–15.

Safriyani, E.; Hasmeda, M.; Munandar; Sulaiman, F.; Holidi, K. K. The Role of *Azolla* on Improving Nitrogen Efficiency in Rice Cultivation. *Iran. J. Plant Physiol.* **2020,** *10,* 3095–3102.

Saikia, N.; Sapcota, D.; Hazarika, R. Effect of Feeding *Azolla* (*Azolla pinnata*) Meal to Broilers: A Field Study in Assam. *Indian J. Poult. Sci.* **2014,** *49,* 113–114.

Saito, K.; Vandamme, E.; Johnson, J. M.; Tanaka, A.; Senthilkumar, K.; Dieng, I.; Wopereis, M. C. S. Yield-Limiting Macronutrients for Rice in Sub-Saharan Africa. *Geoderma* **2019,** *338,* 546–554.

Samad, F. A. A.; Idris, L. H.; Abu, H. H.; Goh, Y. M.; Loh, T. C. Effects of *Azolla* spp. as Feed Ingredient on the Growth Performance and Nutrient Digestibility of Broiler Chicken. *J. Anim. Physiol. Anim. Nutr.* **2020,** 1–8.

Samal, K. C.; Behera, L.; Sahoo, J. P. In *Azolla Biofertilizer - The Nature's Miracle Gift for Sustainable Rice Production*, 2020; pp 1–4.

Seth, N.; Pradhan, C. R.; Mishra, S. K.; Pati, P. K.; Panda, S. K. Performance of Vanaraja Chicken Fed fresh *Azolla* as a Protein Substitute. M.V.Sc, Thesis, Orissa University of Agriculture and Technology, Bhubaneswar, 2013.

Shaaban, A.; Elrasoul, A.; Mousa, A. A.; Orabi, S. H.; Mohamed, M. A. E.; Gad-allah, S. M.; Almeer, R.; Abdel-Daim, M. M.; Khalifa, S. A. M.; El-Seedi, H. R.; Eldaim, M. A. A. Antioxidant, Anti-Inflammatory, and Anti-Apoptotic Effects of *Azolla pinnata* ethanolic Extract against Lead-Induced Hepatotoxicity in Rats. *Antioxidant* **2020,** *9,* 1014.

Shamna, T. P.; Peethambaran, P. A.; Jalaludeen, A.; Joseph, L.; Aslam, M. K. M. Broiler Characteristics of Japanese Quails (*Coturnix coturnix japonica*) at Different Levels of Diet Substitution with *Azolla pinnata. Anim. Sci. Rep.* **2013,** *7,* 75–80.

Shukla, M.; Bhattacharyya, A.; Shukla, P. K.; Roy, D.; Yadav, B.; Sirohi, R. Effect of *Azolla* Feeding on the Growth, Feed Conversion Ratio, Blood Biochemical Attributes and Immune Competence Traits of Growing Turkeys. *Vet. World* **2018,** *11,* 459–463.

Sprent, J. I.; Sprent, P. Nitrogen Fixing Organisms: Pure and Applied Aspects. *Nitrogen Fixing Org.* **1990,** *19,* 288.

Subedi, P.; Shrestha, J. Improving Soil Fertility through *Azolla* Application in Low Land Rice: A Review. *Azarian J. Agric.* **2015,** *2,* 35–39.

Subler, S.; Edwards, C. A.; Metzger, J. Comparing composts and vermicomposts. *BioCycle* **1998,** *39,* 63–66.

Subudhi, B. P. R.; Singh, P. K. Nutritive Value of the Water Fern *Azolla pinnata* for Chicks. *Poult. Sci.* **1978,** *57,* 378–380.

Suhag, M. Potential of Biofertilizers to Replace Chemical Fertilizers. *IARJSET.* **2016,** *3,* 163–167.

Tekle-Haimanot, A.; Doku, E. V. Comparison of *Azolla mexicana* and N and P Fertilization on Paddy Taro (*Colocasia esculenta*) Yield. *Trop. Agric.* **1995,** *72,* 70–72.

Umesha, S.; Singh, P. K.; Singh, R. P. Microbial Biotechnology and Sustainable Agriculture. In *Biotechnology for Sustainable Agriculture*; Ram, L. S., Sukanta, M., Eds.; Amsterdam, The Netherlands, 2018; pp 185–205.

Vaishampayan, A. Cyanobacterial Biofertilizers in Rice Agriculture. *Bot. Rev.* **2001,** *67,* 453–516.

Van, H. C.; Lejeune, A. The *Azolla–Anabaena* Symbiosis. *Biol Environ.* **2002,** *102,* 23–26.

Varadharajan, A.; Gnanasekar, R.; Kothandaraman, S. Studies on Feeding Value of *Azolla* in Quails in Relationship to its Carcass Traits. *J. Pharm. Innov.* **2019,** *8,* 1143–1145.

Venkatashwarlu, B. Role of Bio-Fertilizers in Organic Farming: Organic Farming in Rain Fed Agriculture: Central Institute for Dry Land Agriculture, Hyderabad, 2008, Vol. *8*, pp 5–95.

Vlek, P. L. G.; Vielhauer, K. Nutrient Management Strategies in Stressed Environments. In *Stressed Ecosystems and Sustainable Agriculture*; Virmani, S. M., Katyal, J. C., Eswaran, H., Abrol, I. P. Eds.; Oxford and IBH Publishing Co.: New Delhi, India, 1994; pp 203–229.

Wani, F. S.; Ahmad, L.; Ali, T.; Mushtaq, A. Role of Microorganisms in Nutrient Mobilization and Soil Health. *J. Pure Appl. Microbiol.* **2015,** *9*, 1401–1410.

WHO. Health Indicators of Sustainable Agriculture, Food and Nutrition Security in the Context of the Rio+20 UN Conference on Sustainable Development, 2012.

Yao, Y.; Zhang, M.; Tian, Y.; Zhao, M.; Zeng, K.; Zhang, B.; Yin, B. *Azolla* Biofertilizer for Improving Low Nitrogen use Efficiency in an Intensive Rice Cropping System. *Field Crops Res.* **2018,** *216*, 158–164.

Youssef, M.; Eissa, M. Biofertilizers and their Role in Management of Plant Parasitic Nematodes: A Review. *J. Biotechnol. Pharm. Res.* **2014,** *5*, 1–6.

CHAPTER 7

AZOTOBACTER SALINESTRIS FOR IMPROVING SOIL FERTILITY AND CROP PRODUCTIVITY: MECHANISM, ROLE, AND POTENTIAL APPLICATIONS

AMREENA SULTAN[1*],
ABDEL RAHMAN MOHAMMAD AL-TAWAHA[2],
SHAHNAZ MUFTI[1], AFROZA AKHTER[1], SAYED AZRAH INDRABI[1],
GIRMA BIRRU[3], MUHAMMAD AASIM[4],
EKATERINA KOZUHAROVA[5], INSHA JAVEED[1], NAJMU SAKIB[1],
JAVEID AHMED DAR[1], MUNIR TURK[6],
DEVENDRA KUMAR PANDEY[7], DEVARAJAN THANGADURAI[8],
JEYABALAN SANGEETHA[9], VISHNU D. RAJPUT[10],
ABEER JUBILY[11], and IFTIKHAR ALI[12]

[1]*Division of Vegetable Science, SKUAST-K, Shalimar, Srinagar, Jammu and Kashmir, India*

[2]*Department of Biological Sciences, Al Hussein bin Talal University, Maan, Jordan*

[3]*Agroecosystem Management Research Unit, USDA-ARS, NE, USA*

[4]*Department of Plant Protection, Faculty of Agricultural Sciences and Technologies, Sivas University of Science and Technology, Sivas, Turkey*

[5]*Medical University of Sofia, Faculty of Pharmacy, Department of Pharmacognosy, Sofia, Bulgaria*

[6]*Department of Plant Production, Faculty of Agriculture, Jordan University of Science and Technology, Irbid, Jordan*

Microbial Fertilizer Technology for Sustainable Crop Production. Jeyabalan Sangeetha, Abdel Rahman Mohammad Al-Tawaha, & Devarajan Thangadurai (Eds.)

DOI: 10.1201/9781003594826-7

[7]Department of Biotechnology, School of Bioengineering and Biosciences, Lovely Professional University, Phagwara, Punjab, India

[8]Department of Botany, Karnatak University, Dharwad, Karnataka, India

[9]Department of Environmental Science, Central University of Kerala, Kasaragod, Kerala, India

[10]Academy of Biology and Biotechnology, Southern Federal University, Rostov-on-Don, Russia

[11]Centre for Agricultural Research, General Commission for Scientific Agricultural Research, Syria

[12]Center for Plant Sciences and Biodiversity, University of Swat, Charbagh, Pakistan

*Corresponding author.

ABSTRACT

To meet the needs of a rapidly expanding human population, conventional agriculture has become increasingly dependent on chemical fertilizers and pesticides. On the contrary, to maximize crop yield, intensive agriculture relies on the extensive use of nitrogen fertilizers, as well as the addition of other critical elements. Chemical fertilizers are industrially made substances that contain known amounts of nitrogen, phosphorus, and potassium. When they are used, they pollute the air and groundwater by making water bodies more eutrophicated. On the contrary, the use of beneficial microorganisms as a biofertilizer has risen to the top of the priority list in the agriculture sector due to their potential involvement in food safety and long-term crop productivity. *Azotobacter* is one of the many options for increasing soil fertility. It is a free-living diazotrophic N_2 fixer with a variety of beneficial effects on crop growth and productivity. Biofertilizers are made up of N fixers, K solubilizers, phosphorus solubilizers, and phosphorus mobilizers that can be used alone or in combination with fungi. Various types of microbic cultures and inoculants are now available on the market, and thanks to technological advancements, these have quickly increased.

7.1 INTRODUCTION

High inputs, particularly fertilizers, pesticides, and herbicides, have increased yields, despite growing concern about the negative effects of chemical use on soil, water, and air, as noted by experts in numerous nations (Hayyawi et al., 2018; Amanullah et al., 2019; Alu'datt et al., 2019; Hani et al., 2019; Turk et al., 2003a, 2003b, 2003c; Nikus et al., 2004a, 2004b; Tawaha and Turk, 2004; Turk and Tawaha, 2004). Soil health and fertility are being looked at now to make our farming system more environmentally friendly (Al-Tawaha et al., 2005; Tawaha et al., 2005; Assaf et al., 2006; Turk et al., 2006; Al-Kiyyam et al., 2008; Al-Ajlouni et al., 2009; Al-Tawaha et al., 2010; Al-Tawaha et al., 2018a, 2018b). While sustainable production in an organic farming system is based on managing organic matter to maintain soil health and feed nutrients to developing plants, sustainable production in an industrial farming system is dependent on managing organic matter (Turk and Tawaha, 2001; Tawaha and Turk, 2002a, 2002b; Turk and Tawaha, 2002a, 2002b; Tawaha et al., 2003; Abebe et al., 2005; Abera et al., 2005; Al-Tawaha et al., 2005; Elser et al., 2007; Al-Kiyyam et al., 2008; Hameed et al., 2008; Al-Tawaha et al., 2010a, 2010b; Ananthi et al., 2017; Al-Tawaha et al., 2017; Abu Obaid et al., 2018). As a result, policymakers and regulatory authorities dealing with organic agriculture in a number of nations have permitted the use of biofertilizers, which are comparatively less harmful to the environment. As a result, integrated nutrient management has become a widely acknowledged strategy for increasing soil fertility while also protecting the environment. The natural products combination of fertilizers, organic manures, and biofertilizers is available with this technology. Biofertilizer is a natural substance made up of live microorganisms generated from the soil's foundation or cultivation. As a result, they have no negative impact on soil conditions or the environment. Apart from aiding in area organic processes and phosphoric solubilization, they also help in nitrogen fixation by activating plant growth hormones, resulting in superior nutrient uptake and increased tolerance to drought, wet stress, and diseases. As every gram of carrier of biofertilizers includes a minimum of ten million viable cells of a selected strain, a modest low dose of biofertilizer is adequate to yield exciting results. Biofertilizers are products that comprise a carrier base (solid or liquid) of living microbe preparations that are agriculturally advantageous for nitrogen fixation, phosphorus solubilization, or nutrient mobilization to increase soil and/or crop yield. In scientific jargon, these preparations are known as microorganism inoculants. Biofertilizers, an environmentally acceptable power source, play an important role in reducing inorganic manure levels while simultaneously increasing harvest yields, in addition to maintaining soil fertility. The replacement of nutrients lost due to crop removal with synthetic manures alone is not sustainable in the long run, as their regular usage has damaged the soil's health and efficiency. Biomanures can help to reduce the contribution of inorganic composts by up to 25% while achieving the same or better results. *Azotobacter*, *Aspergillus* sp., *Rhizobium*, *Sinorhizobium*, and *Mycorrhizae* are examples of biofertilizers that are beneficial to crop production. Beijerinck began his investigation with *Azotobacter chroococcum* and *Azotobacter agilis* (1901). The most famous *Azotobacter* species, *A. chroococcum*, of the Azotobacteriaceae family, was discovered in European soils in 1901. *Azotobacter nigricans*, *Azotobacter paspali*, *Azotobacter dobereiner*, *Azotobacter armenicus*, and *Azotobacter salinestris* are examples of alternative *Azotobacter* species. Later on, *Azotobacter beijernckii* was discovered inside the soil and rhizosphere.

7.2 DISTRIBUTION AND TAXONOMY OF *AZOTOBACTER*

In neutral and slightly basic soils, *Azotobacter* species are abundant, but not in acidic soils. Despite the frigid environment, short season, and relatively low pH scale values of those soils, they can be found in Arctic and Antarctic soils. It can live for up to 24 years in dry soils in the form of cysts (Johnstone, 1967). Many species of the genus *Azotobacter* are found in aquatic habitats, such as H_2O and briny marshes, but various are associated with plants and located in the rhizosphere, where they have many connections with the plants. Some strains have also been discovered in earthworm cocoons (*Eisenia fetida*). Martinus-Beijerinck, a Dutch microbiologist and plant scientist who was one of the founding fathers of environmental biology, identified the *Azotobacter* genus in 1901 (Beijerinck, 1901). He chooses and describes the *A. chroococcum* species, which is the principal aerobic, free-living nitrogen fixer. Lipman described *Azotobacter vinelandii* in 1903, and a year later, *A. beijerinckii*, which he named after Beijerinck. The genus *Azotobacter armeniacus* was first described by Thompson and Skerman in 1981. Page and Shivprasad (1991) described a sodium-dependent microaerophilic and air-tolerant type *A. salinestris*. According to a phylogenetic analysis, *A. vinelandii* is related to the bacterium *Pseudomonas aeruginosa* (Rediers et al., 2004). The genera *Azotobacter*, *Azomonas*, and *Pseudomonas* are thought to be related and may be synonyms (Young and Park, 2007). The genus *Azotobacter* has six species, including *A. chroococcum*, which is found in many soils across the world (Mahato et al., 2009). The genus *Azotobacter* has been extensively investigated in saprophytes, particularly nodular organisms (Horner et al., 1942). Some species, particularly aerobic bioagents from the genus *Azotobacter*, are classified as a large group of nonparasitic soil-inhibiting diazotrophic microbes. The taxonomical categorization of *Azotobacter* is as follows:

- Domain: Bacteria
- Kingdom: Bacteria
- Phylum: Proteobacteria
- Class: Gammproteobacteria
- Order: Pseudomonas
- Family: Azotobaceriaceae
- Genus: *Azotobacter*

It belongs to the first group of heterotrophic, nonsymbiotic N-fixing bacteria that are inhibited by alkali or neutral soil. These microorganisms come in a variety of types and are Gram-negative. These are huge ovoid development cells that are rod-shaped or circular, with a diameter of 1.5–2 m or more. Typically, these cells are dispersed or even unexpected, producing chains of varying lengths in microscopic configurations. These cells are mobile in modern media due to the numerous types of flagella attached to the body surface, but over time, they become less skillful, more round, and form a thick covering of body substances. Glycine, an amino acid included in the supplement medium chemical molecule, affects the state of the cell. Their presence is conveyed in a unique way, one by one, in matching or unanticipated groups, and for some reason, constrained by an uncertain length. Among the biofertilizers, it has some exceptional features. They require a large number of nitrogenase chemicals (Joerger and Bishop, 1988). They do not make endospores, but they do form sores and nearly circular bacteria that form thick-walled blisters (methods for a biogenetic replication under reasonable conditions) (Salhia, 2013). Changes in the nutrient concentration in the medium, as well as the addition of organic compounds such as ethyl alcohol, *n*-butanol, or -hydroxybutyrate, cause cyst formation. Chemical variables also play a role in cyst development, which occurs amid metabolic shifts, organic process fluctuations, respiration, and macromolecule biogenesis (Sadoff, 1975). The *Azotobacter* cysts are spherical in shape and encompass the so-called "central body," which is a reduced version of vegetative cells with numerous vacuoles, and hence the "bi-layered shell." The shell's inner half has a fibrous structure called intine, and the outside half has a multilateral crystalline structure called exine (Page and Sadoff, 1975). A few chelation agents also isolate the cytoplasm during a particularly active condition (Parker and Socolofsky, 1968). Cluster phenol, which is made up of long noncyclic chains and aromatic rings, is the main component of the outer shell. Its population is often low in uncultivated soils near crop plants' rhizospheres. Jensen's N-free medium, which is commonly used for *Azotobacter* mass multiplication, grows well at an optimum temperature of 20–30°C and grows well in neutral to alkaline soil (pH 5–7.5), but does not survive once the cation concentration falls below six, and thus is absent in acidic soil. This bacterium has been discovered in the rhizosphere of crops such as maize, rice, sugarcane, bajra, vegetables, and plantation crops (Arun, 2007), and is thus referred to as rhizobacteria (Hecht-Buchholz, 1998). Once enough organic matter is available, they operate better among the foundation part of the crop nonsymbiotically. They are found in the root cortex and leaf sheath parenchymatous cells together. In nonlegume crops, *Azotobacter* is used (Singh and Dutta, 2006). They will illustrate a variety of variables that have an impact on plant growth (Tippannavar and Reddy, 1989). The Azotobacteria, when united and classed as plant-growth-promoting Rhizobacteria (PGPR), are supposed to push plant growth directly or indirectly. These rhizobacteria get their food and energy from the organic matter that donates nitrogen to the soil and root exudates, and they fix nitrogen gas (Maryenko, 1964). These nonspecific PGPR linked to N fixation are critical for ecology and play a key role in agricultural soil fertility (Figure 7.1).

7.3 CULTURE OF *AZOTOBACTER*

It is a genus of free-living diatrophic microorganisms/bacteria with a cyst as its resting stage. It can be found in neutral to alkaline soils, as well as in the aquatic environment and on some plants. It has a variety of metabolic capabilities, including atmospheric nitrogen fixation via ammonia conversion, followed by protein synthesis. Their unique system of three nitrogenase enzymes makes these bacteria very interesting to scientists, who may conduct extensive research to better understand nitrogen fixation and its role in agriculture. Soil-dwelling house diazotrophs such as *Azotobacter* are especially beneficial for cereals and cash crops such as paddy, wheat, bajra, jowar, maize, mustard, cotton, cumin, sugarcane, tobacco, castor, vegetables, and horticulture crops.

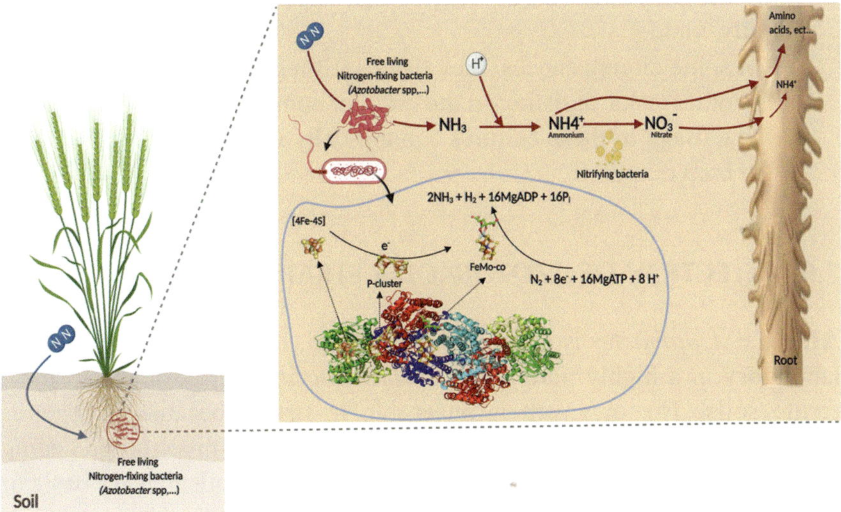

FIGURE 7.1 Mechanism of nonsymbiotic fixation of atmospheric nitrogen by *Azotobacter* sp.

Source: Adapted from Aasfar et al. (2021), https://creativecommons.org/licenses/by/4.0/

7.4 CHARACTERISTICS OF *AZOTOBACTER*

It could be a genus of soil-dwelling bacteria with the following characteristics:

- They are bacilli.
- They're Gram-negative bacteria.
- They're obligate aerobes, which means they cannot live without food.

- They are going to fix N (unlike another nitrogen-fixing bacterium, that goes along with the roots of plants, *Azotobacter* species are independent).

7.5 ADVANTAGES OF *AZOTOBACTER*

- The most common *Azotobacter* strain is used to fix 20–25 kg of nitrogen from the atmosphere per hectare.
- Some of the growth-promoting chemicals released by these cultures are employed to boost seed germination, plant growth, and, eventually, yield.
- In some circumstances, they have antifungal properties in addition to other tasks, assisting in the indirect control of fungal illnesses.
- The usage of these cultures can typically result in a 10–15% boost in crop output.

7.6 SELECTION OF *AZOTOBACTER* STRAINS

After the *Azotobacter* is isolated from the soil, it is tested for purity in a laboratory in a highly pure type. *A. chroococcum* can be found in most fertile soils. The organism is anaerobic in nature because it requires oxygen to grow. The malinin chemical was used to give the preceding culture a dark hue. The organism is mainly found in soils that are basic or neutral. The ability of different *Azotobacter* strains to fix nitrogen varies depending on the pH of the soil, the crop, and the soil atmosphere. As a result, strains' ability to fix nitrogen is frequently studied. As a result, to obtain the best strains of *Azotobacter*, we must undertake a variety of tests or studies in a glass house, in stuffed pots, and in the field, all under the supervision of plant pathologists/microbiologists, some of which are listed below.

7.6.1 AZETYLENE REDUCTION TEST

Various strains of *Azotobacter* in pure form are cultured in separate conical flasks in the laboratory. After that, the flasks are shaken for 50–72 h. After the bacteria have grown to their full potential in the medium, 10–15 mL of broth is transferred to a bottle. About 10 mL of acetylene gas is added to the present bottle, which is then sealed with a sanitized cork borer and left in the shed for 2–4 h to allow the catalyst enzyme (nitrogenase) to react with the acetylene gas. Acetylene is converted to ethylene during this time. The chromatographic analysis is used to determine the proportion of each gas. Additional ethylene gas is produced by the strain with more nitrogenase enzyme. This strain will, of course, be designated for continued usage.

7.6.2 POT CULTURE EXPERIMENT

After testing a variety of strains in the lab and identifying cost-effective strains, the pot culture experiment is the next step. In this test, earthen pots are thoroughly cleaned and disinfected with a common laboratory disinfectant before being filled with a uniform quantity of sterilized garden soil.

The strains that proved to be the most cost-effective during the acetylene reduction test are chosen and multiplied in pure form. To inoculate the seeds, the broth is combined with liquid and inoculated. After that, the seeds are dried in a shed and planted in pots. For comparison, appropriate inoculated control plants are retained. Plants are watered only when necessary and allowed to grow for 45–60 days. There are differences in height, nitrogen content of the plant and soil, and plant dry weight between inoculated and uninoculated plants. For field experiments, low-cost strains are used.

7.6.3 FIELD TEST

Strains that have been determined to be effective in glasshouse settings must go through a field test, which is the most essential test from the farmer's perspective. When applied in the field, strains found to be effective in glass house and acetylene testing must compete for nutrients with wild plants. *Azotobacter* strains that are effective are treated to certain seeds and seeded. For comparison, an adequate control is kept. Except for seed inoculation with effective *Azotobacter* strains, all other variables remain constant.

After noting down the maturity yield figures, a comparison is made. These tests are performed three to four times in different locations over the course of three to four years. Based on this information, an effective strain of *Azotobacter* is chosen and employed for large-scale production. Such strains must be kept in a refrigerator or under cold storage conditions. To be on the safe side, many strains should be employed in the manufacturing of biofertilizers.

7.7 MECHANISMS OF PLANT GROWTH PROMOTION

PGPR influences plant growth promotion by the alteration of the whole organism community in the rhizosphere niche through the gathering of various substances. Generally, PGPR promotes plant growth directly by either enhancing resource acquisition (nitrogen, phosphorus, and essential plant nutrients) or modulating plant hormone levels, or indirectly by lowering the restrictive effects of various pathogens on plant growth and development among the biocontrol agents.

7.7.1 DIRECT MECHANISMS

7.7.1.1 NITROGEN FIXATION

N plays a vital role in plant growth and productivity. Since the plants do not have any ability to synthesize the region N_2, it regenerates into plant-utilizable forms by biological N_2 fixation (BNF) that changes N to ammonia by N fixing microorganisms using a complex catalyst system observed as a catalyst (Kim and Rees, 1994). Biological nitrogen fixation occurs at mild temperatures by N-fixing microorganisms that are present everywhere in nature (Raymond et al., 2004). Nitrogen is a fundamental component for all spheres of life, and it is the most essential supplement for plant blast and yield productiveness.

Nitrogen-fixing organisms are typically classified as: (1) dependent/ symbiotic N_2-fixing microorganisms as well as members of the Rhizobiaceae that form mutualistic relation with leguminous plants (e.g., *Rhizobia*) (Ahemad and Khan, 2012) and nonleguminous trees (e.g., *Frankia*) and (2) nonsymbiotic (free living, associative, and endophytes) gas fixing forms such as eubacteria (*Rhizobium*, *Nostoc*, and *Frankia*), *Gluconoacetobacter diazotrophicus* and *Azocarus*, etc., (Bhattacharyya and Jha, 2012). The plant-growing rhizobacteria broadly introduced as symbionts are *Rhizobium*, *Mesorhizobium* with leguminous plants, and *Frankia* with nonleguminous trees and bushes. Then, nonsymbiotic nitrogen fixation is completed by free-living diazotrophs, and this can enrich nonvegetable plant development, for example, radish and rice. Nonadvantageous nitrogen enhancing rhizospheric microscopic entities having a place with genera including *Azoarcus*, *Azotobacter*, *Acetobacter*, in both symbiotic and free-living systems, "*nif*" genes are found for nitrogen fixation (Reed et al., 2011). The enzyme nitrogenase (*nif*) genes indulge auxiliary qualities associated with the actuation of the Fe protein, press molybdenum cofactor biosynthesis, electron gift, and judicious qualities required for the combination and capacity of the enzyme.

7.7.1.2 PHOSPHATE SOLUBILIZATION

Phosphorus is the essential nutrient of plants like N. It is essentially required for fundamental processes such as photosynthesis, ion exchange, flag transduction, organic compound formation, and plant development. The standard phosphate solubilization used by plant improvement progressing rhizobacteria includes (1) mineral dissolving blends, for example, normal destructive anions, protons, hydroxyl particles, and CO_2; (2) flexibility of extracellular mixes (biochemical phosphate mineralization); and (3) the entry of phosphate at the season of substrate diminishment (natural phosphate mineralization) (Sharma, 2013). Life forms combined with phosphate-solubilizing movement, frequently named phosphate-solubilizing microorganisms (PSMs), may give the absorbable type of P to the plants and subsequently a practical substitute for compound phosphatic bioresources (Khan et al., 2006). Of the different PSMs occupying the rhizosphere, PSMs are referred to as promising biofertilizers (Figure 7.2).

Phosphate solubilizing bacteria bind into the genera *Bacillus*, *Burkholderia*, *Microbacterium*, *Rhodococcus*, and *Serratia* have changed the mentality of agriculturists as soil inoculums to ascertain good plant health and yield.

7.7.1.3 POTASSIUM SOLUBILISATION

K is at number third in terms of essentiality. It is a macronutrient for plant growth. The availability of solvent K within the media is typically low, and over 90% of K inside the soil exists as insoluble rocks and silicate minerals (Parmar and Sindhu, 2013). Without sufficient potassium, the plants will have modified their root systems gradually, produce little seeds, and have lower yields. This underscored the pursuit to locate an option locally available wellspring of potassium for plant take up and to keep up potassium status in soils for managing crop production (Kumar and Dubey, 2012).

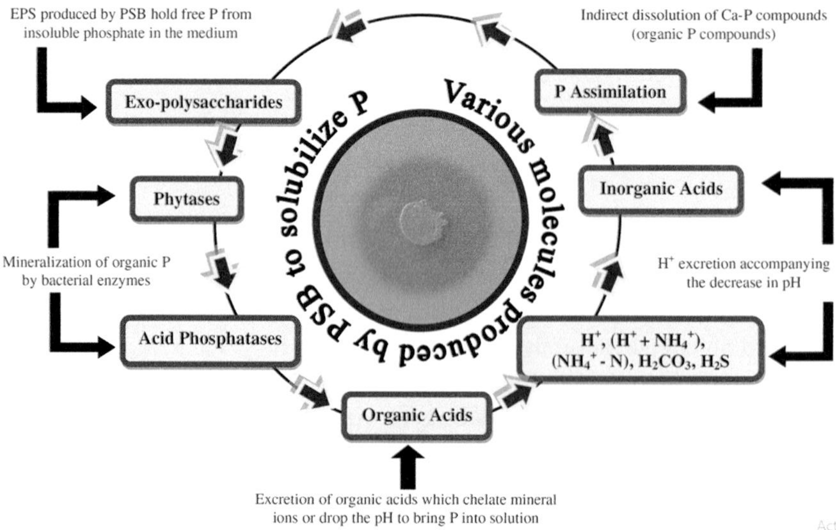

FIGURE 7.2 Various organic/inorganic substances produced by PSB responsible for phosphate solubilization in soil.

Source: Reproduced with permission from Ahemad and Kibret (2014), Copyright © Elsevier (2014).

7.8 GROWTH HORMONE PRODUCTION

Under laboratory conditions, the existence of those plant growth-stimulating substances is consolidated by the sector of demonstrations accomplished with several crops. The bacteria genus *Azotobacter* is rumored to produce cytokinins, auxins, and GA-like substances that are encountered to be directly related to enhancing crop growth and development (Wani et al., 2013). These hormones stem to the rhizosphere or root surface and exploit magnificent effects on the expansion of the upper plants that are growing within the overlapped areas. Barakart and

Gabr (1998) and Akram et al. (2016) ascertained that plant dry weight of various crops, namely, tomato, chili, and maize, was consequently higher on the appliance of *A. chroococcum* in comparison to untreated plants.

7.9 SIDEROPHORE PRODUCTION

Iron could be a very crucial nutrient for pretty much all varieties of life. Commonly, bacterium obtains iron along the secretion of low-molecular mass of Hame-binding chelators noted as siderophores that have tremendous relationship constants for sophisticating iron complexes. The majority of siderophores are soluble and might be bifurcated into intracellular siderophores and extracellular siderophores. In each Gram-positive and Gram-negative PGPR, iron (Fe^{3+}) in Fe^{3+} siderophore complicated on microorganism membrane is reduced to Fe^{2+} that is additionally discharged into the cell from the siderophore via a gating mechanism linking both the membranes that are inner and outer membranes. However, siderophores act as solubilizing bioagents for iron from various minerals or organic compounds underneath conditions of iron prohibition (Indiragandhi et al., 2008). Not solely Heme, siderophores conjointly kind stable complexes with alternative significant heavy metals that are of global concern, such as Cd, Pb, Al, Ga, In, Cu, and Zn, yet like radionuclides together with U and Np (Neubauer et al., 2000).

Azotobacter spp. have the capability to produce siderophores so that they bind to the offered type of iron Fe^{+3} within the soil, making it inaccessible to pathogens and protecting the plant health; equally, hydrogen cyanide production was recorded to be higher in traits of *Azotobacter* (77.00%). *Azotobacter* secretes antibiotic metabolites with a structure like anisomycin, which may be a documented antifungal antibiotic compound. *Azotobacter* spp. can compete with pathogens for nutrients. A number of the soil and air-borne pathogens that are managed by *Azotobacter* within the soil and on the leaf surface include *Fusarium*, *Rhizoctonia*, *Microfomina*, *Colletotrichum*, *Diplodia*, *Batryiodiplodia*, *Cephalosporium*, *Aspergillus*, and *Alternaria*.

7.10 PHYTOHORMONE PRODUCTION

The PGPR synthesis of the plant product indole-3-acetic acid, or indole acetic acid (IAA), has been well-known for a long time by microorganisms.

IAA synthesized by the Rhizobacteria intervenes with the numerous plant biological processes as a result of the endogenous reservoir of plant indole acetic acid could also be modified by getting IAA that has been produced by soil bacterium (Spaepen et al., 2007). Since IAA acts as a reverse signal molecule that touches gene expression in various bioagents and plays a vital role in rhizobacteria–plant interchange. In addition, downregulation of IAA as the signal is related to the plant defense mechanisms contra to a number of plant pathogenic bacterium as has shown an increased status of plants to the bacterial pathogen by exogenous application of IAA or IAA made by the pathogen (Spaepen et al., 2011).

The bacterial IAA will increase root area and stretch and therefore provides the plant with a larger approach to soil nutrients. Also, rhizobacterial IAA slackens the cell walls associated, thereby aiding an increase in the quantity of root excretion that provides extra nutrients to reinforce the expansion of root zone bacterium (Glick, 2012), Therefore, rhizobacterial IAA is known as an associate effector molecule in plant–microorganism association both in pathological and phytostimulation process. 1-Aminocyclopropane-1-carboxylate (ACC) Deaminase PGPR that have the accelerator, ACC deaminase promotes crop growth and development by lowering the ethylene gas levels, causing salt tolerance and inhibiting drought stress in plants. Presently, bioagent strains showing ACC deaminase activity are known in an exceedingly big selection of genera such as *Acinetobacter*, bacteria genera such as *Bacillus*, *Burkholderia*, *Enterobacter*, *Rhizobacteria* take up the ethylene gas precursor ACC and change it into 2-oxobutanoate and ammonia.

7.11 ROLE OF *AZOTOBACTER* IN VEGETABLE PRODUCTION

Mahato et al. (2009) evaluated the response of biofertilizer and inorganic fertilizer on the germination and growth of *Lycopersicon esculentum*. Nitrogen was used as inorganic fertilizer and *Azotobacter* was used as biofertilizer. The germination was observed higher in treatment 3 (soil + *Azotobacter*) as compared to other treatments. It was concluded that *Azotobacter* as a biofertilizer reported better than inorganic fertilizers in relation to seed germination and all plant growth parameters described in Table 7.1. *Azotobacter* makes availableness of bound nutrients such as carbon, nitrogen, phosphorus, and sulfur by accelerating the mineralization

Plant height, number of branches per plant, and yield attributes were found most when eggplants were grown in soils treated jointly with varying levels of fertilizers, PSB, *Azospirillum*, and *Azotobacter*. Physiological functions such as photosynthesis and transpiration rate, stomatal resistance, internal CO_2 concentration, and leaf temperature at the flowering stage were highest at 100% RDF applied together with PSB, *Azospirillum* and *Azotobacter*. Inoculation increased root yield by 20%, and repeated inoculation caused a 23% increase. The increase in the yield of crystal sugar obtained when inoculation was 21%, whereas continual inoculation caused a rise of 22–23% (Flores et al., 2013). Applying FYM 50% + *Azospirillum* 50% and *Azotobacter* 100% at an optimum level is an effective way of pushing growth, yield, and quality of cabbage. It redoubled height, leaves, head size, spread, and yield per hectare at the side of higher quality heads in terms of TSS, acidity, and vitamin C (Egamberdieva et al., 2011).

TABLE 7.2 Effect of *Azotobacter* on Crop Yield.

Crop	Increase in yield over yield obtained with chemical fertilizers (%)
Potato	13
Carrot	16
Cauliflower	40
Tomato	2–24
Sugarcane	9–24

Source: Adapted from Bhattacherjee and Dey (2014), https://creativecommons.org/licenses/by/4.0/

7.12 ROLE OF *AZOTOBACTER* IN PLANT DISEASE MANAGEMENT

Maheshwari et al. (2012) proved the strain TRA2 of *A. chroococcum* is often associated with the isolate of wheat rhizosphere and appeared vigorous opposing movement against root decay organisms, *Macrophomina phaseolina* and *Fusarium oxysporum*. Numerous components are included backing the management system utilized by the bacterium for the management of host infections. They might grasp the generation of siderophores, microbial-inhibiting substances, tab toxins, and furthermore, the extension hormones such as auxins, gibberellins, and cytokinins. In any case, no minute component is controlled and fully responsible for the disease inhibition and very a strategy may be utilized by the bacterium depending upon the rhizobacterial strain, natural situations, virulent pathogen concerned, and also the target of the host. These ways utilized by the bacterium are incontestable to give greater resistance toward the assault of virulent plant pathogens. Verma et al. (2001) conducted an exploration to check the *in vitro* generation of antifungal substances by totally distinctive strains of *A. chroococcum* and found that exclusively 37% of the overall strains were prepared to inhibit the extension of *Rhizoctonia solani*. *Azotobacter* spp. possess the adaptability to supply iron chelating substances that attach to the accessible kind of iron Fe^{+3} within the rootzone, subsequently hindering the pathogens from iron availableness and defensive the plant health. *Azotobacter* is reportable to supply an antibiotic substance that possesses a comparative structure as that of anisomycin, which is a well-developed fungicidal antibiotic. Beyond doubt, *Azotobacter* ensures the availability of supplements such as P, N, and S by the quick mineralization of natural wastes in soil and reducing the uptake of heavy metals. *Azotobacter* plays frequently a substitute for chemical compost, as a result of which, it gives N in substances such as nitrate and amino acids whereas not except for higher doses that may be one among the potential choices of inorganic N supply (e.g., urea). It also makes a difference in supporting plant development and yield, indeed, fairly in case of poor phosphate substance soil; moreover, it aids in taking up large scale, and beyond any doubt, little supplements encourage higher consumption of plant root wastes.

7.13 *AZOTOBACTER* IN NUTRIENT CYCLING

Azotobacter increases the availability of certain nutrients such as carbon, nitrogen, phosphorus, and sulfur in soil by accelerating the mineralization of organic wastes and preventing heavy metal uptake. *Azotobacter* is a highly important alternative to chemical fertilizers since it gives nitrogen in the form of ammonia, nitrate, and amino acids without overdosing, which is one of the potential inorganic N supply alternatives (e.g., urea). It also aids in the uptake of macro and certain minor nutrients, allowing for better usage of plant root exudates, even when the soil has a low phosphate level.

7.14 VARIOUS FORMULATIONS

7.14.1 INOCULUM FORMULATIONS

The presence of a competent bacterial strain will validate an agent's success or failure. When it comes to formulation manufacturing, an organism that is performing well in the lab may not be able to produce similar results in the field. Once an inoculant composition that works *in situ* has been developed, it should be fine-tuned to allow for end-user advancement (Stephens and Rask, 2000).

7.14.2 TALC FORMULATIONS

Talc, also known as metal silicate ($Mg_3Si_4O_{10}(OH)_2$), is an inactive mineral found in soap-rock industries that is used as a carrier for formulation development. Under nursery and field circumstances, seed treatment, soil application, and seed plant plunge with talc-based Pf1 bioformulation successfully reduced sheath spoil infection on rice plants.

7.14.3 PRESS MUD FORMULATIONS

It is a micronutrient-rich by-product of the sugar industry that could reduce the use of synthetic fertilizers. It creates the ideal environment for bacteria to try nitrogen fixation and phosphate solubilization. The manure is manufactured to meet requirements such as being free of all diseases, dangerous bacteria, and weeds, as well as being simple to handle, pack, and transport.

7.14.4 VERMICULITE FORMULATIONS

It might be a display layer silicate mineral [$(Si_3Al)Mg_3(OH)_2O_{10}MgO_5nH_2O$] and considered as a prospective carrier, especially if their generation approach incorporates the work of specific elite strains. Adding N-fixing and P-solubilizing bacteria to a vermicompost, for example, will increase the amount of N and P available inside the final product. It has characteristics such as a large enough zone for microbic multiplication, good air circulation due to its multi-lamellate construction, and it is easily accessible.

7.14.5 PEAT FORMULATIONS

These are the determination carriers, and they are the first to be used in the rhizobia inoculation exchange. Peat is widely available and has a long history of field testing. Thus, it is commonly used as a carrier for PGPR, particularly rhizobia inoculants. The most often used way for inoculating grain vegetables with rhizobia is to use peat inoculants attached to the seed as a suspension (e.g., *Bradyrhizobium* spp., *Mesorhizobium* spp., *Rhizobium* spp., etc.).

7.14.6 LIGNATE FORMULATIONS

Alginate is the most often used material for encapsulating microbic cells. It is a common chemical compound made up of D-mannuronic acid and L-glucuronic acid that's obtained from different bacteria (*Pseudomonas* and *Azotobacter*). It has a number of advantages over free cell formulations, including protecting the bacteria from biotic and abiotic stresses, as well as limiting the effects of harmful compounds.

7.15 MOLECULAR APPROACHES TO IMPROVE BIOFERTILIZATION PROPERTIES OF *AZOTOBACTER*

While improving the nutritional properties of *Azotobacter* spp. as biofertilizers to restore the component level (Gauri et al., 2012), it's critical to consider esteem successful technique that will allow the cheaper supply of biofertilizers to the farming trade. When evaluating *Azotobacter*'s commercial generation, it is critical to emphasize its cultural and nutritional features to strengthen its development in fermentation, with a focus on boosting its capacity as a biofertilizer (Gomare et al., 2013). Because of their beneficial properties and extensive scope to produce new substances, bacterial agents, and polymers made by them have sparked molecular and economic interest. These molecules will be used in the most severe amount of persuasive tonic for soil and plant health. Because of its ability to supply fundamental organic thirteen particles, such as polyhydroxybuterate, exopolysaccharides (EPSs), and iron-chelating compounds, *A. vinelandii* is extremely important in molecular applications. The nitrogen fixation capacity of *A. vinelandii* would be greatly increased with the help of arranged composing either expansion or deletion of focused on. However, the targeted figure control is distributed in such a way that the urea from antimicrobial metabolites is recycled into the finished product (Barney et al., 2015). In light of the importance of EPSs and unique metabolites produced inside the foundation of microscopic organisms in organic soil.

7.16 FUTURE PROSPECTS AND POSSIBILITIES IN COMMERCIALIZATION OF *AZOTOBACTER*

Azotobacter is one of the most viable options for use as a biofertilizer for eco-friendly and property crop production because of its ability to improve plant health through nitrogen fixation, development hormone production, phosphate solubilization, disease management, and recovery of higher soil wellbeing. Understanding and regulating these important features of *Azotobacter* could be key to long-term crop change experiments (Kyaw et al., 2019). However, there's a squeezing ought to execute a parcel of considerations connected to improved screening approaches, characterization of plant growth progressing and antibacterial substances from bacterial segregate, and demonstration of the molecular premise of procedures involved. Furthermore, additional research into the potential of *Azotobacter* in increasing soil ripeness is necessary, as is the use of cutting-edge soil hereditary qualities, among other things.

7.17 ADVANTAGES OF *AZOTOBACTER*

i) *Azotobacter* is capable of nitrogen fixation, phosphate solubilization, phosphate mobilization, and rhizobacteria advancement.

ii) The preparation includes certain living organisms that will benefit the soil. The packaging's reasonableness ensures a longer shelf life and, as a result, the environment's and the user's safety.
iii) They will mobilize nutrients that help biological activity in soils take place.
iv) Increasing the amount of adjusted nutrients helps to maintain plant health. As a result of the pleasant structure given to the soil, root development is promoted.
v) Food is provided, and the growth of beneficial microbes and soil worms is encouraged.
vi) The amount of organic matter in the soil is higher than average.
vii) Encourages the formation of mycorrhizal associations, which increases the amount of phosphorus (P) available in the soil.
viii) Assists in the killing of soil-borne diseases, allowing for the ongoing supply of micronutrients to the soil.
ix) *Azotobacter* has a role in maintaining constant nitrogen (N) and phosphorus (P) contents.
x) Enhances the ability of nutrients to exchange within the soil.

7.18 CONCLUSION

Azotobacter spp. are a widespread, nonsymbiotic, heterotrophic bacteria that may fix nitrogen, phosphorus, and sulfur in soil by the fast mineralization of organic exudates while also preventing the intake of essential elements. PGPR are bacteria that harmonize growth substances to improve plant development and growth while also inhibiting plant pathogenic development by producing inhibitors. It also affects nutrient uptake and creates a few biological molecules such as amino acids, among other things. *Azotobacter* promotes seed germination, but it also makes a difference by increasing nutrient comfort and replenishing soil fertility for improved crop response. It is an extremely important organic component of the nutrient management framework, owing to its critical role in soil property. In any way, enough biological component knowledge gained through the administration of biological N_2 fixation frameworks of bacteria in crop fields is critical for biofertilizer production innovation development and selection. Additional research is needed to assess the possibility and investigate the role of *Azotobacter* in soil fertility.

KEYWORDS

- ***Azotobacter***
- **biofertilizer**
- **chemical fertilizer**
- **crop productivity**
- **phosphorus solubilizer**
- **soil fertility**

REFERENCES

Aasfar, A.; Bargaz, A.; Yaakoubi. K.; Hilali, A.; Bennis, I.; Zeroual, Y.; Meftah, K. I. Nitrogen Fixing *Azotobacter* Species as Potential Soil Biological Enhancers for Crop Nutrition and Yield Stability. *Front. Microbiol.* **2021,** *12*, 628379. doi: 10.3389/fmicb. 2021.628379.

Abebe, G.; Hattar, B.; Al-Tawah, A. M. Nutrient Availability as Affected by Manure Application to Cowpea (*Vigna unguiculata* L. Walp.) on Calacarious Soils. *J. Agric. Soc. Sci.*, **2005,** *1* (1), 1–6.

Abera, T.; Feyisa, D.; Yusuf, H.; Nikus, O.; Al-Tawaha, A. R. Grain Yield of Maize as Affected by Biogass Slurry and NP Fertilizer Rate at Bako, Western Oromiya, Ethopia. *Biosci. Res.* **2005,** *2*, 31–38.

Ahemad, M.; Kibret, M. Mechanism and Applications of Plant Growth Promoting Rhizobacteria: Current Perspective. *J. King Saud Univ. Sci.* **2014,** *26* (1), 1–20. doi: 10.1016/j.jksus.2013.05.001

Ahemad, M.; Khan, M. S. Effects of Pesticides on Plant Growth Promoting Traits of *Mesorhizobium* Strain MRC4. *J. Saudi Soc. Agric. Sci.* **2012,** *11*, 63–71.

Akram, M.; Rizvi, R.; Sumbul, A.; Ansari, R. A.; Mahmood, I. Potential Role of Bioinoculants and Organic Matter for the Management of Root-Knot Nematode Infesting Chickpea. *Cogent Food Agric*. **2016,** *2* (1), 1183457.

Al-Ajlouni, M. M.; Al-Ghzawi, A. L. A.; Al-Tawaha, A. R.; et al. Crop Rotation and Fertilization Effect on Barley Yield Grown in Arid Conditions. *J. Food Agric. Environ.* **2009,** *88* (3), 869–872.

Al-Juthery, H. W.; Habeeb, K. H.; Jawad, F.; Altaee, K.; AL-Taey, D. K. A.; Al-Tawaha, A. R. M. Effect of Foliar Application of Different Sources of Nano-Fertilizers on Growth and Yield of Wheat. *Biosci. Res.* **2010,** *15* (4), 3988–3997.

Al-Kiyyam, M. A.; Turk, M.; Al-Mahmoud, M.; Al-Tawaha, A. R. Effect of Plant Density and Nitrogen Rate on Herbage Yields of Marjoram under Mediterranean Conditions. *J. Am.-Eurasian J. Agric. Environ. Sci.* **2008,** *3*, 153–158.

Al-Tawaha, A. R.; Al-Karaki, G.; Al-Tawaha, A. R.; Sirajuddin, S. N.; Makhadmeh, I.; Wahab, P. E. M.; Youssef, R. A.; Al Sultan, W.; Massadeh, A. Effect of Water Flow Rate on Quantity and Quality of Lettuce (*Lactuca sativa* L.) in Nutrient Film Technique (NFT) under Hydroponics Conditions. *Bulg. J. Agric. Sci.* **2018a,** *24* (5), 791–798.

Al-Tawaha, A. R.; Al-Tawaha, A. R.; Alu'datt, M.; Al-Ghzawi, A. L.; Wedyan, M.; Al-Obaidy, S. D. A.; Al-Ramamneh, E. A. D. Effects of Soil Type and Rainwater Harvesting Treatments in the Growth, Productivity and Morphological Trains of Barley Plants Cultivated in Semi-Arid Environment. *Aust. J. Crop Sci.* **2018b,** *12* (6), 975–979.

Al-Tawaha, A. R.; McNeil, D. L.; Yadav, S. S.; Turk, M.; Ajlouni, M.; Abu-Darwish, M. S.; Al-Ghzawi, A. L. A.; Al-udatt, M.; Aladaileh, S. Integrated Legume Crops Production and Management Technology. In *Climate Change and Management of Cool Season Grain Legume Crops*; Yadav, S. S., McNeil, D. L., Robert, R., Sharanagouda, A. P., Eds.; Springer: Dordrecht, 2010b; pp 325–349.

Al-Tawaha, A. R. M.; Nidal, O. Use of Sorghum and Maize Allelopathic Properties to Inhibit Germination and Growth of Wild Barley (*Hordeum spontaneum*). *Not. Bot. Horti Agrobot. Cluj-Na.* **2010a,** *38* (3), 124–127.

Al-Tawaha, A. R.; Turk, M. A.; Abu-Zaitoon, Y. M.; Aladaileh, S. H.; Al-Rawashdeh, I. M.; Alnaimat, S.; Al-Tawaha, A. R. M.; Aludatt, M. H.; Wedyan, M. Plants Adaptation to Drought Environment. Plants Adaptation to Drought Environment. *Bulg. J. Agric. Sci.* **2017,** *23* (3), 381–388.

Al-Tawaha, A. M.; Turk, M.; Hameed, K.; Assaf, T.; Tahhan, R.; Al-Jamali, A. F. Effects of Application of Olive Mill by-Products on Lentil Yield and their Symbiosis with Mycorrhizal Fungi under Arid Conditions. *Crop Res.* **2010,** *40* (3), 66–75.

Al Zoubi, O. M.; Al-Tawaha, A. R. Allelopathic Effect of Beetroot (*Beta vulgaris* L.) on Germination and Growth *Zea mays* and *Vigna umbellata*. *Int. J. Bot. Stud.* **2019,** *4* (4), 47–51.

Alu'datt, M. H.; Rababah, T.; Alhamad, M. N.; Al-Tawaha, A.; Al-Tawaha, A. R.; Gammoh, S.; Ereifej, K. I.; Al-Karaki, G.; Hamasha, H. R.; Tranchant, C. C.; Kubow, S. Herbal Yield, Nutritive Composition, Phenolic Contents and Antioxidant Activity of Purslane (*Portulaca oleracea* L.) Grown in Different Soilless Media in a Closed System. *Ind. Crops Prod.* **2019,** *141*, 111746.

Ananthi, T.; Amanullah, M. M.; Al-Tawaha, A. R. M. S. A Review on Maize-Legume Intercropping for Enhancing the Productivity and Soil Fertility for Sustainable Agriculture in India. *Adv. Environ. Biol.* **2017,** *11* (5), 49–63.

Arun, K. S. Bio-fertilizers for Sustainable Agriculture. Agrobios Publishers: Jodhpur, 2007; pp 76–77.

Barakat, M. A. S.; Gabr, S. M. Effect of Different Biofertilizer Types and Nitrogen Fertilizer Levels on Tomato Plants. *ALEXJA* **1998,** *43*, 149–160.

Baral, B. R.; Adhikari, P. Effect of *Azotobacter* on Growth and Yield of Maize. *SAARC J. Agric.* **2013,** *11* (2), 141–147.

Barney, B. M.; Eberhart, L. J.; Ohlert, J. M.; Knutson, C. M.; Plunkett, M. H. Gene Deletions Resulting in Increased Nitrogen Release by *Azotobacter vinelandii*: Application of a Novel Nitrogen Biosensor. *Appl. Environ. Microbiol.* **2015,** *81* (13), 4316–4328.

Beijerinck, M. W. Über ologonitrophile mikroben. Zentralbl. Bakteriol. Parasitenkd. Infektionskr. *I. I. Abt.,* **1901,** *7*, 561–582.

Bhat, M. I.; Rashid, A.; Faisul-ur- Rasool; Mahdi, S. S.; Haq, S. A.; Bhat, R. A. Effect of *Rhizobium* and Vesicular Arbuscular Mycorrhizae Fungi on Green Gram (*Vigna radiata* L. Wilczek) under Temperate Conditions. *Res. J. Agric. Sci.* **2010,** *1*, 113–118.

Bhattacherjee, R.; Dey, U. A Way Towards Organic Farming: A Review. *Afr. J. Microbiol. Res.* **2014,** *8* (24), 2332–2342.

Bhattacharyya, P. N.; Jha, D. K. Plant Growth-Promoting Rhizobacteria (PGPR): Emergence in Agriculture. *World J. Microbiol. Biotechnol.* **2012,** *28* (4), 1327–1350.

Brar, S. K.; Sarma. S. J.; Chaabouni, E. Shelf-Life of Biofertilizers: An Accord between Formulations and Genetics. *J. Biofertilizers Biopestic.* **2012,** *3*, 109.

Chen, Y. P.; Rekha, P. D.; Arunshen, A. B.; Lai, W. A.; Young, C. C. Phosphate Solubilizing Bacteria from Subtropical Soil and their Tricalcium Phosphate Solubilizing Abilities. *Appl. Soil Ecol.* **2006,** *34*, 33–41.

Döberainer, J. *Azotobacter paspali* sp. nov., uma bacteria fixadora de nitrogeniona rhizosfera de Paspalum. *Pesquisa Agropecuaria Brasileira* **1966,** *1*, 357–365.

Egamberdieva, D.; Kucharova, Z.; Davranov, K.; Berg, G.; Makarova, N.; Azarova, T.; Chebotar, V.; Tikhonovich, I.; Kamilova, F.; Validov, Z. V.; Lugtenberg, B. Bacteria Able to Control Foot and Root Rot and to Promote Growth of Cucumber in Salinated Soils. *Biol. Fertil. Soils* **2011,** *47* (2), 197–205.

Flores-Félix, J. D.; Menéndez, E.; Rivera, L. P.; Marcos, G. M.; Martínez-Hidalgo, P.; Mateos, P. F.; Martínez-Molina, E.; Velázquez, M. E.; García-Fraile, P.; Rivas, R. Use of *Rhizobium leguminosarum* as a Potential Biofertilizer for *Lactuca sativa* and *Daucus carota* Crops. *J. Plant Nutr. Soil Sci.* **2013,** *176*, 876–882.

Gauri, S. S.; Mandal, S. M.; Pati, B. R. Impact of *Azotobacter* Exopolysaccharides on Sustainable Agriculture. *Appl. Microbiol. Biotechnol.* **2012,** *95*, 331–338.

Glick, B. R. Plant Growth-Promoting Bacteria: Mechanisms and Applications. *Scientifica* **2012,** *2012*, 1–15.

Gomare, K. S.; Mese, M.; Shetkar, Y. Isolation of *Azotobacter* and Cost Effective Production of Biofertilizer. *Indian J. App. Res.* **2013,** *3* (5), 54–56.

Hakeem, K. R.; Akhtar, M. S.; Abdullah, S. N. A. *Plant, Soil and Microbes – Vol. 1, Implications in Crop Science*; Springer International Publishing: Cham, Switzerland, 2016.

Han, H. S.; Lee, K. D. Plant Growth Promoting Rhizobacteria Effect on Antioxidant Status Photosynthesis, Mineral Uptake and Growth of Lettuce Undersoil Salinity. *Res. J. Agric. Biol. Sci.* **2004,** *1*, 210–215.

Hecht, C. B. The Apoplast-Habitat of Endophytic Dinitrogen-Fixing Bacteria and their Significance for the Nitrogen Nutrition of Non-Leguminous Plants. *J. Plant Nutr. Soil Sci.* **1998,** *161*, 509–520.

Hewedy, A. M. Influence of Single and Multi-Bacterial Fertilizer on the Growth and Fruit Yield of Tomato. *Egypt. J. Appl. Sci.* **1999,** *14* (7), 508–523.

Horner, C. K.; Burk, D.; Allison, F. E.; Sherman, M. S. Nitrogen Fixation by *Azotobacter* as Influenced by Molybdenum and Vanadium. *J. Agric. Res.* **1942,** *65*, 173–193.

Indiragandhi, P.; Anandham, R. M.; Adhaiyan, M.; Sa, T. M. Characterization of Plant Growth-Promoting Traits of Bacteria Isolated from Larval Guts of Diamondback Moth *Plutella xylostella* (Lepidoptera: Plutellidae). *Curr. Microbiol.* **2008,** *56* (4), 327–333.

Jnawali, A. D.; Ojha, R. B.; Marahatta, S. Role of *Azotobacter* in Soil Fertility and Sustainability – A Review. *Adv. Plants Agric. Res*. **2015,** *2* (6), 1–5.

Joerger, R. D.; Bishop, P. E. Bacterial Alternative Nitrogen Fixing Systems. *Crit. Rev. Microbiol.* **1988,** *16*, 1–14.

Johnstone, D. B. Isolation of *Azotobacter insignis* from Fresh Water. *Ecology* **1967,** *48* (4), 671–672.

Khalequzzaman, K. M.; Hossain, I. Pathology: 1. Bangladesh Agricultural University, Mymensingh, Bangladesh, 2009.

Kim, J.; Rees, D. C. Nitrogenase and Biological Nitrogen Fixation. *Biochemistry* **1994,** *33*, 389–397.

Krassilnikov, N. A. Opred Zlitabaktzrij i aktinomycetov. Moskva, 1949.

Kumar, P.; Dubey, R. C.; Maheshwari, D. K. *Bacillus* Strains Isolated from Rhizosphere Showed Plant Growth Promoting and Antagonistic Activity against Phytopathogens. *Microbiol. Res.* **2012,** *167* (8), 493–499.

Kyaw, E. P.; Soe, M. M.; San San Yu, Z. K. L.; Lynn, T. M. Study on Plant Growth Promoting Activities of *Azotobacter* Isolates for Sustainable Agriculture in Myanmar. *J. Biotech. Biores.* **2019,** *1* (5), 1–6.

Levai, L.; Szilvia, V.; Nora, B.; Eva, G. Can Wood Ash and Biofertilizer Play a Role in Organic Agriculture? *Agronomski Glasnic.* **2008,** *3*, 263–271.

Lipman, J. G. Report on the New Jersey Agricultural Experiment Station, 1903, Vol *24*, pp 217–285.

Lipman, J. G. Report on the New Jersey Agricultural Experiment Station, 1904, Vol *25*, pp 237–289.

Mahato, P.; Anoop, B.; Chauhan, J. S. Effect of *Azotobacter* and Nitrogen on Seed Germination and Early Seedling Growth in Tomato. *Researcher* **2009,** *1* (4), 62–66.

Mahdi, S. S.; Hassan, G. L.; Samoon, S. A.; Rather, H. A.; Dar, S. A.; Zehra, B. Biofertilizers in Organic Agriculture. *J. Phytol.* **2010,** *2* (10), 42–54.

Maheshwari, D. K.; Dubey, R. C.; Aeron, A.; Kumar, B.; Kumar, S.; Tewari, S.; Arora, N. K. Integrated Approach for Disease Management and Growth Enhancement of *Sesamum indicum* L. Utilizing *Azotobacter chroococcum* TRA2 and Chemical Fertilizer. *World J. Microb. Biot.* **2012,** *28* (10), 3015–3024.

Maryenko, V. G. Zavisimost lurozaja kukuruzyl balansa azota *Aztobacter chroococcum* V usbvijah monobakterialznoj kultury. *Dokt ISHA* **1964,** *99*, 399–406.

Neubauer, U.; Furrer, G.; Kayser, A.; Schulin, R. Siderophores, NTA, and Citrate: Potential Soil Amendments to Enhance Heavy Metal Mobility in Phytoremediation. *Int. J. Phytoremediation* **2000,** *2* (4), 353–368.

Nikus, O.; Al-Tawaha, A. M.; Turk, M. A. Effect of Manure Supplemented with Phosphate Fertilizer on the Fodder Yield and Quality of Two Sorghum Cultivars (*Sorghum bicolor* L.). *Biosci. Res.* **2004a,** *1*, 1–7.

Nikus, O.; Al-Tawaha, A. M.; Turk, M. A. Yield Response of Sorghum (*Sorghum bicolor* L.) to Manufacture Supplemented with Phosphate Fertilizer Under Semi-Arid Mediterranean Conditions. *Int. J. Agric. Biol.* **2004b,** *6*, 889–893.

Page, W. J.; Sadoff, H. L. Relationship between Calcium and Uronic Acids in the Encystment of *Azotobacter vinelandii*. *J. Bacteriol.* **1975,** *122* (1), 145–151.

Page, W. J.; Shivprasad, S. *Azotobacter salinestris* sp. nov., A Sodium-Dependent, Microaerophili, and Aeroadaptive Nitrogen-Fixing Bacterium. *Ann. Microbiol.* **1991,** *51*, 145–158.

Parker, L.T.; Socolofsky, M. D. Central Body of the *Azotobacter* Cyst. *J. Bacteriol.* **1968,** *91* (1), 297– 303.

Partha, N.; Sivasubramanian, V. Recovery of Chemicals from Pressmud a Sugar Industry Waste. *Indian Chem. Eng.* **2006,** *48*, 161–163.

Raymond, J.; Siefert, J. L.; Staples Blankenship, C. R. The Natural History of Nitrogen Fixation. *Mol. Biol. Evol.* **2004,** *21* (3), 541–554.

Sadoff, H. L. Encystment and Germination in *Azotobacter vinelandii*. *Microbiol. Rev.* 1975, *39* (4), 516–539.

Salhia, B. The effect of *Azotobacter chrococcumas* Nitrogen Biofertilizer on the Growth and Yield of *Cucumis sativus*. Deanery of Higher Education Faculty of Science, Master of Biological Sciences Botany: The Islamic University, Gaza, 2013.

Schmidt, W. Mechanisms and Regulation of Reduction-Based Iron Uptake in Plant. *New Phytol.* **1999,** *141*, 1–2.

Sharma, K.; Dak, G.; Agrawal, A.; Bhatnagar, M.; Sharma, R. Effect of Phosphate Solubilizing Bacteria on the Germination of *Cicer arietinum* Seeds and Seedling Growth. *J. Herbal Med. Toxicol.* **2007,** *1* (1), 61–63.

Singh, M. S.; Dutta, S. Mustard and Rapeseed Response to *Azotobacter* – A Review. *Agric. Rev.* **2006,** *27* (3), 232–234.

Spaepen, S.; Vanderleyden, J.; Remans, R. Indole-3-Acetic Acid in Microbial and Microorganism-Plant Signaling. *FEMS Microbiol. Rev.* **2007,** *31*, 425–448.

Spaepen, S.; Vanderleyden, J. Auxin and Plant-Microbe Interactions. *Cold Spring Harb. Perspect. Biol.,* **2011,** *3* (4), a001438.

Stephens, J. H. G.; Rask, H. M. Inoculant Production and Formulation. *Field Crops Res.* **2000,** *65*, 249–258.

Subedi, R.; Khanal, A.; Aryal, K.; Chhetri, L. B. B.; Kandel, B. P. Response of *Azotobacter* in Cauliflower (*Brassica oleracea* L. var. Botrytis) Production at Lamjung, Nepal. *Acta Sci. Malays.* **2019,** *3* (1), 17–20.

Tawaha, A. M.; Singh, V. P.; Turk, M. A. A Review on Growth, Yield Components and Yield of Barley as Influenced by Genotypes, Herbicides and Fertilizer Application. *Res. Crop* **2003,** *4* (1), 1–9.

Tawaha, A. M.; Turk, M. A. Lentil (*Lens culinaris* Medic.) Productivity as Influenced by Rate and Method of Phosphate Placement in a Mediterranean Environment. *Acta Agron. Hung.* **2002a,** *50* (2), 197–201.

Tawaha, A. M.; Turk, M. A. Field Pea Seeding Management for Semi-Arid Mediterranean Conditions. *J. Agron. Crop Sci.* **2004,** *190*, 86–92.

Tawaha, A. R. M.; Turk, M. A.; Lee, K. D.; Zheng, W. J.; Ababneh, M.; Abebe, G.; Musallam, I. W. Impact of Fertilizer and Herbicide Application on Performance of Ten Barley Genotypes Grown in Northeastern Part of Jordan. *Int. J. Agric. Biol.* **2005,** *7* (2), 162–166.

Thompson, J. P.; Skerman, V. B. D. Validation List No 6. *Int. J. Syst. Bacteriol.* **1981,** 31.

Tippanavar, C. M.; Reddy, T. K. *Azotobacter* in Root Stem and Leaf Tissues of Cells of *Triticum aestivum* and *Triticum durum* L. *Curr. Sci.* **1989,** *58*, 1342–1343.

Turk, M. A.; Hmaeed, K. M.; Aqeel, A. M.; Tawaha, A. M. Nutritional Status of Durum Wheat Grown in Soil Supplemented with Olive Mill by-Products. *Agrochimica* **2003c,** *47* (5-6), 209–219.

Turk, M. A.; Lee, K. D.; Tawaha, A. M. Inhibitory Effects of Aqueous Extracts of Black Mustard on Germination and Growth of Radish. *Res. J. Agric. Biol. Sci.* **2005,** *1* (3), 227–231.

Turk, M. A.; Tawaha, A. M. Onion (*Allium cepa* L.) as Influenced by Rate and Method of Phosphorus Placement. *Crop Res.* **2002a,** *23* (1), 105–107.

Turk, M. A.; Tawaha, A. M. Allelopathic Effect of Black Mustard (*Brassica nigra* L.) on Germination and Growth of Wild Oat (*Avena fatua* L.). *Crop Prot.* **2003a,** *22* (4), 673.

Turk, M. A.; Tawaha, A. M. Allelopathic Effect of Black Mustard (*Brassica nigra* L.) on Germination and Growth of Wild Barley (*Hordeum spontaneum*). *J. Agron. Crop Sci.* **2003b,** *189* (5), 298–303.

Turk, M. A.; Tawaha, A. M. In *Effect of Variable Sowing Ratios and Sowing Rates of Bitter Vetch on the Herbage Yield of Oat-Bitter Vetch Mixed Cropping*, Proceedings 7th International Oat Conference; Peltonen-Sainio, P., Topi-Hulmi, M., Eds; MTT, 2004.

Turk, M. A.; Tawaha, A. M.; Samara, N. Effects of Seedling Rate and Date, and Phosphorous Application on Growth and Yield of Carbon Vetch (*Vicia narbonensis*). *Agronomie* **2003b,** *23*, 1–4.

Turk, M. A.; Tawaha, A. M.; Shatnawi, M. Lentil (*Lens culinaris* Medik) Response to Plant Density, Sowing Date, Phosphorous Fertilization and Ethephon Application in the Absence of Moisture Stress. *J. Agron. Crop Sci.* **2003a,** *189* (1), 1–6.

Verma, S.; Kumar, V.; Narula, N.; Merbach, W. In Vitro produktion von antimikrobiellen substanzen durch *Azotobacter chroococcum* isolate/mutanten. *Zeitschrift für Pflanzenkrankheiten und Pflanzenschutz.* **2001,** 152–165.

Wani, S. A.; Chand, S.; Ali, T. Potential Use of *Azotobacter chroococcum* in Crop Production: An Overview. *Curr. Agri. Res. J.* **2013,** *1* (1), 35–38.

Young, J. M.; Park, D. C. Probable Synonymy of the Nitrogen-Fixing Genus *Azotobacter* and the Genus *Pseudomonas*. *Int. J. Syst. Evolution. Microbiol.* **2007,** *57* (12), 2894–2901.

CHAPTER 8

POTENTIAL USE OF MYCORRHIZAL FUNGI IN AGRICULTURAL ECOSYSTEMS: AN OVERVIEW

KHALDOUN OTHMAN AL SANE[1], ABDEL RAHMAN MOHAMMAD AL-TAWAHA[2*], ROGHAYEH KARIMIRAD[3], ABHIJIT DEY[4], SIDDHARTHA PATI[5], ABDEL RAZZAQ AL-TAWAHA[6], PUTERI E. MEGAT WAHAB[7], RAZZAGH KARIMIRAD[8], R. Z. SAYYED[9], DEVARAJAN THANGADURAI[10], JEYABALAN SANGEETHA[11], HIBA ALATRASH[12], DEVENDRA KUMAR PANDEY[13], IMRAN[14], AMANULLAH[14], SHAH KHALID[14], EKATERINA KOZUHAROVA[15], SABAA JAWAD ABDULKADHIM[16], SAMIA KHANUM[17], and IFTIKHAR ALI[18]

[1]*Directorate of Plant Biodiversity and Plant Genetic Resources, National Agricultural Research Center (NARC), Jordan*

[2]*Department of Biological Sciences, Al Hussein bin Talal University, Maan, Jordan*

[3]*Department of Horticultural Sciences, Faculty of Agriculture and Natural Resources, University of Mohaghegh Ardabili, Ardabīl, Iran*

[4]*Department of Life Sciences, Presidency University, Kolkata, India*

[5]*NatNov Bioscience Private Limited, Balasore, Odisha, India*

[6]*National Agricultural Research Center (NARC), Baq'a 19381, Jordan*

[7]*Department of Crop Science, Faculty of Agriculture, University Putra Malaysia, Serdang, Selangor, Malaysia*

[8]*Department of Animal Science, Lorestan University, Khorramabad, Iran*

Microbial Fertilizer Technology for Sustainable Crop Production. Jeyabalan Sangeetha, Abdel Rahman Mohammad Al-Tawaha, & Devarajan Thangadurai (Eds.)

DOI: 10.1201/9781003594826-8

[9]*Department of Microbiology, PSGVP Mandal's Arts, Science and Commerce College, Shahada, Maharashtra, India*

[10]*Department of Botany, Karnatak University, Dharwad, Karnataka, India*

[11]*Department of Environmental Science, Central University of Kerala, Kasaragod, Kerala, India*

[12]*University of Helsinki, Helsinki, Finland*

[13]*Department of Biotechnology, School of Bioengineering and Biosciences, Lovely Professional University, Phagwara, Punjab, India*

[14]*Department of Agronomy, The University of Agriculture, Peshawar, Pakistan*

[15]*Medical University of Sofia, Faculty of Pharmacy, Department of Pharmacognosy, Sofia, Bulgaria*

[16]*Al Furat Al-Awsat Technical University, Al-Mussaib Technical College, Iraq*

[17]*Department of Botany, University of the Punjab, Lahore, Pakistan*

[18]*Center for Plant Sciences and Biodiversity, University of Swat, Charbagh, Pakistan*

**Corresponding author.*

ABSTRACT

Arbuscular mycorrhiza (AM) is the most frequent type of symbiotic relationship between plants and microorganisms on the planet. Plant nutrition, stress resistance and tolerance, soil structure, and fertility are all improved by AM fungi, which can be found in the majority of natural habitats. They also play an important role in a variety of ecological services, such as improving plant nutrition, stress resistance and tolerance, soil structure, and fertility. For their role in sustainable agriculture, AM fungi interact with a wide range of crop plants, including cereals, vegetables, and fruit trees. Mycorrhizal fungi associated with plant roots boost nutrient absorption, notably phosphorus absorption, and so promote crop and tree growth. Ectomycorrhizae are found in most economically important tree species of temperate regions of the world, and in some tropical trees. It is for this reason that these symbiotic relationships are critical in crop and biomass production. This is why they are attracting a great deal of interest in the agricultural and forestry industries.

8.1 INTRODUCTION

The purpose of integrated soil fertility management is to increase crop yield while also enhancing nutrient efficiency (Hayyawi et al., 2018; Amanullah et al., 2019; Hani et al., 2019). The introduction of grain legumes, which improve soil fertility by promoting biological nitrogen fixation, as well as synthetic fertilizers can help achieve this (Turk et al., 2003a, 2003b; Aludatt et al., 2019). Cover crops, which add organic matter to the soil and promote a healthy, fertile soil (Turk et al., 2003c; Tawaha and Turk 2004; Turk and Tawaha, 2004), green manure or the cultivation of legumes, which fix nitrogen from the air through the process of biological nitrogen fixation (Al-Tawaha et al., 2005; Tawaha et al., 2005; Assaf et al., 2006; Turk et al., 2006). Soil fertility requires nutrient cycling between organic and inorganic forms. By decomposing organic waste (plant and animal dung), microbes release inorganic nutrients into the soil solution (Al-Kiyyam et al., 2008; Al-Ajlouni et al., 2009; Al-Tawaha et al., 2010; Al-Tawaha et al., 2018a, 2018b). Soil microbes may be able to alter those nutrients further. When microbes compete with plants for nutrients, they immobilize the nutrients in their biomass, making them unavailable to plants. Soil fertility refers to a soil's ability to support plant growth by providing essential plant nutrients and desirable chemical, physical, and biological properties. Plants need nitrogen, phosphorus, potassium, sulfur, calcium, and magnesium. Iron, manganese, molybdenum, zinc, and selenium are micronutrients (Turk and Tawaha, 2001; Tawaha and Turk, 2002a, 2002b; Tawaha et al., 2003; Nikus et al., 2004a, 2004b; Abebe et al., 2005; Abera et al., 2005; Al-Tawaha et al., 2005; Al-Kiyyam et al., 2008; Al-Tawaha et al., 2010). Organic farming involves boosting the production of foods that are free of harmful pollutants. It entails the application of biofertilizers and biopesticides, which improve the nutrient quality of the crop while also controlling any pest or pathogen that may be present. hand the contrary, biofertilizers are microorganisms that improve the soil's nutritional quality by consuming nutrients from the soil. Beneficial microorganisms such as bacteria, fungi, and algae are examples of beneficial microorganisms that contribute to the improvement of soil fertility.

The word *mycorrhiza* (singular) is a combination of two classical Greek words for "mushroom" and "root" (Abbasi et al., 2015). In a mycorrhizal association, the fungal hyphae of an underground mycelium come in contact with plant roots in a mutual–benefit relationship in which mycorrhizae form mutualistic interaction with the roots of the majority of plant species (Azcbn-Aguilar and Barea, 1997; Bhandari and Garg, 2017; Shuab et al., 2017; Barceló et al., 2019; Bennett and Classen, 2020). Extraradical mycorrhizal fungi mycelia are usually the "hidden half" of the symbiotic relationship with plants (Futai et al., 2008). Most plant species develop symbiosis with mycorrhizal fungi (Smith and Read, 2008; Siddiqui and Pichtel, 2008; Aggarwal et al., 2011; Bidartondo et al., 2011). Hence, plant growth and performance are improved by the acquisition of mineral nutrients from the soil (Al-Karaki et al., 1998; Dighton et al., 2016; Bhandari and Garg, 2017). Actually, the onset of symbiosis and the beneficial mutualism between mycorrhizal fungi and the host plant has been well reviewed (Smith and Read, 2008; Bidartondo et al., 2011; Wyatt et al., 2014; Chen et al., 2018). In this relation, the fungus receives some of the carbohydrates from the host plant and the plant obtains its requirements of inorganic mineral nutrients and micro-nutrients from the mycelium (Giasson et al., 2008; Chen et al., 2018).

In addition, plant disease protection, vigor, and survivability of the seedlings and overall soil quality can be highly improved by mycorrhizal fungi association (Coleman et al., 2004; Futaiet al., 2008; Siddiqui and Pichtel, 2008). However, much more research work has to be done to explore the relative significance of mycorrhizae in nutrient uptake, plant growth, and defense (Dighton et al., 2016). Actually, the colonization of land by plants relied on vital biological evolutions, among which was the symbiotic relationship with fungi to enhance nutrient uptake (Bidartondo et al., 2011). Besides, the hyphal network does not only contribute to the uptake of mineral nutrients by its host plant and thus promote plant growth, but the fungus also represents a rather large carbon sink in the soil (Miller and Zastrow, 2000; Coleman et al., 2004), increase the survival of the trees in the native habitats, minimize nutrient leaching from the soil and, therefore, leading to nutrient accumulation in the soil and reducing the risk of groundwater pollution (Chen et al., 2018).

The mechanisms proposed to explain the enhanced nutrient uptake include increased exploration of the soil; increased phosphorous (P) translocation into plants via the arbuscules; efficient transfer of P to plant roots; and increased storage of absorbed P; modification of rhizosphere (root surroundings); and efficient utilization of phosphorous within plants (Aggarwal et al., 2011). Further, the effects of AMF on growth and yield and on the absorption of micro-nutrients by plants from agricultural and metal-contaminated soils have been documented (Giasson et al., 2008). Indeed, considerable progress in the investigation of plant-fungus interaction has been achieved mainly through the increased use of biochemical and molecular techniques (Sawers et al., 2008). When both partners co-occur, they actually interact (Montesinos-Navarro et al., 2019). Every individual may communicate with multiple partners simultaneously; multiple mycorrhizal fungi may colonize a single host plant, and each of these fungal individuals may possibly interact with multiple host plants (Wyatt et al., 2014). In many of these associations, about 10–30% of the food produced by the plant through photosynthesis transfers to the fungi. Furthermore, the functions of these mycorrhizal fungi range from stress alleviation to bioremediation in soils affected by heavy metals (Giasson et al., 2008; Aggarwal et al., 2011; Garg et al., 2017).

8.2 TYPES OF MYCORRHIZAL FUNGI

There are three most common types of mycorrhizal fungi recorded worldwide (Barceló et al., 2019; Bennett and Classen, 2020; http://mycorrhizas.info/#intro), namely, arbuscular mycorrhiza (AM), ericoid mycorrhizal (ErM) and ectomycorrhizal (EcM) fungi. However, there are also ectendomycorrhizae which exhibit some structural similarities to both ecto- and endo-mycorrhizal fungi (Siddiqui and Pichtel, 2008). Physically, these different types establish different forms of mycorrhizae, with different arrangements or associated microscopic structures of fungal hyphae. Mycorrhizal fungi can form structures either on the outside (ectomychorrhizae) or inside (endomycorrhizae) of plant roots. Ectomycorrhizal fungi form a net-like structure called the mantle (or sheath) which covers around the plant roots and encloses the rootlet (Futai et al., 2008; Smith and Read, 2008). Hyphae radiate outwards from the mantle into the substrate and grow into the spaces between root cells without penetrating the root cells (Smith and Read, 2008). To our knowledge, over 6000 species of fungi are involved in ectomycorrhizae and they are predominantly associated with forest trees. Several investigations have reported ectomycorrhizal associations in *Pinus halepensis* L. (Querejeta et al., 1998), *Quercus coccifera* (Caravaca et al., 2005), *Acacia holosericea* (Duponnois et al., 2005), *Pinus sylvestris* (Wallander et al., 2000), *Araucaria angustifolia* L. (Pereira et al., 2020), *Ceratonia siliqua* L. (Boutasknit et al., 2020) and much more examples.

On the other hand, there are endomycorrhizae in which the fungal hyphae penetrate root cells and send out their hyphae for a few centimeters into the neighboring soil and are involved in facilitating the nutrient acquisition, particularly phosphate ions and form complicated branching, shrub-like arbuscles within the host plant cells and bladder-like vesicles at later stages may develop as well (Coleman et al., 2004; Bidartondo et al., 2011; Bhandari and Garg, 2017). Furthermore, this type is the most widespread type of mycorrhizae, and the most ancient (Panwaret al., 2008; Shuab et al., 2017). They originated 350–450 million years ago with key roles in ecosystem functioning at various scales and are thought to play a crucial role in the colonization by the plants on the earth (Dighton et al., 2016; Field and Pressel, 2018). AM fungi are obligate symbionts that depend on their host plants to complete their life cycle (Montesinos-Navarro et al., 2019). Indeed, the role and importance of fungal hyphae for the high affinity for phosphorus and the ability for nutrient acquisition are well known (Al-Karaki et al., 1998; Nichols, 2008; Panwar et al., 2008; Shuab et al., 2017). Several studies have shown significant increases in plant height, leaf area, stem diameter, chlorophyll concentration, and the percentage of N, P, and K mineral nutrients in the leaves (Khalifah et al., 2016; Chen et al., 2018). Inoculation of mycorrhizal fungi has significantly increased plant height, leaf area, dry weight, percentage of N, P, K, and other minerals in leaves of maize compared to no biofertilization (Khalifah et al., 2016) (Figure 8.1).

A wide spectrum of endomycorrhizal inoculations on different plant species has been thoroughly investigated (Siddiqui and Pichtel, 2008; Shaaban and Wahbi, 2009; Aggarwal et al., 2011; Chen et al., 2018). Examples included wheat (*Triticum durum* Desf.) (Al-Karaki et al., 1998), *Aquilaria filaria* (Turjaman et al., 2006), *Cupressus atlantica* (Ouahmane et al., 2007), lentil (Shaaban and Wahbi, 2009), *Eucalyptus tereticornis* (Karthikeyan and Krishnakumar, 2012), *Zea mays* (Zhu et al., 2012), argan tree (*Argania spinosa*) (Sellal et al., 2017), Carob (*C. siliqua* L.) (Boutasknit et al., 2020), and numerous other plant species.

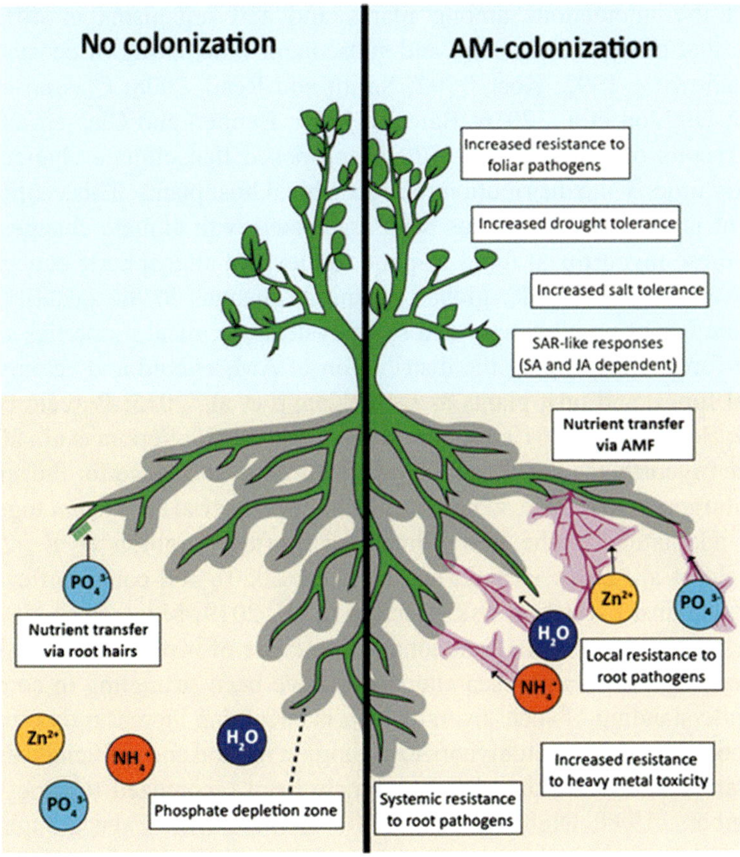

FIGURE 8.1 Positive effects of arbuscular mycorrhizal (AM) colonization. The hyphal network of arbuscular mycorrhizal fungi (AMF) extends beyond the depletion zone (grey), accessing a greater area of soil for phosphate uptake. A mycorrhizal-phosphate depletion zone will also eventually form around AM hyphae (purple). Other nutrients that have enhanced assimilation in AM roots include nitrogen (ammonium) and zinc. Benefits from colonization include tolerance to many abiotic and biotic stresses through induction of systemic acquired resistance (SAR).

Source: Adapted from Jacott et al. (2017), https://creativecommons.org/licenses/by/4.0/

8.3 MYCORRHIZAL FUNGI: CLIMATE CHANGE AND ECOLOGICAL CONSIDERATIONS

There is a growing awareness of the significant environmental roles of mycorrhizal fungi (Dighton et al., 2016). Climate change can, considerably, affect the interactions among plants and soil organisms in different ways that affect the structure and subsequent functioning of ecosystems (Bujakiewicz, 1992; Kost, 1992; Smith and Read, 2008; Classen et al., 2015; Dighton et al., 2016; Barceló, 2019; Bennett and Classen, 2020). The results of Barceló et al. (2019) proposed that climate change can greatly modify the distribution of mycorrhizal host plants. This ecological role of mycorrhizal fungi has been associated with climate changes and how these mycorrhizal fungi respond to elevated atmospheric concentrations of carbon dioxide, global warming, changes in the quantity and distribution of precipitation, and soil physical–chemical properties as the main forces determining the distribution of AM, ericoid and ectomycorrhizal fungi, and host plants as well (Knapp et al., 2015; Barceló et al., 2019; Begum et al., 2019; Bennett and Classen, 2020; Pereira et al., 2020). These mycorrhizae are expected to provide time and space for the spatial distribution of mycorrhizal host plants in the terrestrial ecosystem together with adaptation to the new climatic scenarios (Dighton et al., 2016; Barceló et al., 2019; Bennett and Classen, 2020), and could buffer host plants against extinction risk (Barceló et al., 2019; Montesinos-Navarro et al., 2019). Indeed, the functional importance of mycorrhizae has been addressed by so many researchers who have been struggling to come to an understanding of such diversity as it is becoming increasingly obvious that communities of ectomycorrhizal fungi in natural and artificial habitats are far more diverse than has previously been recognized (Cairney and Chambers, 1999; Dighton et al., 2016; Geml, 2018). Even though, we currently have limited knowledge about their biology and physiology.

These fungi, which are closely related to the flow of carbon and other essential elements in the biosphere, may have greater significance in combination with revegetation and phytoremediation projects (Miller and Gastrow, 2000; Al-Karaki, 2013; Bhat et al., 2017; Shuab et al., 2017). The vital role of AMF in ecological restoration services has been well studied (Smith and Read, 2008; Shuab et al., 2017). Further, mycorrhizae are essential for plant revegetation programs in sand dunes of the coastal regions and could be a critical component in strategies to protect exposed sandy soils against soil erosion and improve their fertility level (Maremmani et al., 2003; Rodríguez-Echeverría et al., 2008; Jobim et al., 2016; Chen et al., 2018). These areas are characterized by diverse biotic and abiotic components under the effects of anthropogenic activities, including loss of natural habitats (Jobim et al., 2016; Boeraeve et al., 2019).

Thus, mycorrhizal research, phytoremediation, and renaturation of degraded locations are widely adopted (Shuab et al., 2017; Chen et al., 2018). In this context, trees may be used to revegetate abandoned areas to create ecosystem services or wildlife areas (Smith and Read, 2008; Chen et al., 2018) and enhance the protection of biodiversity (Castellanos et al., 2005; Geml, 2018). In addition, mycorrhizae are also important in desert ecosystems subjected to soil erosion (Aide and Grau, 2005; Castellanos et al., 2005).

Indeed, natural restoration can help reduce soil erosion and floods and at a parallel line provide habitat for many living organisms (Aide and Grau, 2005; Jobim et al., 2016).The mycelial tissues of mycorrhizal fungi, being natural root symbionts, are more resistant to abiotic factors than the root itself, and this may compensate for the reduced root growth (Begum et al., 2019). Several studies revealed that AM increased plant tolerance to harsh environmental conditions including drought (Al-Karaki et al., 1998; Bhat et al., 2017; Shuab et al., 2017; Boutasknit et al., 2020), soil salinity (Chandrasekaran et al., 2016; Shuab et al., 2017; Begum et al., 2019) and heavy metal toxicity (Solís-Domínguez et al., 2011; Garg et al., 2017; Begum et al., 2019). Actually, the potential of this symbiosis in improving crop production is widely recognized. Even though it is not widely implemented in agricultural systems (Pagano and Cabello, 2011) (Table 8.1).

In addition to their nutritional role in a symbiotic relationship with plant roots, several reports indicated that a well-aggregated soil structure by mycorrhizae could enhance root penetration, control soil erosion, enhance water conservation capacity and water use efficiency, and regulate the entire water relations network (Auge, 2001; Nichols, 2008; Smith and Read, 2008; Zhu et al., 2012; Bennett and Classen, 2020). The importance of mycorrhizal fungi as a stabilizing agent in the formation and maintenance of soil and enhancing soil structure by stimulating the formation of water-stable soil aggregates has been pointed out (Miller and Gastrow, 2000; Nichols, 2008; Begum et al., 2019). Across a wide range of soils, these contributions are largely dependent on the textural properties and whether the structure of the soil is hierarchical in construction, along with many other variables (Miller and Gastrow, 2000). Besides, AM fungi could also prevent soil erosion and enhance soil nutrient value and fertility, which are key elements for secondary succession and the whole structure of plant community (Dighton et al., 2016; Shuab et al., 2017).

TABLE 8.1 Influence of AM Fungi on Plant under Salinity Stress Conditions.

Plant species	Effect of AM fungi
Cucumis sativus	Enhanced nutrient uptake, stand establishment rate, and photosynthetic rate.
	Elevated levels of antioxidant enzymes (superoxide dismutase [SOD], peroxidase [POD], and catalase [CAT]) and lower uptake of Na^+.
	Increased fruit fresh and dry weights, proline and electrolyte leakage, root and shoot dry weights, photosynthesis, and stomata conductance.
Trigonella foenum-graecum	Enhanced growth, proline, and antioxidant enzymes and phosphatase.
Triticum aestivum	Increased leaf area, stomatal density, stomatal conductance (g_s) and maximum rate of g_s to water vapor (g_{smax}), and leaf water potential.
	Increased total phenol, proline content, leaf area, shoot and root dry weights, and lengths.
	Increased plant dry weight; stomatal conductance, density, size, and aperture; higher soluble sugar concentration; and K^+:Na^+ ratio, C assimilation, and N uptake.
Triticum durum	Increased aboveground and root biomass, N uptake and aboveground N, and stability of plasma membranes.
Solanum lycopersicum	Increased plant metabolic plasticity.
Medicago sativa	Increased photosynthesis-related performance, such as net photosynthesis, stomatal conductance, the maximum quantum yield (F_v/F_m), and effective quantum yield (ΦPSII) of PSII photochemistry.
	Increased K^+, K^+:Na^+, and Ca^{+2}:Na^+.
Arundo donax	Increase in proline accumulation and a trend toward higher isoprene emission and higher H_2O_2.
Zea mays	Increased maize salt tolerance by decreasing salt concentrations, increasing soil macroaggregate proportions, soil bacterial diversity, maize mineral uptake, and photosynthesis; Enhanced K selected transportation from the root to the shoot and reduced the shoot malondialdehyde (MDA) content to promote maize photosynthesis and transpiration.
	Increased osmolyte accumulation, water potential, and photosynthetic activity.
Chrysanthemum morifolium	Increased root length, shoot, and root dry weight, total dry weight, and root N concentration.

TABLE 8.1 *(Continued)*

Plant species	Effect of AM fungi
Ricinus communis	Increased the net photosynthetic rate, stomatal conductance, transpiration rate, soluble protein, and proline, and decreased intercellular CO_2 concentration and MDA.
Ocimum basilicum	Increased chlorophyll content and water use efficiency K, P and Ca, K/Na, and Ca/Na.
Jatropha curcas	Growth rate and photosynthetic activity.
Capsicum annuum	Increased total fresh and dry weight, chlorophyll a and b concentrations, antioxidant content expressed as total soluble phenols and proline concentrations, relative water content, and decreased membrane permeability.
Cajanus cajan	Improved plant biomass and reduced oxidative burst by strengthening antioxidant enzymatic activities.
Leymus chinensis	Increased plant biomass, photosynthetic parameters, and contents of photosynthetic pigments.
Robinia pseudoacacia	Improved the net photosynthetic rate, quantum efficiency of photosystem II photochemistry, and K^+ content and relative water content, but Na^+ content and shoot/root ratio of Na^+; Increased chlorophyll content, net photosynthetic rate, intercellular CO_2 concentration, stomatal conductance and transpiration rate, carbon content, calorific value, carbon, and energy accumulations.
S. lycopersicum	Increased K^+ concentration, K^+:Na^+ ratio cytokinins, indoleacetic acid, and abscisic acid concentrations.
	Enhanced chlorophyll content.
	Increased antioxidant activity.
Poncirus trifoliata	Increased plant height, stem diameter, leaf number, shoot biomass and root biomass, length, surface area, and volume.
Citrus aurantium	Increased shoot and root dry matter, root infections, spore production, and concentrations of N, P, K, Zn, Mn, and Cu in plant tissues.
Cicer arietinum	Enhanced nodulation and nitrogen fixation.
	Improved growth and productivity, Si uptake.
Oryza sativa	Decreased Na^+ root-to-shoot distribution and an increase of Na^+ accumulation in rice roots.
Vitis vinifera	Increased root growth, root and shoot biomass, leaf area and improved leaf Ψw, $\Psi \pi$, P_n, and chlorophyll content, K^+:Na^+ ratio, and P_n.

TABLE 8.1 *(Continued)*

Plant species	Effect of AM fungi
Acacia gerrardii	Increased osmoprotectants such as glycine, betaine, and proline and reduced lipid peroxidation.
Gossypium hirsutum	Promoted leaf proline and K^+:Na^+.
Lupinus termis	Increased proline, glycine betaine, and sugars and increased the activities of key antioxidant enzymes such as SOD, CAT, ascorbate peroxidase (APX), and POD.
O. basilicum	Enhanced the biomass production and accumulation of nutrients, reducing sugars, total soluble carbohydrates, photosynthetic pigments, proline, and protein by reducing Na^+.
Vigna radiata	Improved plant growth.
Cucurbita pepo	Enhanced relative water content.
Sesbania grandiflora	Decreased potassium.
Sesbania aegyptiaca	Increased chlorophyll content and potassium.
Acacia nilotica	Increased potassium.
Trifolium alexandrinum	Increased plant biomass.
Glycine max	Increased CAT activity.
C. cajan	Higher osmolyte accumulation.
Phragmites australis	Enhanced osmolyte accumulation.
Ocimum basilicum	Increased proline accumulation.
Piper nigrum	Enhanced chlorophyll content.
Citrus sinensis	Increased potassium and calcium uptake.
Lactuca sativa	Increased growth and mineral content.

Source: Reproduced with permission and modified slightly from Abdelhamid et al. (2019), Copyright © 2019 Springer Nature.

8.4 POTENTIAL USE OF MYCORRHIZAL FUNGI IN RESTORATION, REFORESTATION, AND LANDSCAPING PROJECTS

Restoration practices such as enhancing soil properties and expanding vegetation cover can promote methods for restoring soil productivity together with sustainability (Abbasi et al., 2015; Shuab et al., 2017). Ecological restoration, now, has been identified as one of the major challenges in environmental and economic policy in the world (Castellanos et al., 2005; Aradottir and Hagen, 2013; Jacobs et al., 2015). Practically, restoring at least 15% of the world's depleted habitats is one of the 20 Aichi targets of the United Nations Convention on Biological Diversity (CBD, 2010). Even though restoration knowledge and practices so far indicate that many restoration projects have achieved modest success or have completely failed (Thomas et al., 2014; Shuab et al., 2017). Such failure could be due to loss/lack of mycorrhizal propagules following degradation of the vegetation cover. Mycorrhizae could be excellent tools in the restoration process as they are responsible for making soil nutrients accessible to plants, fixing nitrogen, and enhancing phosphorous uptake.

In addition, ecosystem functioning may be modified through land restoration, which results in many biological changes, including changes in microbial biomass and decomposition of organic matter (Querejeta et al., 1998; Potthoff et al., 2006; Smith and Read, 2008). Reforestation facilitates the conservation of forest ecosystem functions and services, including enhanced soil fertility, leading to increased productivity and/or preservation of the environment (Pagano and Cabello, 2011; Dighton et al., 2016). Furthermore, Mycorrhizal fungi can also make a significant contribution to forest ecosystems by enhancing biomass and biodiversity and by creating a tree network through which nutrients can be transported (Futai et al., 2008). However, in order to perform effective reforestation projects, it is important first to improve soil quality and the ability of plant species to cope with harsh environments (Shuab et al., 2017). In a study conducted by Menkis et al. (2005), they have suggested that abundant mycorrhizae of conifer seedlings in forest nurseries may have a positive effect on the vitality and growth of outgrown seedlings in reforestation programs.

Moreover, the combination of selected ectomycorrhizal fungi and tree species has been used in a successful reforestation project to significantly improve productivity and reduce the cost of abandoned minelands program

(AML) recovery projects (Cordell et al., 2002). The growth of host plants at the seedling stage was found to be improved by ecto-mycorrhizal fungi (Futai et al., 2008). Surely, the extra-radical mycelia are the key organs for nutrient uptake in many woody plants, frequently connecting seedlings to mature trees. Biogeochemical cycling and the composition of plant populations in the forest ecosystem are also affected by these mycorrhizae (Futai et al., 2008). In a research conducted on *C. atlantica* (Ouahmane et al., 2007), only AM fungi isolated from the natural habitat were used in order to increase the chances of engaging fungi with superior adaptability to drought and to the host tree and to avoid introducing new AM fungal species with erratic effects on the local environment. In another example, the growth of the argan trees (*A. spinosa*) has been enhanced by the presence of endomycorrhizal fungi at the level of the culture substrate (Sellal et al., 2017). Their work aimed at studying the effect of inoculation of native endomycorrhizal fungi on the growth and development of argan trees. In practice, reforestation of argan trees requires young seedlings to be raised in nurseries and then planted and adapted to the dry climate of the original natural habitat of these plants (Sellal et al., 2017). In the same context, *Araucaria* forests in Brazil were indeed a source of high diversity for AMF (Pereira et al., 2020). In most of the regions studied, they suggested a greater reliance on AMF in actively growing primary forests (Figure 8.2).

Actually, the response of seedlings stage to mycorrhization has been observed by other authors (Turjaman et al., 2006). It was observed that AM fungi increased the early growth of two nontimber forest species (*Dyerapolyphylla*) and *A. filaria* under greenhouse conditions. Their results revealed that AM fungi can accelerate the establishment of the planting stocks by promoting their conservation ecologically and sustaining the economic production of these nontimber forest products (Turjaman et al., 2006) and they are considered to be sensitive ecological indicators for specific factors in a forest ecosystem and react strongly to climate changes (Bujakiewicz, 1992; Kost, 1992; Smith and Read, 2008; Dighton et al., 2016). In another study, seedlings of *Desmoncus orthacanthos*, along with inoculation of AM fungi, have shown a threefold increase in their survival in the field (Ramos-Zapata et al., 2006) compared to doubled seedling survival and significantly increased establishment in research conducted on *E. tereticornis* (Karthikeyan and Krishnakumar, 2012). Furthermore, the selection of the inoculating equipment and the most appropriate plant species and mycorrhizal fungi species for planting sites have to be combined in a comprehensive approach for efficient practical remediation and/or forestation (Cordell et al., 2002). Nonetheless, a thorough understanding of the complex interconnected cellular processes and arbuscular mycorhizal fungi (AMF)-mediated mechanisms controlling plant survival and tolerance to abiotic stress factors would enhance its practical application as a powerful strategy for stress bio-improvement (Boeraeve et al., 2019; Boutasknit et al., 2020). Moreover, phylogenetic restoration in mycorrhizal-plant interactions increases the opportunities for community stability in natural landscapes (Montesinos-Navarro et al., 2019).

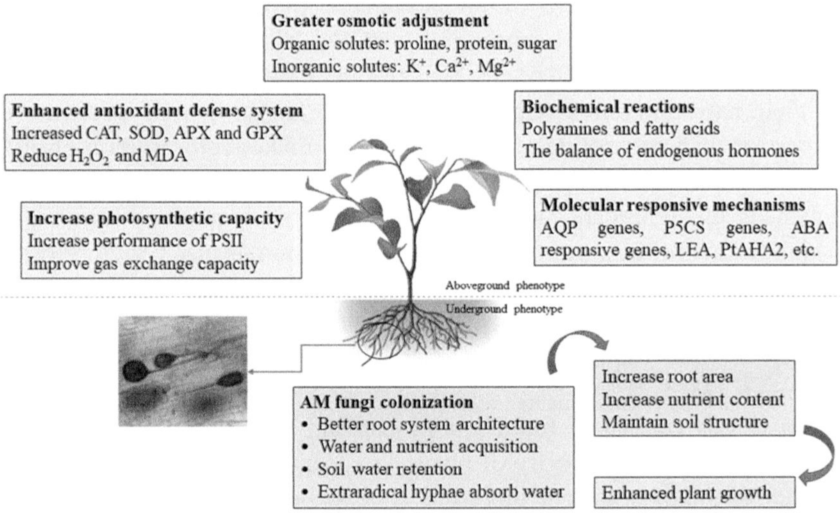

FIGURE 8.2 A schematic diagram regarding the underlying mechanisms of AM fungi enhancing plant drought tolerance. AM fungi affect the morphology, physiological activities, and molecular regulation of host plants through direct or indirect interactions in response to drought stress and thus help host plants regulate and maintain various processes to deal with the harmful effects of drought on plant.

Source: Adapted from Cheng et al. (2021), https://creativecommons.org/licenses/by/4.0/

From the practical point of view, the most suitable natural habitats for examining interactions between macrofungi and plant communities are deciduous forests (Bujakiewicz, 1992) and coniferous forests (Kost, 1992). Nevertheless, the complex structure of forests creates serious difficulties in the identification and classification of different fungal groups within forest ecosystems. The conceptual differences observed in the fungi are largely consistent with the ecologically described forest types, which indicate the importance of such habitat types as a framework for a better insight of the environmental differences that influence the dynamics of the fungal community at the landscape level (Geml, 2018). The data presented in his study provide unparalleled insights into the biodiversity and landscape-level population dynamics of the Pannonian forests. Indeed, the function of macrofungi rising in the soil is essential and significant in the development of forests. Furthermore, the richness of AMF has increased with increasing forest area and increasing soil pH but decreased with nitrate content in the soil (Boeraeve et al., 2019). Besides, the community composition of mycorrhizae was affected by plant-available phosphorus. AMF species richness in mycorrhizal community was also discussed (Menkis et al., 2005; Moreira et al., 2016). In fact, the increased survival of the trees in their native habitats (Chen et al., 2018) appears to be even more important than promoting growth, because it makes reforestation efforts more sustainable (Giasson et al., 2008). Moreover, reforestation could be enhanced by significant biophysical alterations such as cloud formation, which further reflects sunlight and forest protection, is also a vital priority for controlling greenhouse gas emissions (Pagano and Cabello, 2011). However, we have limited knowledge about the potential role and impact of richness and community composition of fungi at the landscape level (Geml, 2018).

In fact, mycorrhizal fungi are widespread in the landscape and are associated with the majority of plants in various ecosystems (Smith and Read, 2008). In addition, there are clear experimental and innovative evidence that mycorrhizal fungi influence a number of community traits and ecosystem functions. Unfortunately, large areas of natural vegetation have been cleared and/or replaced by other land uses due to human interference and anthropogenic activities (Boeraeve et al., 2019). This has resulted in pronounced changes in landscape structure worldwide, and increasingly fragmented landscapes consisting of small fragments of well-conserved habitats scattered within an intensively managed matrix. In a research conducted on fungal diversity and community structure by Geml (2018), these were shown to be strongly affected by vegetation type and environmental factors.

8.5 CONCLUSION AND FUTURE PERSPECTIVES

Our understanding of the function of mycorrhizae at the community level is still fragile (Smith and Read, 2008). While there are many gaps in our understanding of how climate changes would affect mycorrhizal fungi, we also highlight numerous opportunities for investigating plant and mycorrhizae responses to and potential modifications, especially upon climate changes (Bennett and Classen, 2020). Actually, there are some future research lines, opportunities, and challenges with respect to mycorrhizal fungi (Panwar et al., 2008; Chen et al., 2018). In this context, producers of AM fungal products rely on self-imposed quality standards to ensure best practices in production (Vosatka et al., 2008), and are focusing their efforts on developing relevant case studies and field trials to illustrate and emphasize the benefits of mycorrhizal fungi in agriculture and horticulture (Chen et al., 2018; Begum et al., 2019). Because plants and microbial strains vary in the efficacy of the symbiotic relationship, the selection of suitable symbiotic consortia is needed to be of particular value, ensuring better chances for plant survival (Panwaret al., 2008). Despite its growing trend, the current market for mycorrhizal products seems to be beyond its full potential. Challenges and threats for AM fungal products, regardless of the technical issues, include political and regulatory restrictions, quality assurance and product value, and customer awareness and acceptance (Chen et al., 2018). Such systems cause limitations and create barriers in the world market for AM fungal products (Vosatka et al., 2008; Chen et al., 2018). In this context, the European Parliament is currently assessing the establishment of a fair EU market for biostimulants (Chen et al., 2018). Basically, the quality and efficiency of mycorrhizae products are still areas that require further consideration (Vosatka et al., 2008). Actually, the proper dose (i.e., density of propagules) for a given market is not yet formally established and leaves space for promoters to establish these standards. Furthermore, those restrictions open a wide horizon of opportunities for the research community to support producers and consumers in determining what the minimum standards of treatment should be (Vosatka et al., 2008). In the same context, Wyatt et al. (2014), proposed a theory called "the biological market theory" and claimed that its strength is to make networks of mutual trading connections tractable by treating them as if partners were trading goods within a central market.

Several future trends and opportunities have been suggested (Leyval et al., 2002; Panwar et al., 2008; Sawers et al., 2008; Vosatka et al., 2008; Shaaban and Wahbi, 2009; Garg et al., 2017; Chen et al., 2018; Geml, 2018; Barceló et al., 2019; Begum et al., 2019; Montesinos-Navarro et al., 2019; Bennett and Classen, 2020).

KEYWORDS

- **arbuscular mycorrhiza**
- **ecosystems**
- **ectomycorrhizae**
- **micro-nutrients**
- **phosphorus absorption**

REFERENCES

Abbasi, H.; Akhtar, A.; Sharf, R. Vesicular Arbuscular Mycorrhizal (VAM) Fungi: A Tool for Sustainable Agriculture. *Am. J. Plant Nutr. Fert. Technol.* **2015,** *5* (2), 40–49.

Abdelhamid, M. T.; El-Masry, R. R.; Darwish, D. S.; Abdalla, M. M. F.; Oba, S.; Ragab, R. The Mechanisms Involved in Improving the Tolerance of Plants to Salt Stress using Arbuscular Mycorrhizal Fungi. In *Microorganisms in Saline Environments: Strategies and Functions. Soil Biology*; Giri, B., Varma, A., Eds.; Springer: Cham, 2019; Vol. *56*. https://doi.org/10.1007/978-3-030-18975-4_13.

Abebe, G.; Hattar, B.; Al-Tawaha, A. M. Nutrient Availability as Affected by Manure Application to Cowpea (*Vigna unguiculata* L. Walp.) on Calacarious Soils. *J. Agric. Soc. Sci.* **2005,** *1* (1), 1–6.

Abera, T.; Feyisa, D.; Yusuf, H.; Nikus, O.; Al-Tawaha, A. R. Grain Yield of Maize as Affected by Biogas Slurry and NP Fertilizer Rate at Bako, Western Oromiya, Ethopia. *Biosci. Res.* **2005,** *2*, 31–38.

Aggarwal, A.; Kadian, N.; Tanwar, A.; Yadav, A.; Gupta, K. Role of Arbuscular Mycorrhizal Fungi (AMF) in Global Sustainable Development. *J. Appl. Nat. Sci.* **2011,** *3*, 340–351. https://doi.org/10.31018/jans.v3i2.211.

Aide, T. M.; Grau, H. R. In *Implications of Rural-Urban Migration for Forest Restoration in Latin America*, Oral Presentation at the World Conference on Ecological Restoration, Zaragoza, Spain, Europe, Sept 12–18, 2005.

Al-Ajlouni, M. M.; Al-Ghzawi, A. L. A.; Al-Tawaha, A. R.; et al. Crop Rotation and Fertilization Effect on Barley Yield Grown in Arid Conditions. *J. Food Agric. Environ.* **2010,** *88* (3), 869–872.

Al-Juthery, H. W.; Habeeb, K. H.; Jawad, F.; Altaee, K.; AL-Taey, D. K. A.; Al-Tawaha, A. R. M. Effect of Foliar Application of Different Sources of Nano-Fertilizers on Growth and Yield of Wheat. *Biosci. Res.* **2010,** *15* (4), 3988–3997.

Al-Karaki, G. N. The Role of Mycorrhiza in the Reclamation of Degraded Lands in Arid Environments. In *Developments in Soil Classification, Land Use Planning and Policy Implications: Innovative Thinking of Soil Inventory for Landuse Planning and Management of Land Resources*; Shahid, S. A., Taha, F. K., Abdelfattah, M, A., Eds.; Springer Science and Business Media: Dordrecht, pp 823–836.

Al-Karaki, G. N.; Al-Raddad, A.; Clark, R. B. Water Stress and Mycorrhizal Isolate Effects on Growth and Nutrient Acquisition of Wheat. *J. Plant Nutr.* **1998,** *21* (5), 891–902. doi: 10.1080/01904169809365451

Al-Kiyyam, M. A.; Turk, M.; Al-Mahmoud, M.; Al-Tawaha, A. R. Effect of Plant Density and Nitrogen Rate on Herbage Yields of Marjoram under Mediterranean Conditions. *J. Am.-Eurasian J. Agric. Environ. Sci.* **2008,** *3*, 153–158.

Al-Tawaha, A. R.; Al-Karaki, G.; Al-Tawaha, A. R.; Sirajuddin, S. N.; Makhadmeh, I.; Wahab, P. E. M.; Youssef, R. A.; Al Sultan, W.; Massadeh, A. effect of Water Flow Rate on Quantity and Quality of Lettuce (*Lactuca sativa* L.) in Nutrient Film Technique (NFT) under Hydroponics Conditions. *Bulg. J. Agric. Sci.* **2018a,** *245*,791–798.

Al-Tawaha, A. R.; Al-Tawaha, A. R.; Alu'datt, M.; Al-Ghzawi, A. L.; Wedyan, M.; Al-Obaidy, S. D. A.; Al-Ramamneh, E. A. D. Effects of Soil Type and Rainwater Harvesting Treatments in the Growth, Productivity and Morphological Trains of Barley Plants Cultivated in Semi-Arid Environment. *Aust. J. Crop Sci.* **2018b,** *12* (06), 975–979.

Al-Tawaha, A. R.; McNeil, D. L.; Yadav, S. S.; Turk, M.; Ajlouni, M.; Abu-Darwish, M. S.; Al-Ghzawi, A. L. A.; Al-udatt, M.; Aladaileh, S. Integrated Legume Crops Production and Management Technology. In *Climate Change and Management of Cool Season Grain Legume Crops*; Yadav, S. S., McNeil, D. L., Redden, R., Patil, S. A., Eds.; Springer: Dordrecht, 2010; pp 325–349.

Al-Tawaha, A. R. M.; Nidal, O. Use of Sorghum and Maize Allelopathic Properties to Inhibit Germination and Growth of Wild Barley (*Hordeum spontaneum*). *Not. Bot. Horti Agrobot. Cluj-Na.* **2010,** *38* (3), 124–127.

Al-Tawaha, A. R.; Turk, M. A.; Abu-Zaitoon, Y. M.; Aladaileh, S. H.; Al-Rawashdeh, I. M.; Alnaimat, S.; Al-Tawaha, A. R. M.; Alu'datt, M. H.; Wedyan, M. Plants Adaptation to Drought Environment. *Bulg. J. Agric. Sci.* **2017,** *23* (3), 381–388.

Al-Tawaha, A. M.; Turk, M.; Hameed, K.; Assaf, T.; Tahhan, R.; Al-Jamali, A. F. Effects of Application of Olive Mill By-Products on Lentil Yield and their Symbiosis with Mycorrhizal Fungi under Arid Conditions. *Crop Res.* **2010,** *40* (3), 66–75.

Aludatt, M. H.; Rababah, T.; Alhamad, M. N.; Al-Tawaha, A.; Al-Tawaha, A. R.; Gammoh, S.; Ereifej, K. I.; Al-Karaki, G.; Hamasha, H. R.; Tranchant, C. C.; Kubow, S. Herbal Yield, Nutritive Composition, Phenolic Contents and Antioxidant Activity of Purslane (*Portulaca oleracea* L.) Grown in Different Soilless Media in a Closed System. *Ind. Crops Prod.* **2019,** *141*, 111746.

Ananthi, T.; Amanullah, M. M.; Al-Tawaha, A. R. M. S. A Review on Maize-Legume Intercropping for Enhancing the Productivity and Soil Fertility for Sustainable Agriculture in India. *Adv. Environ. Biol.* **2017,** *11* (5), 49–63.

Aradottir, A. L.; Hagen, D. Ecological Restoration: Approaches and Impacts on Vegetation, Soils and Society. *Adv. Agron.* **2013,** *120*, 173–222.

Auge, R. M. Water Relations, Drought and Vesicular-Arbuscular Mycorrhizal Symbiosis. *Mycorrhiza* **2001,** *11*, 3–42.

Azcbn-Aguilar, C.; Barea, J. M. Applying Mycorrhiza Biotechnology to Horticulture: Significance and Potentials. *Sci. Horticulturae* **1997,** *68*, 1–24.

Barceló, M.; van Bodegom, P. M.; Soudzilovskaia, N. A. Climate Drives the Spatial Distribution of Mycorrhizal Host Plants in Terrestrial Ecosystems. *J. Ecol.* **2019,** *107*, 2564– 2573. https://doi.org/10.1111/1365-2745.13275.

Begum, N.; Qin, C.; Ahanger, M. A.; Raza, S.; Khan, M. I.; Ashraf, M.; Ahmed, N.; Zhang, L. Role of Arbuscular Mycorrhizal Fungi in Plant Growth Regulation: Implications in Abiotic Stress Tolerance. *Front. Plant Sci.* **2019,** *10*, 1068. doi: 10.3389/fpls.2019.01068.

Bennett, A. E.; Classen, A. T. Climate Change Influences Mycorrhizal Fungal-Plant Interactions, but Conclusions are Limited by Geographical Study Bias. *Ecology* **2020,** *e02978*. https://doi.org/10.1002/ecy.2978.

Bhandari, P.; Garg, N. Dynamics of Arbuscular Mycorrhizal Symbiosis and its Role in Nutrient Acquisition: An Overview. In *Mycorrhiza - Nutrient Uptake, Biocontrol, Ecorestoration*; Varma, A., Prasad, R., Tuteja, N., Eds.; Springer International Publishing, 2017.

Bidartondo, M. I.; Read, D. J.; Trappe, J. M.; Merckx, V.; Ligrone, R.; Duckett, J. G. The Dawn of Symbiosis between Plants and Fungi. *Biol. Lett.* **2011,** *7*, 574–577.

Boeraeve, M.; Honnay, O.; Jacquemyn, H. Local Abiotic Conditions are more Important than Landscape Context for Structuring Arbuscular Mycorrhizal Fungal Communities in the Roots of a Forest Herb. *Oecologia* **2019,** *190*, 149–157.

Boutasknit, A.; Baslam, M.; Ait El Mokhtar, M.; Anli, M.; Ben-Laouane, R.; Douira, A.; El Modafar, C.; Mitsui, T.; Wahbi, S.; Meddich, A. Arbuscular Mycorrhizal Fungi Mediate Drought Tolerance and Recovery in Two Contrasting Carob (*Ceratonia siliqua* L.) Ecotypes by Regulating Stomatal, Water Relations and (In) Organic Adjustments. *Plants* **2020,** *9*, 80. doi: 10.3390/plants9010080.

Bujakiewicz, A. Macrofungi on Soil in Deciduous Forests. In *Fungi in Vegetation Science*; Winterhoff, W., Ed.; Springer, 1992.

Caravaca, F.; Alguacil, M. M.; Barea, J. M.; Roldán, A. Survival of Inocula and Native AM Fungi Species Associated with Shrubs in a Degraded Mediterranean Ecosystem. *Soil Biol. Biochem.* **2005,** *37*, 227–233.

CBD. Aichi Biodiversity Targets. [Online] 2010. http://www.cbd.int/sp/targets/ (accessed Feb 20, 2020)

Chen, M.; Arato, M.; Borghi, L.; Nouri, E.; Reinhardt, D. Beneficial Services of Arbuscular Mycorrhizal Fungi - From Ecology to Application. *Front. Plant Sci.*, **2018,** *9*, 1270.

Cheng, S.; Zou, Y. N.; Kuča, K.; Hashem, A.; Abd Allah, E. F.; Wu, Q. S. Elucidating the Mechanisms Underlying Enhanced Drought Tolerance in Plants Mediated by Arbuscular Mycorrhizal Fungi. *Front. Microbiol.* **2021,** *12*, 809473. doi: 10.3389/fmicb.2021.809473.

Coleman, D. C.; Crossley, D. A.; Hendrix, J.; Hendrix, P. F. In *Fundamentals of Soil Ecology*; Elsevier Inc., 2004.

Cordell, C. E.; Mans, L. F.; Marx, D. H. In *Mycorrhizal Fungi and Trees - A Successful Reforestation Alternative for Mineland Reclamation*, In *National Proceedings: Forest and Conservation Nursery Associations - 1999, 2000, and 2001*, Dumroese, R. K., Riley, L. E., Landis, T. D., Eds.; USDA, USA; 2002; pp 206–212.

Dighton, J. In *Fungi in Ecosystem Processes*; CRC Press: Florida, USA, 2016.

Duponnois, R.; Colombet, A.; Hien, V.; Jean, T. The Mycorrhizal Fungus *Glomus intraradices* and Rock Phosphate Amendment Influence Plant Growth and Microbial Activity in the Rhizosphere of *Acacia holosericea. Soil Biol. Biochem.* **2005,** *37*. doi: 1460-1468. 10.1016/j.soilbio.2004.09.016.

Futai, K.; Taniguchi, T.; Kataoka, R. Ectomycorrhizae and their Importance in Forest Ecosystems. In *Mycorrhizae: Sustainable Agriculture and Forestry*; Siddiqui, Z. A., Akhtar, M. S., Futai, K., Eds.; Springer, 2008; pp 241–285.

Garg, N.; Singh, S.; Kashyap, L. Arbuscular Mycorrhizal Fungi and Heavy Metal Tolerance in Plants: An Insight into Physiological and Molecular Mechanisms. In *Mycorrhiza-Nutrient Uptake, Biocontrol, Ecorestoration*; Varma, A., Prasad, R.,Tuteja, N., Eds.; Springer International Publishing AG, 2017.

Geml, J. Soil Fungal Communities Reflect Aspect-Driven Environmental Structuring and Vegetation Types in a Pannonian Forest Landscape. *Fungal Ecol.* **2018,** *39*, 63–79. https://doi.org/10.1016/j.funeco.2018.12.005

Giasson, P.; Karam, A.; Jaouich, A. Arbuscular Mycorrhizae and Alleviation of Soil Stresses on Plant Growth. In *Mycorrhizae: Sustainable Agriculture and Forestry*; Siddiqui, Z. A., Akhtar, M. S., Futai, K., Eds.; Springer, 2008; pp 99–134.

Jacob, D. F.; Oliet, J. A.; Aronson, J.; Bolte, A.; Bullock, J. M.; Donos, P. J. Restoring Forests: What Constitutes Success in the Twenty-First Century? *New For.* **2015,** *46*, 601–614.

Jacott, C. N.; Murray, J. D.; Ridout, C. J. Trade-Offs in Arbuscular Mycorrhizal Symbiosis: Disease Resistance, Growth Responses and Perspectives for Crop Breeding. *Agronomy* **2017,** *7*, 75. https://doi.org/10.3390/agronomy7040075

Jobim, K.; Silva, I. R.; Assis, D. M. A.; Vieira, L. C.; Silva, D. K. A.; Oehl, F.; Maia, L. C.; Goto, B. T. Arbuscular Mycorrhizae in Coastal Areas. In *Recent Advances on Mycorrhizal Fungi*; Pagano, M. C., Ed., 2016. doi: 10.1007/978-3-319-24355-9_9

Karthikeyan, A.; Krishnakumar, N. Reforestation of Bauxite Mine Spoils with *Eucalyptus tereticornis* Sm. Seedlings Inoculated with Arbuscular Mycorrhizal Fungi. *Ann Res.* **2012,** *55*, 207–216.

Khalefah, K. M.; Hamada, A. A.; Khalaf, A. A.; Ameen, T. A. The Effect of Interaction Between *Mycorrhiza* (*G. mosseae*), *Trichoderma* (*T. harzianum*) and Phosphate Fertilization on Growth of Maize. *Tikrit. J. Agric. Sci.* **2016,** *16* (4), 1813–1646.

Kost, G. Macrofungi on Soil in Coniferous Forests. In *Fungi in Vegetation Science*; Winterhoff, W., Ed.; Springer, 1992; pp 79–112.

Menkis, A.; Vasiliauskas, R.; Taylor, A. F. S.; Stenlid, J.; Finlay, R. Fungal Communities in Mycorrhizal Roots of Conifer Seedlings in Forest Nurseries under Different Cultivation Systems, Assessed by Morphotyping, Direct Sequencing and Mycelial Isolation. *Mycorrhiza* **2005,** *16*, 33–41. https://doi.org/10.1007/s00572-005-0011-z

Miller, R. M.; Jastrow, J. D. Mycorrhizal Fungi Influence Soil Structure. In *Arbuscular Mycorrhizas: Physiology and Function*; Kapulnik, Y., Douds, D. D., Eds.; Kluwer Academic Publishers, 2000; pp 3–18.

Montesinos-Navarro, A.; Díaz, G.; Torres, P.; Caravaca, F.; Roldán, A. Phylogenetic Rewiring in Mycorrhizal–Plant Interaction Networks Increases Community Stability in Naturally Fragmented Landscapes. *Commun. Biol.* **2019,** *2*, 452. https://doi.org/10.1038/s42003-019-0700-3.

Moreira, M.; Zucchi, M.; Gomes, J. E.; Tsai, S. M.; Alves-Pereira, A.; Cardoso, J. B. N. *Araucaria angustifolia* above Ground Roots Presented High Arbuscular Mycorrhizal Fungal Colonization and Diversity in the Brazilian Atlantic Forest. *Pedosphere* **2016,** *26* (4), 561–566.

Nichols, K. A. Indirect Contributions of AM Fungi and Soil Aggregation to Plant Growth and Protection. In *Mycorrhizae: Sustainable Agriculture and Forestry*; Siddiqui, Z. A., Akhtar, M. S., Futai, K., Eds.; Springer, 2008; pp 177–194.

Nikus, O.; Al-Tawaha, A. M.; Turk, M. A. Effect of Manure Supplemented with Phosphate Fertilizer on the Fodder Yield and Quality of Two Sorghum Cultivars (*Sorghum bicolor* L.). *Biosci. Res.* **2004a,** *1*, 1–7.

Nikus, O.; Al-Tawaha, A. M.; Turk, M. A. Yield Response of Sorghum (*Sorghum bicolor* L.) to Manufacture Supplemented with Phosphate Fertilizer under Semi-Arid Mediterranean Conditions. *Int. J. Agric. Biol.* **2004b,** *6*, 889–893.

Ouahmane, L.; Hafidi, M.; Thioulou; Duponnois, R. Improvement of *Cupressus atlantica* Gaussen. Growth by Inoculation with Native Arbuscular Mycorrhizal Fungi. *J. Appl. Microbiol.* **2007,** *103*, 683–690,

Pagano, M.; Cabello, M. Mycorrhizal Interactions for Reforestation: Constraints to Dryland Agroforest in Brazil. *ISRN Ecol.* **2011,** *13*, 890850. doi: 10.5402/2011/890850

Panwar, J.; Yadav, R. S.; Yadav, B. K.; Tarafdar, J. C. Arbuscular Mycorrhizae: A Dynamic Microsymbiont for Sustainable Agriculture. In *Mycorrhizae: Sustainable Agriculture and Forestry*; Siddiqui, Z. A., Akhtar, M. S., Futai, K., Eds.; 2008; pp 159–176.

Pereira, J.; Vasconcellos, R.; Pereira, A.; Stürmer, S.; Silva, A.; Baretta, D.; Bonfim, J.; Cardoso, E. Reforestation Processes, Seasonality and Soil Characteristics Influence Arbuscular Mycorrhizal Fungi Dynamics in *Araucaria angustifolia* Forest. *Forest Ecol. Manage.* **2020,** *460*. doi: 10.1016/j.foreco.2020.117899.

Potthoff, M.; Steenwerth, K. L.; Jackson, L. A.; Drenovsky, R. E.; Scow, K. M.; Joergensen, R. G. Soil Microbial Community Composition as Affected by Restoration Practices in California Grassland. *Soil Biol Biochem.* **2006,** *38*, 1851–1860.

Querejeta, J.; Roldán, A.; Albaladejo, J.; Castillo, V. The Role of Mycorrhizae, Site Preparation, and Organic Amendment in the Afforestation of a Semi-Arid Mediterranean Site with *Pinus halepensis*. *For. Sci.* **1998,** *44* (2), 203–211.

Ramos-Zapata, J. A.; Orellana, R.; Allen, E. B. Establishment of *Desmoncus orthacanthos martius* (Arecaceae): Effect of Inoculation with Arbuscular Mycorrhizae. *Rev. Biol. Trop.* **2006,** *54*, 65–72.

Sawers, R. J. H.; Yang, S.; Gutjahr, C.; Paszkowski, U. The Molecular Components of Nutrient Exchange in Arbuscular Mycorrhizal Interactions. In *Mycorrhizae: Sustainable Agriculture and Forestry*; Siddiqui, Z. A., Akhtar, M. S., Futai, K., Eds., 2008; pp 37–60.

Sellal, Z.; Ouazzani, T. A.; Mouden, N.; Ouarraqi, El. M.; Selmaoui, K.; Dahmani, J.; Benkirane, R.; Modafar, C. El.; Douira, A. Effect of an Endomycorrhizal Inoculum on the Growth of Argan Tree Plants. *Intl. J. Env. Agric. Biotech.* **2017,** *2* (2), 928–939.

Shaaban, A.; Wahbi, A. The Response of Two Lentil Varieties and One Wheat Variety to Phosphate Fertilization and Inoculation with Mycorrhiza. *Arab J. Arid Environ.* **2009,** *2* (2), 4–14.

Shuab, R.; Lone, R.; Ahmad, J.; Reshi, Z. Arbuscular Mycorrhizal Fungi: A Potential Tool for Restoration of Degraded Land. In *Mycorrhiza - Nutrient Uptake, Biocontrol,*

Ecorestoration; Varma, A., Prasad, R., Tuteja, N., Eds.; Springer International Publishing, 2017.

Siddiqui, Z.; Pichtel, J. Mycorrhizae: An Overview. In *Mycorrhizae: Sustainable Agriculture and Forestry*; Siddiqui, Z. A., Akhtar, M. S.; Futai, K., Eds.; Springer, 2008; pp 1–35.

Smith, S. E.; Read, D. J. In *Mycorrhizal Symbiosis*, 3rd Edition; Academic Press: New York, 2008.

Tawaha, A. M.; Turk, M. A. Lentil (*Lens culinaris* Medic.) Productivity as Influenced by Rate and Method of Phosphate Placement in a Mediterranean Environment. *Acta Agron. Hung.* 2002a, *50* (2), 197–201.

Tawaha, A. M.; Turk, M. A. Field Pea Seeding Management for Semi-Arid Mediterranean Conditions. *J. Agron. Crop Sci.* 2004, *190*, 86–92.

Tawaha, A. M.; Singh, V. P.; Turk, M. A. A Review on Growth, Yield Components and Yield of Barley as Influenced by Genotypes, Herbicides and Fertilizer Application. *Res. Crop* **2003,** *4* (1), 1–9.

Tawaha, A. R. M.; Turk, M. A.; Lee, K. D.; Zheng, W. J.; Ababneh, M.; Abebe, G.; Musallam, I. W. Impact of Fertilizer and Herbicide Application on Performance of Ten Barley Genotypes grown in Northeastern Part of Jordan. *Int. J. Agric. Biol.* **2005,** *7* (2), 162–166.

Thomas, E.; Jalonen, R.; Loo, J.; Boshier, D.; Gallo, L.; Cavers, S. Genetic Considerations in Ecosystem Restoration using Native Tree Species. *For. Ecol. Manag.* **2014,** *333*, 66–75.

Turjaman, M.; Tamai, Y.; Santoso, E.; Osaki, M.; Tawaraya, K. Arbuscular Mycorrhizal Fungi Increased Early Growth of Two Nontimber Forest Product Species *Dyerapolyphylla* and *Aquilaria filaria* under Greenhouse Conditions. *Mycorrhiza* **2006,** *16* (7), 459–464.

Turk, M. A.; Tawaha, A. M. Onion (*Allium cepa* L.) as Influenced by Rate and Method of Phosphorus Placement. *Crop Res.* **2002a,** *23* (1), 105–107.

Turk, M. A.; Tawaha, A. M. Allelopathic Effect of Black Mustard (*Brassica nigra* L.) on Germination and Growth of Wild Oat (*Avena fatua* L.). *Crop Prot.* **2003a,** *22* (4), 673.

Turk, M. A.; Tawaha, A. M. Allelopathic Effect of Black Mustard (*Brassica nigra* L.) on Germination and Growth of Wild Barley (*Hordeum spontaneum*). *J. Agron. Crop Sci.* **2003b,** *189* (5), 298–303.

Turk, M. A.; Tawaha, A. M. In *Effect of Variable Sowing Ratios and Sowing Rates of Bitter Vetch on the Herbage Yield of Oat-Bitter Vetch Mixed Cropping*, Proceedings 7th International Oat Conference/Pirjo Peltonen-Sainio and Mari Topi-Hulmi, Eds.; MTT, 2004.

Turk, M. A.; Hmaeed, K. M.; Aqeel, A. M.; Tawaha, A. M. Nutritional Status of Durum Wheat Grown in Soil Supplemented with Olive Mill by-Products. *Agrochimica* **2003c,** *47* (5-6), 209–219.

Turk, M. A.; Lee, K. D.; Tawaha, A. M. Inhibitory Effects of Aqueous Extracts of Black Mustard on Germination and Growth of Radish. *Res. J. Agric. Biol. Sci.* **2005,** *1* (3), 227–231.

Turk, M. A.; Tawaha, A. M.; Samara, N. Effects of Seedling Rate, Date and Phosphorous Application on Growth and Yield of Carbon Vetch (*Vicia narbonensis*). *Agronomie* **2003b,** *23* (4), 1–4.

Turk, M. A.; Tawaha, A. M.; Shatnawi, M. Lentil (*Lens culinaris* Medik) Response to Plant Density, Sowing Date, Phosphorous Fertilization and Ethephon Application in the Absence of Moisture Stress. *J. Agron. Crop Sci.* **2003a,** *189* (1), 1–6.

Vosatka, M.; Albrechtova, J.; Patten, R. The International Market Development for Mycorrhizal Technology. In *Mycorrhiza*; Varma, A., Ed.; Springer, 2008.

Wallander, H. Uptake of P from Apatite by *Pinus sylvestris* Seedlings Colonised by Different Ectomycorrhizal Fungi. *Plant Soil* **2000,** *218*, 249–256.

Wyatt, G. A. K.; Kiers, E. T.; Gardner, A.; West, S. A. A Biological Market Analysis of the Plant-Mycorrhizal Symbiosis. *Evolution* **2014,** *68*, 2603–2618.

Zhu, X. C.; Song, F. B.; Liu, S. Q.; Liu, T. D.; Zhou, X. Arbuscular Mycorrhizae Improves Photosynthesis and Water Status of *Zea mays* L. Under Drought Stress. *Plant Soil Environ.* **2012,** *58*, 186–191.

CHAPTER 9

AGRICULTURAL APPLICATIONS OF THE ARBUSCULAR MYCORRHIZAL FUNGI

IMRAN[1], ABDEL RAHMAN MOHAMMAD AL-TAWAHA[2*], AMANULLAH[1], ROGHAYEH KARIMIRAD[3], ABHIJIT DEY[4], SIDDHARTHA PATI[5], ABDEL RAZZAQ AL-TAWAHA[6], PUTERI E. MEGAT WAHAB[7], MUHAMMAD AASIM[8], R. Z. SAYYED[9], DEVARAJAN THANGADURAI[10], JEYABALAN SANGEETHA[11], HIBA ALATRASH[12], DEVENDRA KUMAR PANDEY[13], MUNIR TURK[14], TAHA A. AL-ISSA[14], EKATERINA KOZUHAROVA[15], IFTIKHAR ALI[16], and SAMIA KHANUM[17]

[1]*Department of Agronomy, The University of Agriculture, Peshawar, Pakistan*

[2]*Department of Biological Sciences, Al Hussein bin Talal University, Maan, Jordan*

[3]*Department of Horticultural Sciences, Faculty of Agriculture and Natural Resources, University of Mohaghegh Ardabili, Ardabīl, Iran*

[4]*Department of Life Sciences, Presidency University, Kolkata, India*

[5]*NatNov Bioscience Private Limited, Balasore, India*

[6]*National Agricultural Research Center (NARC), Baq'a 19381, Jordan*

[7]*Department of Crop Science, Faculty of Agriculture, University Putra Malaysia, Serdang, Selangor, Malaysia*

[8]*Department of Plant Protection, Faculty of Agricultural Sciences and Technologies, Sivas University of Science and Technology, Sivas, Turkey*

Microbial Fertilizer Technology for Sustainable Crop Production. Jeyabalan Sangeetha, Abdel Rahman Mohammad Al-Tawaha, & Devarajan Thangadurai (Eds.)

DOI: 10.1201/9781003594826-9

[9]*Department of Microbiology, PSGVP Mandal's Arts, Science and Commerce College, Shahada, Maharashtra, India*

[10]*Department of Botany, Karnatak University, Dharwad, Karnataka, India*

[11]*Department of Environmental Science, Central University of Kerala, Kasaragod, Kerala, India*

[12]*University of Helsinki, Helsinki, Finland*

[13]*Department of Biotechnology, School of Bioengineering and Biosciences, Lovely Professional University, Phagwara, Punjab, India*

[14]*Department of Plant Production, Faculty of Agriculture, Jordan University of Science and Technology, Irbid, Jordan*

[15]*Medical University of Sofia, Faculty of Pharmacy, Department of Pharmacognosy, Sofia, Bulgaria*

[16]*Center for Plant Sciences and Biodiversity, University of Swat, Charbagh, Pakistan*

[17]*Department of Botany, University of the Punjab, Lahore, Pakistan*

**Corresponding author.*

ABSTRACT

Companies are increasingly creating fungal inoculants and offering them as biofertilizers to promote plant growth and soil safety. However, as proven in a new study, there is little or no proof of their efficacy and possible ecological consequences. Arbuscular mycorrhizal fungi (AMF) are naturally occurring microorganisms that create symbiotic associations with plant roots, assisting them in water and nutrient intake, and enhancing plant performance. Despite the rising usage of commercial AMF inoculants by growers worldwide, research has revealed inconsistent success. Although some plants have responded favorably to inoculation, in the majority of cases, there has been no change or even a decline in plant performance. Ectomycorrhizal and endomycorrhizal fungi are both frequent mycorrhizas that play a critical role in the growth of plants. Mycorrhizal technology can be used in agricultural and horticultural crops, as well as forestry, to improve nutrient utilization, mitigate environmental and ecological issues by reducing chemical inputs manage disease by reducing biotic stress caused by pathogenic fungi, and more effectively manage land use. While inoculating plants with mycorrhiza is a frequent practice, developing inoculants with consistent and dependable action in field circumstances remains a barrier to their widespread use. The capability of the inoculum production technology and the formulation carrier are critical to their rapid adoption. This chapter discusses the commercial viability of mycorrhizal fungus as a tool for enhancing plant growth and productivity.

9.1 INTRODUCTION

In today's agricultural world, microbial-based biofertilizers are regarded as one of the most significant agricultural ingredients, as they increase crop production while also contributing to the development of sustainable agricultural ecosystems (Imran et al., 2020a). Integrated Soil Fertility Management (ISFM) is a collection of methods on smallholder farms that include cropping, fertilizers, organic resources, and other amendments aimed at increasing productivity and input efficiency. Several studies have been published on this topic (Hayyawi et al., 2018; Amanullah et al., 2019; Hani et al., 2019). Incorporating legumes into crop rotations has the potential to significantly increase the available nitrogen supply for cereals at a cheap cost. Additionally, legumes can improve soil quality, porosity, and structure, as well as influence certain microorganism populations in the rhizosphere to benefit subsequent crops (Aludatt et al., 2019; Turk et al., 2003a; Turk et al., 2003b). Increased soil fertility can be achieved through the use of cover crops, which add organic matter to the soil and promote a healthy fertile soil (Turk et al., 2003c; Nikus et al., 2004a; Nikus et al., 2004b; Tawaha and Turk, 2004; Turk and Tawaha, 2004); the use of green manure or the cultivation of legumes, which fix nitrogen from the air via the biological nitrogen fixation process; and the use of micro-dose fertilizers (Al-Tawaha et al., 2005; Lee et al., 2005; Sulpanjan et al., 2005; Tawaha et al., 2005; Assaf et al., 2006; Turk et al., 2006). Fertile soil requires a constant exchange of nutrients between organic and inorganic forms. Mineralization happens when microbes degrade organic substances in the soil solution, such as plant and animal dung, to liberate inorganic nutrients (Al-Kiyyam et al., 2008; Al-Ajlouni et al., 2009; Al-Tawaha et al., 2010; Al-Tawaha et al., 2018a; Al-Tawaha et al., 2018b). Additional modifications to those nutrients may be achievable through the action of soil microbes. When microbes compete with plants for nutrients, they immobilize those resources inside the microbial biomass, rendering them unavailable to plants. In agriculture, soil fertility refers to a soil's ability to support plant growth by providing essential plant nutrients as well as favorable chemical, physical, and biological characteristics that promote plant growth. Plant nutrition contains the macronutrients nitrogen, phosphorus, and potassium, as well as sulfur, calcium, and magnesium. Boron, chlorine, copper, iron, manganese, molybdenum, zinc, and selenium are the most abundant micronutrients (Turk and Tawaha, 2001; Tawaha and Turk 2002a; Tawaha and Turk, 2002b; Tawaha et al., 2003; Abebe et al., 2005; Abera et al., 2005; Al-Tawaha et al., 2005; Elser et al., 2007; Al-Kiyyam et al., 2008; Hameed et al., 2008; Al-Tawaha et al., 2010) (Figure 9.1).

A heterogeneous mixture of microbiological-based bio-products whose biological availability is critical for the acceleration and improvement of the biological processes of complex plant-microbe soil continuity are included in this component (Ahmed et al., 2008; Aise et al., 2011; Akpalu et al., 2014). To produce effective biofertilizers, different soil microorganisms (particularly bacteria and fungi) that display PGP characteristics might be employed (Ali et al., 2019; Karim and Imran, 2019). It is generally agreed that microbial bioformulation can be divided into four categories: (1) non-fermentable bacteria, (2) mobilizing/activating P microorganisms, (3) composting microorganisms, and (4) the use of biopesticides. Always keep in mind that, in addition to the important function for which they were selected, microbial groups that exhibit other PGP features (e.g., phytohormones, siderophores, amino acids, and polysaccharides) that may contribute to further crop development should be considered as well (Ankomah et al., 1996; Anchal et al., 1997; Alam et al., 2007; Akter et al., 2013; Imran et al., 2018). When it comes to the growth and development of essential crops, the beneficial effects of microbial-based biofertilizers can be restricted to a single nutrient, such as nitrogen (thanks to N-fixing bacteria), but they can also be limited to a number of nutrients (e.g., due to arbuscular mycorrhizal fungi [AMF]) (Ayoub et al., 1994; Assiouty and Sedera, 2005; Arancon et al., 2005; Ansari and Sukhraj, 2010; Argaw, 2012; Imran et al., 2017a; Imran, et al., 2017b). Moreover, the development of microorganism consortiums, which is a polymicrobial complex including multiple microbial strains belonging to distinct functional classes, has been shown to significantly increase plant growth, productivity,

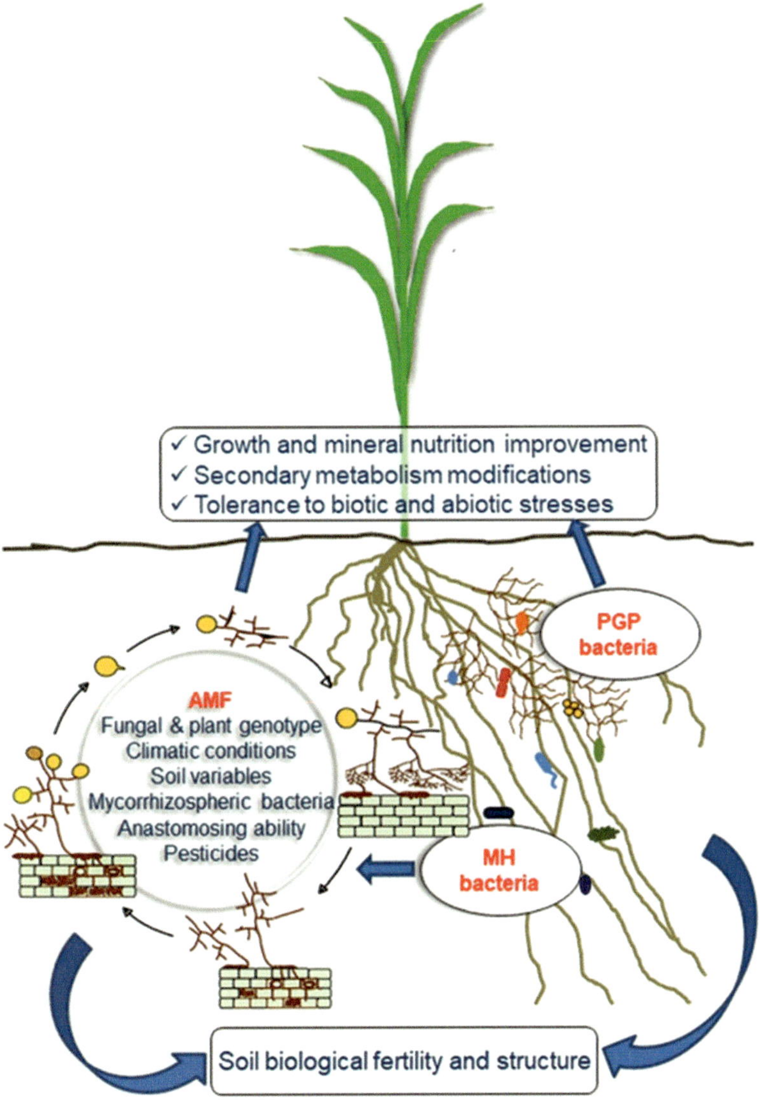

FIGURE 9.1 The impacts of arbuscular mycorrhizal fungi (AMF) and beneficial bacteria on plant performance and soil fertility. On the left: a visual representation of the AMF life cycle and factors affecting the different AMF developmental stages; on the right: mycorrhizal helper (MH) and plant growth promoting (PGP) bacteria synergistically interacting with AMF.

Source: Adapted from Giovannini et al. (2020), https://creativecommons.org/licenses/by/4.0/

and the overall health of agricultural systems (Bardan, 2003; Azarmi et al., 2009; Aziz et al., 2016). The effectiveness of a range of approaches for generating successful polymicrobial formulations is determined by the candidate strain's ability to be active, complementary, and coherent (Bellore and Mall, 1975; Bekere et al., 2012; Begum et al., 2015). These favorable effects on crops have also been demonstrated in conjunction with *Rhizobium* spp. In addition, there is little and contradictory knowledge of how AMF inoculants develop and disseminate (Imran, 2019). Many field tests only analyzed the structure for a short period of time. Even more concerning are the unknown environmental consequences of employing fungal inoculants (Imran and Amanullah, 2018; Imran 2018d; Imran, 2018e.). The difficulty is that we don't know how these fungi spread, but we do know that they may travel through soil, air, and even migrating birds. Thus, if an AMF inoculant is able to spread beyond its intended zone and become invasive, it has the potential to negatively disrupt local fungal communities and have devastating effects on plant variety and ecosystem function. This is a clear example of industrial practices that are not in line with contemporary scientific knowledge. The current usage of AMF inoculation is, at best, a danger and, at worst, a hazard to the environment (Imran et al., 2015a; Imran, 2017). Long-term inoculation of how these commercial biofertilizers are developed and how they can alter existing fungal and plant communities is urgently needed, which can subsequently help to design optimal practices (Imran, 2019).

9.2 HISTORY OF SOIL MICROBES COMMERCIALIZATION

There is a long history of commercialization of microbial biofertilizers under the name *Rhizobium*-based bioinklut, also known as "nitrogen." For more than a decade, experts from academic and private research institutes have investigated the potential of microorganisms such as fungi, bacteria (*Azotobacter*), and then cyanobacteria to promote plant growth (Figure 9.2).

Several decades later, Russia and Eastern European countries were dominated by two largely microbiological products, "*Azotobacterin*" and "*Fasobacterin*," which included *Azotobacter crococum* (NF) and *Bacillus megaterium* (PSB), respectively. In 1958, it was claimed that around 10 million hectares in Russia were treated with these microbial formulations and that there was some increase in crop yield for various crops. Farmers transported fertile soils from one field to another before agriculturally beneficial microorganisms, particularly AMF, were discovered, positively increasing crop productivity. The involvement of some soil microorganisms in this phenomenon is not new; it first occurred in early 1903. As a result, a large number of P solubilizing microorganisms (PSMs) live alongside helpful bacteria and fungi. This opportunity arose in close contact with the plants. The goal of these strong tools is to improve the use of phosphors in agriculture to establish sustainable agriculture. A microbial inoculant based on P-solubilizing *B. megaterium* (i.e., fibrobacterin) was effectively employed in Russia a few years later. The introduction of efficient products targeted at increasing nutrient efficiency.

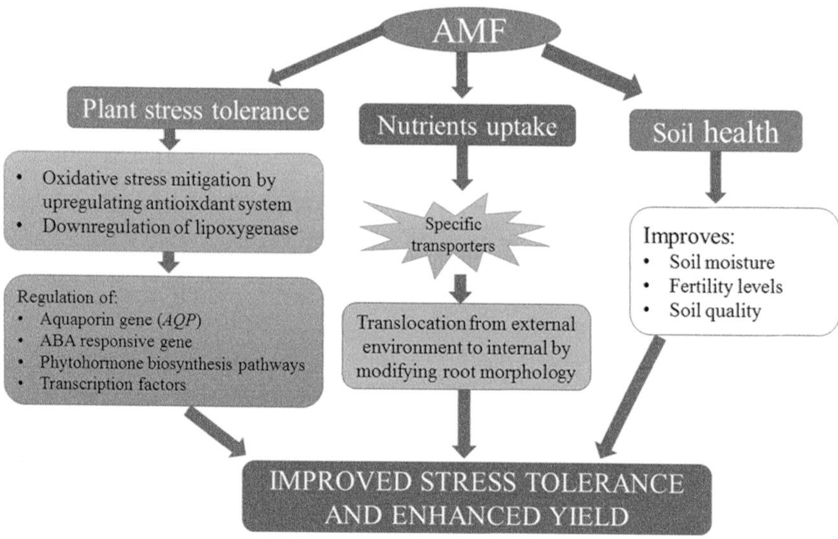

FIGURE 9.2 Mycorrhizal functions to regulate various processes in the ecosystem and plant growth promotion under abiotic stress conditions.

Source: Adapted from Begum et al. (2019).

9.3 MYCORRHIZA COMMERCIAL USE: A BUSINESS AND CONSUMER DESCRIPTION

Despite the fact that mycorrhizal fungi are found in 85% of all plant families and are responsible for the majority of crop growth, the widespread presence of these fungi in virtually all soils indicates the urgent need for inoculation with these natural connecting fungi in agricultural production systems (Berg and Lynd, 1985; Imran et al., 2019; Imran et al., 2020b). In recent years, advanced scientific understanding of mycorrhizal symbioses has increased the commercial potential for the application of mycorrhizal technology in agriculture, horticulture, plant production, forestry landscaping, phytoremediation in disturbed sites, and other aspects of the plant industry.

To produce mycorrhizal inoculum, professional firms must have the necessary biotechnological knowledge and capacity to counterbalance the acceptable, ethical, educational, and marketable requirements that are associated with this technique (Carsky et al., 2001; Brady, 2002; Bhattacharyya et al., 2010; Imran et al., 2018; Imran, 2018a; Imran 2018b; Karim and Imran, 2019).

There is a large variety of formulation types available, both liquid and solid, that can be customized to meet specific requirements. There are a multitude of sources for smart and high-quality commercial mycorrhizal inoculum now that it is available (Chen et al., 2008; Chiezey and Odunze, 2012). Consortium inoculum, which contains mixtures of species of mycorrhizal fungus, consistently produces the most favorable reaction. Mycorrhizal inoculum is available in the following forms: granular, powder, liquid, and tablet (Imran and Khan, 2017; Imran et al., 2017a). Dust, granules, and wettable powders are all examples of dry products. Dusts range in particle size from 5 to 20 mm and contain around 10% of an organism's weight. Granules are distinct masses ranging in size from 5–10 mm^3, pellets are larger than 10 mm^3, and briquettes are big blocks ranging in size from 5 g to a few cubic centimeters. These products may boost plant biomass output by using an inert carrier such as carbon dioxide, lignite, clay minerals (perlite, vermiculite, and bentonite), starch polymers, dry fertilizers, and soil–plant leftovers.

9.4 INTERACTION OF PLANTS AND AMF, SYMBIOSIS FOR NUTRITIONAL SECURITY

Mycorrhizal fungi offer enormous potential for use in agriculture when employed in conjunction with a set of conditions that aid plants in obtaining nourishment via symbiosis (Cassman et al., 1981; Carvalho et al., 2011; Imran et al., 2015a; Imran 2017; Imran et al., 2017b). Despite this, potential crop inoculation is rarely observed, and mycorrhizas are intentionally introduced into only a few businesses. Regardless of the potential boost in yield, the commercialization of mycorrhizas has not yet occurred. On the contrary, forestry is one of the few industries that have unequivocally recognized mycorrhizas' importance in plant growth (Imran et al., 2019; Imran et al., 2020b). The majority of commercial lumber comes from trees that develop ectomycorrhizal connections, whereas alien trees form mycorrhizal symbioses.

9.5 APPLICATION OF AMF

The fast-changing biotechnological environment, which is affected by globalization, antagonism, financial pressures, and the advancement of innovative technologies, has had an impact on the industry. It will be at the center of interest in developed countries as demand for organic food continues to grow. Growth and pathogenic resistance increases could be achieved by inoculating soil with mycorrhizae rather than by using inorganic fertilizers, pesticides, and fungicides (Cassman et al., 1981; Chauhan et al., 1992; Chaturvedi, 2006). Similarly, in developing countries, once inoculating soil with appropriate fungal isolates could eliminate the need for recurrent applications of expensive fertilizers that farmers cannot willingly afford (Imran et al., 2015a; Imran et al., 2015b). Until the process of translating this idea into a successful venture, barriers are emerging due to a lack of diffusion of knowledge, vibrant consulting services, and hope generators. As a result, it has only landed with a linear reassignment of technology or intellectual property.

Mycorrhizal fungi may be one option that can immediately improve productivity, whereas crops without AMF application may reduce biomass production in natural and managed ecosystems while preserving the natural resource base and lowering fertilizer and energy costs to restore economic efficiency and environmental safety. Better mycorrhizal fungus strains might potentially boost the growth of practically all agronomic crops if they were developed. Endomycorrhizae and ectomycorrhizal are both commercially produced on a modest scale. Some expert scientists have proposed and developed mycorrhizal inoculum generation devices in a gel carrier that are extremely concentrated and devoid of microbial contamination. Breakthroughs like this are required for the practical application of mycorrhizal fungi. The lack of large-scale field tests under typical agroecology, adequate economics of mycorrhizal fungi technology, and a growing lack of willingness on the part of farmers to switch from an energy-intensive system to a new, but cheaper, energy-conservative system using mycorrhizal fungi appear to be the most significant barriers to commercialization of mycorrhizal fungi.

9.6 CONCLUSION

Plant development and productivity are hampered by abiotic stressors. As a result of climate change and agricultural malpractices, such as the overuse of fertilizers and pesticides, the impacts of abiotic pressures on crop output have been exacerbated, and the ecosystem itself has been damaged. When it comes to increasing agricultural yield, there is an urgent need for environmentally acceptable management strategies, such as the application of AMF. AMF are also referred to as "biofertilizers" in some circles. Furthermore, it is widely considered that the inoculation of AMF confers tolerance on host plants to a variety of stressful conditions such as heat, salinity, drought, metals, and severe temperatures. Host plants may benefit from AMF since it can both support them in the upregulation of tolerance mechanisms and prevent the downregulation of critical metabolic pathways. Due to the fact that they are natural root symbionts, AMF give vital plant inorganic nutrients to their host plants, hence increasing growth and yield under both stress and non-stressed conditions. The use of AMF as a biofertilizer has the potential to improve plants' ability to adapt to changing environmental conditions. As a result, additional study concentrating on the increase of crop quality and productivity through the use of AMF is required.

KEYWORDS

- **arbuscular mycorrhizal**
- **biofertilizers**
- **biotic stress**
- **ectomycorrhizal**

REFERENCES

Abebe, G.; Hattar, B.; Al-Tawah, A. M. Nutrient Availability as Affected by Manure Application to Cowpea (*Vigna unguiculata* L. Walp.) on Calacarious Soils. *J. Agric. Social Sci.* **2005,** *1*(1), 1–6.

Abera, T.; Feyisa, D.; Yusuf, H.; Nikus, O.; Al-Tawaha, A. R. Grain Yield of Maize as Affected by Biogass Slurry and NP Fertilizer Rate at Bako, Western Oromiya, Ethopia. *Biosci. Res.* **2005,** *2*, 31–38.

Ahmed, M. F.; Kennedy, I. R.; Choudhury, A. T. M. A.; Kecskés, M. L.; Deaker, R. Phosphorus Adsorption in Some Australian Soils and Influence of Bacteria on the Desorption of Phosphorus. *Commun. Soil Sci. Plant Anal.* **2008,** *39*, 1269–1294.

Aise, D.; Erdal, S.; Hasanand, A.; Ahment, M. Effects of Different Water, Phosphorus and Magnesium Doses on the Quality and Yield Factors of Soybean in Harran Plain Conditions. *Int. J. Phys. Sci.* **2011,** *6*(6), 1484–1495.

Akpalu, M. M.; Siewobr, H.; Oppong, D.; Akpalu, S. E. Phosphorus Application and Rhizobia Inoculation on Growth and Yield of Soyabean (*Glycine max* L. Merrill). *Am. J. Exp. Agric.* **2014,** *4*(6), 674–685.

Akter, F. N. I.; Shamsuddoha, A. T. M.; Bhuiyan, M. S. I.; Shilpi, S. Effect of Phosphorus and Sulphur on Growth and Yield of Soybean (*Glycine max* L.). *Int. J. Bioresource Stress Manag.* **2013,** *4*(4), 555–560.

Al-Ajlouni, M. M.; Al-Ghzawi, A. L. A.; Al-Tawaha, A. R. Crop rotation and Fertilization Effect on Barley Yield Grown in Arid Conditions. *J. Food Agric. Environ.* **2009,** *88*(3), 869-872.

Alam, M. N.; Jahan, M. S.; Ali, M. K.; Ashraf, M. A.; Islam, M. K. Effect of Vermicompost and Chemical Fertilizers on Growth, Yield and Yield Components of Potato in Barind Soils of Bangladesh. *J. Appl. Sci. Res.* **2007,** *3*, 1879–1888.

Ali, I.; Khan, A. A.; Imran, Inamullah, Khan, A.; Asim, M.; Ali, I.; Zib, B.; Khan, I.; Rab, A.; Sadiq, G. Humic Acids and Nitrogen Levels Optimizing Productivity of Green Gram (*Vigna radiata* L.). *Russ. Agric. Sci.* **2019,** *45*(1), 43–47.

Al-Juthery, H. W.; Habeeb, K. H.; Jawad, F.; Altaee, K.; AL-Taey, D. K. A.; Al-Tawaha, A. R. M. Effect of Foliar Application of Different Sources of Nano-Fertilizers on Growth and Yield of Wheat. *Biosci. Res.* **2010,** *15*(4), 3988–3997.

Al-Kiyyam, M. A.; Turk, M.; Al-Mahmoud, M.; Al-Tawaha, A. R. Effect of Plant Density and Nitrogen Rate on Herbage Yields of Marjoram Under Mediterranean Conditions. *J. Am. Eurasian J. Agric. Environ. Sci.* **2008,** *3*, 153–158.

Al-Tawaha, A. M.; Turk, M.; Hameed, K.; Assaf, T.; Tahhan, R.; Al-Jamali, A. F. Effects of Application of Olive Mill By-Products on Lentil Yield and their Symbiosis with Mycorrhizal Fungi Under Arid Conditions. *Crop Res.* **2010,** *40*(3), 66–75.

AL-Tawaha, A. R. M.; Nidal, O. Use of Sorghum and Maize Allelopathic Properties to Inhibit Germination and Growth of Wild Barley (*Hordeum spontaneum*). *Not. Bot. Horti. Agrobo. Cluj-Napoca*, **2010,** *38*(3), 124–127.

Al-Tawaha, A. R.; Al-Karaki, G.; Al-Tawaha, A. R.; Sirajuddin, S. N.; Makhadmeh, I.; Wahab, P. E. M.; Youssef, R. A.; Al Sultan, W.; Massadeh, A. Effect of Water Flow Rate on Quantity and Quality of Lettuce (*Lactuca sativa* L.) in Nutrient Film Technique (NFT) Under Hydroponics Conditions. *Bulg J. Agric. Sci.* **2018a,** *245*, 791–798.

Al-Tawaha, A. R.; Al-Tawaha, A. R.; Aludatt, M.; Al-Ghzawi, A. L.; Wedyan, M.; Al-Obaidy, S. D. A.; Al-Ramamneh, E. A. D. Effects of Soil Type and Rainwater Harvesting Treatments in the Growth, Productivity and Morphological Traits of Barley Plants Cultivated in Semi-Arid Environment. *Aust. J. Crop Sci.* **2018b,** *12*(6), 975–979.

Al-Tawaha, A. R.; McNeil, D. L.; Yadav, S. S.; Turk, M.; Ajlouni, M.; Abu-Darwish, M. S.; Al-Ghzawi, A. L. A.; Al-udatt, M.; Aladaileh, S. Integrated Legume Crops Production and Management Technology. In *Climate Change and Management of Cool Season Grain Legume Crops*; Yadav, S. S., David. L., McNeil, R. R., Sharanagouda, A. P., Eds.; Springer: Dordrecht, 2010; pp 325–349.

Al-Tawaha, A. R.; Turk, M. A.; Abu-Zaitoon, Y. M.; Aladaileh, S. H.; Al-Rawashdeh, I. M.; Alnaimat, S.; Al-Tawaha, A. R. M.; Aludatt, M. H.; Wedyan, M. Plants Adaptation to Drought Environment. *Bulg. J. Agric. Sci.* **2017,** *23*(3), 381–388.

Alu'datt, M. H.; Rababah, T.; Alhamad, M. N.; Al-Tawaha, A.; Al-Tawaha, A. R.; Gammoh, S.; Ereifej, K. I.; Al-Karaki, G.; Hamasha, H. R.; Tranchant, C. C.; Kubow, S. Herbal Yield, Nutritive Composition, Phenolic Contents and Antioxidant Activity of Purslane (*Portulaca oleracea* L.) Grown in Different Soilless Media in a Closed System. *Indus. Crops Prod.* **2019,** *141*, 111746.

Ananthi, T.; Amanullah, M. M.; Al-Tawaha, A. R. M. S. A Review on Maize-Legume Intercropping for Enhancing the Productivity and Soil Fertility for Sustainable Agriculture in India. *Adv. Environ. Biol.* **2017,** *11*(5) 49–63.

Anchal, D.; Kharwara, P. C.; Rana, S. S.; Dass, A. Response of Gram Varieties to Sowing Dates and Phosphorus Levels Under on Farm Condition. *Himachal J. Agric. Res.* **1997,** *23*, 112–115.

Ankomah, A. B.; Zapata, F.; Hardarson, G. Yield, Nodulation and N_2 Fixation by Cowpea Cultivars at Different Phosphorus Levels. *Biol. Fertil. Soils* **1996,** *22*, 10–15.

Ansari, A. A.; Sukhraj, K. K. Effect of Vermiwash and Vermicompost on Soil Parameters and Productivity of Okra in Guyana. *Afr. J. Agric. Res.* **2010,** *5*, 1794–1798.

Arancon, N. Q.; Edwards, C. A.; Bierman, P.; Metzger, J. D.; Lucht, C. Effects of Vermin Composts Produced From Cattle Manure, Food Waste and Paper Waste on the Growth and Yield of Peppers in the Field. *Pedobiologia* **2005,** *49*, 297–306.

Argaw, A. Evaluation of Co-inoculation of *Bradyrhizobium Japonicum* and Phosphate Solubilizing *Pseudomonas* spp. Effect on Soybean (*Glycine max* L.) in Assossa Area. *J. Agric. Sci. Technol.* **2012,** *14*, 213–224.

Assiouty, F. M.; Abo-Sedera, S. A. Effect of Bio and Chemical Fertilizers on Seed Production and Quality of Spinach. *Int. J. Agric. Biol.* **2005,** *6*, 947–952.

Ayoub, M.; Guertin, S.; Lussier, S.; Smith, D. L. Timing and Level of Nitrogen Fertilizer Effects on Spring Wheat Yield in Eastern Canada. *Crop Sci.* **1994,** *34*, 748–756.

Azarmi, R.; Giglou, M. T.; Taleshmikail, R. D. Influence of Vermicompost on Soil Chemical and Physical Properties in Tomato Field. *Afr. J. Biotechnol.* **2009,** *7*, 2397–2401.

Aziz, A. L. A.; Ahiabor, B. D. K.; Opoku, A.; Abaidoo, R. C. Contributions of *Rhizobium* Inoculants and Phosphorus Fertilizer to Biological Nitrogen Fixation, Growth and Grain Yield of Three Soybean Varieties on a *Fluvic Luvisol. Am. J. Exp. Agric.* **2016,** *10*(2), 1–11.

Bardan, M. M. Effect of Nitrogenous and Phosphetic Fertilization on Some Economical Characters of Soybean Crawford Cultivar Under Calcareous Soil Conditions. *Egypt. J. Agric. Res.* **2003,** *81*(2), 433–440.

Begum, A.; Islam, M. D. A.; Ahmed, Q. M.; Rahman, M. Effect of Nitrogen and Phosphorus on the Growth and Yield Performance of Soybean. *Res. Agric. Livest. Fish.* **2015,** *2*(1), 35–42.

Begum, N.; Qin, C.; Ahanger, M. A.; Raza, S.; Khan, M. I.; Ashraf, M.; Ahmed, N.; Zhang, L. Role of Arbuscular Mycorrhizal Fungi in Plant Growth Regulation: Implications in Abiotic Stress Tolerance. *Front. Plant Sci.* **2019,** *10*, 1068. doi: 10.3389/fpls.2019.01068.

Bekere, W.; Endalkachew, W.; Tesfu, K. Growth and Nodulation Response of Soybean (*Glycine max* L.) to *Bradyrhizobium* Inoculation and Phosphorus Levels Under Controlled Condition in South Western Ethiopia. *Afric. J. Agric. Res.* **2012,** *7*(30), 4266–4270.

Bellore, S. K.; Mall, L. R. Chlorophyll Content as an Ecological Index of Dry Matter Production. *Indian J. Bot. Soc.* **1975,** *54*(1/2), 75–77.

Berg, R. K.; Lynd, J. Q. Soil Fertility Effects on Growth, Yield and Nodulation and Nitrogenase Activity of Australian Winter Pea. *J. Plant Nutr.* **1985,** *8*, 131–145.

Bhattacharyya, R. P. S. C.; Chandra, S.; Kundu, S.; Supradip, S.; Mina, B. L.; Srivastva, A. K.; Gupta, H. Fertilization Effects on Yield Sustainability and Soil Properties Under Irrigated Wheat-Soybean Rotation of an Indian Himalayan Upper Valley. *Nutr. Cycling Agroecosyst.* **2010,** *86,* 255–268.

Brady. N. Phosphorus and Potassium. In *The Nature and Properties of Soils*. Prentice Hall of India Pvt. Limited: New Delhi, India, 2002; p 352.

Carsky, R. J.; Singh, B. B.; Oyewole, R. Contribution of Early–Season Cowpea to Late Season Maize in the Savanna Zone of West Africa. *Biol. Agric. Hort.* **2001,** *18*, 303–315.

Carvalho, E. R.; Rezende, P. M.; Andrade, M. J. B.; Passos, A. M. A.; Oliveira, J. A. Fertilizante mineral e resíduo orgânico sobre características agronômicas da soja e nutrientes no solo [Mineral Fertilizer and Organic Waste on Soybean Crop and Soil Nutrients]. *Rev. Ciênc. Agro.* **2011,** *42*, 930–939.

Cassman, K, G.; Munns, D. N.; Beck, D, P. Phosphorus Nutrition of *Rhizobium japonicum*: Strain Differences in Phosphate Storage and Utilization. *Soil Sci. Am. J.* **1981,** *45*, 517–520

Cassman, K. G.; Whitney, A. S.; Fox, R. L. Phosphorus Requirements of Soybean and Cowpea as Affected by Mode of Nitrogen Nutrition. *Agric. J.* **1981,** *73*,17–22.

Chaturvedi, I. Effects of Phosphorus Levels Alone or in Combination with Phosphate-Solubilizing Bacteria and Farmyard Manure on Growth, Yield and Nutrient Uptake of Wheat (*Triticum aestivum*). *J. Agric. Soc. Sci.* **2006,** *2*(2), 96–100.

Chauhan, Y. S.; Johanson, C.; Venkataratnam. Effect of Phosphorus Deficiency on Phenology and Yield Components of Short Duration Pigeonpea. *Trop. Agric.* **1992,** *69*, 235–238.

Chela, G. S.; Tiwana, M. S.; Thind, I. S.; Puri, K. P.; Kaur, K. Effect of Bacterial Cultures and Nitrogen Fertility on the Yield and Quality of Maize Fodder (*Zea mays* L.). *Ann. Biol.* **1993,** *9*, 83–86.

Chen, Y. P.; Rekha, P. D.; Arunshen, A. B.; Lai, W. A.; Young, C. C. PSB From Subtropical Soil and their Tri-Calcium Phosphate Solubilizing Abilities. *Appl. Soil Ecol.* **2006,** *34*, 33–41.

Chen, Z.; Shengwu, M.; Lily, L. Studies on Phosphorus Solubilizing Activity of a Strain of *Hosphobacteria* Isolated from Chestnut Type Soil in China. *Bioresource Technol.* **2008,** *99*, 6702–6707.

Chiezey, U. F.; Odunze, A. C. Soybean Response to Application of Poultry Manure and Phosphorus Fertilizer in the Sub-Humid Savanna of Nigeria. *J. Ecol. Nat. Environ.* **2012,** *1*(2), 25–31.

Giovannini, L.; Palla, M.; Agnolucci, M.; Avio, L.; Sbrana, C.; Turrini, A., Giovannetti, M. Arbuscular Mycorrhizal Fungi and Associated Microbiota as Plant Biostimulants: Research Strategies for the Selection of the Best Performing Inocula. *Agronomy* **2020,** *10*(1), 106. https://doi.org/10.3390/agronomy10010106

Imran, Amanullah. Global Impact of Climate Change on Water, Soil Resources and Threat Towards Food Security: Evidence from Pakistan. *Adv. Plants Agric. Res.* **2018,** *8*(5), 350–355. doi: 10.15406/apar.2018.08.00349.

Imran, Khan, A. A. Canola Yield and Quality Enhanced with Sulphur Fertilization. *Russ. Agric. Sci.* **2017,** *43*, 113. doi: 10.3103/S1068367417020100.

Imran. Climate Change is a Real Fact Confronting to Agricultural Productivity. *Int. J. Environ. Sci. Nat. Res.* **2017,** *3*(3), 555613. doi: 10.19080/IJESNR.2017.03.555613.

Imran. Climate Change is Threat Toward Agronomy (Base of Food, Fiber System), and Food Security. *Food Nutr. J.* **2018a,** doi: 10.29011/2575-7091.100060.

Imran. Physiological and Morphological Traits of Agronomic Crops Influenced by Climate Change. *Mod. Concept Dev. Agron.* **2018b,** *1*(4), doi: 10.31031/MCDA.2018.01.000524.

Imran. Ecological Environmental Variability Influence Growth and Yield Potential of Rice Under Northern Climatic Scenario. *Russ. Agric. Sci.* **2018c,** *44*(1), 18–24.

Imran. Phosphorous Fertilization Influenced Weeds Attributes and Phenological Characteristics of Mungbean Cultivars (*Vigna radiata* L.). *Russ. Agric. Sci.* **2018d,** *44*(3), 229–234.

Imran. Organic Matter Amendments Improve Soil Health, Productivity and Profitability of Maize and Soybean. *Ann. Rev. Res.* **2018e,** *1*(3), 555564.

Imran. Advance Production Technologies of Legume Crops. Production Technology. In *Agronomic Crops*; 2019; doi: 10.1007/978-981-32-9151-5_16.

Imran. Plant microbe's Interaction in Agronomic Crops. In *Management Practices*; 2019; doi: 10.1007/978-981-32-9151-8_21.

Imran, Amanullah, Arif, M.; Shah, Z.; Bari, A. Soil Application of *Trichoderma* and Peach (*Prunus persica* L.) Residues Possesses Biocontrol Potential for Weeds and Enhances Growth and Profitability of Soybean *Glycine max*. *Sarhad J. Agric.* **2020a,** *36*(1), 10–20.

Imran, Amanullah, Bari, A.; Ali, R. Peach Sources, Phosphorous and Beneficial Microbes Enhance Productivity of Soybean. *Soy. Res.* **2018,** *16*(2), 39–48.

Imran, Amanullah, Bari, A.; Khan, H.; Ali. R. Climatic Variability and Agronomic Cropping Pattern. In *Production Technology, Agronomic Crops*; **2019;** doi: 10.1007/978-981-32-9151-5_3.

Imran, Amanullah, Khan, A. A.; Bari A.; Uddin, R.; Ali, R.; Ahmed, N.; Ali, A.; Maula, F.; Ahmad, F.; Naveed, S.; Ullah, I. Production Statistics and Modern Technology of Maize Cultivation in Khyber Pakhtunkhwa Pakistan. *Plant Sci. Archiv.* **2019,** *2*(2), 1–12.

Imran, Bari, A.; Ali, R.; Ahmad, N.; Ahmad, Z.; Khattak, M. I.; Ali, A.; Ahmad, F.; Khan, I.; Naveed, S. Traditional Rice Farming Accelerate CH_4 and N_2O Emissions Functioning as a Stronger Contributors of Climate Change. *Int. J. Environ. Sci. Nat. Resour.* **2017b,** *9*(3), 555765. doi: 10.19080/ARTOAJ.2017.09.555765.

Imran, Jamal, N.; Alam A.; Khan, A. A. Grain Yield, Yield Attributes of Wheat and Soil Physio-Chemical Characteristics Influenced by Biochar, Compost and Inorganic Fertilizer Application. *Agric. Res. Tech.* **2017a,** *10*(4), 555795. doi: 10.19080/ARTOAJ.2017.10.555795.

Imran, Khan, A. A.; Ullah, I.; Zada, H.; Ahmad, F.; Shah, S. T.; Usman, A.; Ullah, I. Yield and Yield Attributes of Rapeseed Cultivars as Influence by Sulfur Level Under Swat Valley Conditions. *J. Pure Appl. Biol.* **2015a,** *4*(3), 296–301.

Imran, Naveed, S.; Khan, A. A.; Khattak, I. Impact of Phosphorus Levels and Seed Rates on Growth and Yield of Late Sown Maize on High Elevation in Swat, Pakistan. *Pakistan J. Agric. Res.* 2015b, *28*(4), 406–413.

Karim, S. A.; Imran. Reduction in Substrate Moisture Content Reduce Final Yield of Wheat. *Biomed. J. Sci. Tech. Res.* **2019,** *15*, 11150–11154.

Nikus, O.; Al-Tawaha, A. M.; Turk, M. A. Effect of Manure Supplemented with Phosphate Fertilizer on the Fodder Yield and Quality of Two Sorghum Cultivars (*Sorghum bicolor* L.). *Biosci. Res.* **2004a,** *1*, 1–7.

Nikus, O.; Al-Tawaha, A. M.; Turk, M. A. Yield Response of Sorghum (*Sorghum bicolor* L.) to Manufacture Supplemented with Phosphate Fertilizer Under Semi-Arid Mediterranean Conditions. *Int. J. Agric. Biol.* **2004b,** *6*, 889–893.

Tawaha, A. M.; Turk, M. A. Field Pea Seeding Management for Semi-Arid Mediterranean Conditions. *J. Agron. Crop Sci.* **2004,** *190*, 86–92.

Tawaha, A. M.; Turk, M. A. Lentil (*Lens culinaris* Medic.) Productivity as Influenced by Rate and Method of Phosphate Placement in a Mediterranean Environment. *Acta Agron. Hung.* **2002a,** *50*(2), 197–201.

Tawaha, A. M.; Singh, V. P.; Turk, M. A. A Review on Growth, Yield Components and Yield of Barley as Influenced by Genotypes, Herbicides and Fertilizer Application. *Res. Crop* **2003,** *4*(1), 1–9.

Tawaha, A. R. M.; Turk, M. A.; Lee, K. D.; Zheng, W. J.; Ababneh, M.; Abebe, G.; Musallam, I. W. Impact of Fertilizer and Herbicide Application on Performance of Ten Barley Genotypes Grown in Northeastern Part of Jordan. *Int. J. Agric. Biol.* **2005,** *7*(2), 162–166.

Turk, M. A.; Tawaha, A. M. Onion (*Allium cepa* L.) as Influenced by Rate and Method of Phosphorus Placement. *Crop Res.* **2002a,** *23*(1), 105–107.

Turk, M. A.; Tawaha, A. M. Effect of Variable Sowing *Ratios and Sowing Rates of Bitter Vetch on the Herbage Yield of Oat-Bitter Vetch Mixed Cropping*, Proceedings 7th International Oat Conference, 2004; Peltonen-Sainio, P., Topi-Hulmi, M., Eds.; MTT.

Turk, M. A.; Tawaha, A. M. Allelopathic Effect of Black Mustard (*Brassica nigra* L.) on Germination and Growth of Wild Barley (*Hordeum spontaneum*). *J. Agron. Crop Sci.* **2003b,** *189*(5), 298–303.

Turk, M. A.; Hameed, K. M.; Aqeel, A. M.; Tawaha, A. M. Nutritional Status of Durum Wheat Grown in Soil Supplemented with Olive Mill By-products. *Agrochimica* **2003c,** *47*(5/6), 209–219.

Turk, M. A.; Lee, K. D.; Tawaha, A. M. Inhibitory Effects of Aqueous Extracts of Black Mustard on Germination and Growth of Radish. *Res. J. Agric. Biol. Sci.* **2005,** *1*(3), 227–231.

Turk, M. A.; Tawaha, A. M.; Samara, N. Effects of Seedling Rate and Date, and Phosphorous Application on Growth and Yield of Carbon Vetch (*Vicia narbonensis*). *Agronomie* **2003b,** *23*, 1–4.

Turk, M. A.; Tawaha, A. M.; Shatnawi, M. Lentil (*Lens culinaris* Medik) Response to Plant Density, Sowing Date, Phosporous Fertilization and Ethephon Application in the Absence of Moisture Stress. *J. Agron. Crop Sci.* **2003a,** *189*(1), 1–6.

CHAPTER 10

ROLE OF MYCORRHIZAE IN SOIL NUTRIENT AVAILABILITY AND PLANT NUTRIENT UPTAKE

EDAPPAYIL JANEESHMA[1*], JOS THOMAS PUTHUR[2], ABDEL RAHMAN MOHAMMAD AL-TAWAHA[3], VISHNU D. RAJPUT[4], ABEER JUBILY[5], POOJASHREE NAGAPPA KUMMUR[6], DEVARAJAN THANGADURAI[7], and JEYABALAN SANGEETHA[7]

[1]*Department of Botany, MES KEVEEYAM College, Valanchery, Kerala, India*

[2]*Plant Physiology and Biochemistry Division, Department of Botany, University of Calicut, Calicut, Kerala, India*

[3]*Department of Biological Sciences, Al Hussein Bin Talal University, Maan, Jordan*

[4]*Academy of Biology and Biotechnology, Southern Federal University, Rostov-On-Don, Russia*

[5]*Centre for Agricultural Research, General Commission for Scientific Agricultural Research, Syria*

[6]*Department of Botany, Karnatak University, Dharwad, Karnataka, India*

[7]*Department of Environmental Science, Central University of Kerala, Kasaragod, Kerala, India*

**Corresponding author.*

ABSTRACT

From 400 to 450 million years ago, arbuscular mycorrhizal (AM) connections emerged and were crucial in plant colonization. A mycorrhizal relationship

Microbial Fertilizer Technology for Sustainable Crop Production. Jeyabalan Sangeetha, Abdel Rahman Mohammad Al-Tawaha, & Devarajan Thangadurai (Eds.)

DOI: 10.1201/9781003594826-10

with ubiquitous soil fungi is formed by 80% of all known terrestrial plant species. There is a bidirectional exchange of resources across the mycorrhizal interface that benefits both partners. In addition to supplying phosphorus (P) and nitrogen (N), mycorrhizal fungi also help plants withstand biotic (root infections) and abiotic (drought, salt, heavy metals) stresses. In exchange for the mycorrhizal fungus-assisted nutrient uptake, the host plant gives back its photosynthetically fixed carbon. In addition to structural differences, AM and ectomycorrhizal (ECM) were different in associating plant species. Mycorrhizae are useful for crops that are cultivated in soils with a limited supply of nutrients. It is possible that mycorrhizal connections will result in higher nutrient concentrations in plant tissue, particularly for minerals (such as P and N) that are dependent on diffusion toward the roots for uptake. In order to manage crop nutrient levels, mycorrhizal management is becoming a more viable option, albeit there are still a number of practical issues to overcome in this application. Along with providing nutritional benefits to plants, mycorrhizal relationships with plants have the potential to compete with fungal diseases for the infection of plant roots.

10.1 INTRODUCTION

Extensive application of chemical fertilizers in agricultural lands to improve the soil nutrient status invites acidification, air pollution, leaf scorch of crop plants, eutrophication, and related environmental risks (Janeeshma and Puthur, 2022). At this point, sustainable agricultural approaches become essential to overcome the drawbacks of chemical fertilizers and the mineral depletion of the soil and one of the approaches is utilization of mycorrhizae for this purpose (Labidi et al., 2015). Previous reports have discussed the role of mycorrhizal symbiosis in the mobilization and transport of minerals from the soil to plants (Labidi et al., 2015). AM-mediated enhancement in the mineral absorption was achieved by different methods like enhancing the area of absorption of the root, production, and secretion of organic acids as well as acid phosphatases and the maintenance of consortium with other soil microbiota including nitrogen-fixing bacteria (Weisany et al., 2017). This symbiotic association enhanced the uptake of different micro- and macronutrients including nitrogen (N), phosphorus (P), potassium (K), sulfur (S), iron (Fe), magnesium (Mg), manganese (Mn), calcium (Ca), copper (Cu), zinc (Zn), and boron (B) from the soil (Marschner and Del, 1994; Labidi et al., 2015). *Rhizophagus irregularis* can be considered as one of the most studied model systems to analyze the mycorrhizal mediated nutrient acquisition mechanisms in different host plants and its association boost up the absorption of mineral ions (Labidi et al., 2012; Burghelea et al., 2015; Zhang et al., 2017; Santander et al., 2017). Whereas a mixture of AM fungus (*Septoglomus constrictum*, *Funneliformis geosporum*, *Glomus fuegianum*, *R. irregularis*, and *Glomus* sp.) further enhanced the uptake of P in *Hedysarum coronarium* (Labidi et al., 2015).

The abiotic stress tolerance offered by the mycorrhizal association depends on the elevated rate of mineral acquisition, even under different suboptimal conditions (Mirshad and Puthur, 2016). The symbiosis of *Glomus etunicatum*, *Glomus intraradices* (Daei et al., 2009), *Glomus fasciculatum* (Giri et al., 2007), and *G. macrocarpum* (Giri and Mukerji, 2004) assisted the host plant in overcoming salinity stress by increasing the essential nutrient acquisition and also by preventing excess uptake of Na^+. The enhancement of mineral absorption by mycorrhizal roots was observed even in the elevated $CaCO_3$ concentrations in soil (Clark and Zeto, 2000; Labidi et al., 2012; Mirshad et al., 2014). According to the report of Al-Karaki (2000), *Glomus mosseae* induced the assimilation of P, Zn, Cu, and Fe in tomato plants growing under salt stress. Similar responses were observed under drought stress in *Ziziphus mauritiana* associated with *Gigaspora margarita*, *Glomus constrictum*, *G. fasciculatum*, *G. mosseae*, *Sclerocystis rubiformis*, and *Scutellospora calospora* (Mathur and Vyas, 2000) (Table 10.1).

TABLE 10.1 Improved Nutrient Uptake by Various Endophytic Fungi Which are not Root Associated.

Fungal endophyte	Host plant	Alteration observed in nutrient acquisition
Epichloë festucae	*Festuca rubra* L.	Increases P and N content in shoots
Epichloë sp.	Perennial ryegrass (*Lolium perenne*)	Increases Mn content
Epichloë gansuensis (C. J. Li & Nan.) Schardl.	Drunken horse grass (*Achnatherum inebrians* (Hance) Keng.)	Increases N and P content
Colletotrichum tropicale (Rojus, Rehner & Samuels)	Cacao tree (*Theobroma cacao* L.)	Increases ^{15}N uptake at whole-plant level

Source: Adapted and slightly modified from Sarkar et al. (2021). https://creativecommons.org/licenses/by/4.0/

10.2 PHOSPHORUS (P)

The inevitable role of P in nucleic acid biosynthesis, membrane stability and functioning, biosynthesis of the high-energy molecule, photosynthesis, and other key physiological and developmental functions of plants, enlighten the importance of enhancing the availability of this element in the rhizosphere (Figure 10.1). However, extremely low diffusion rate and complexation with other minerals make the element unavailable to plants (create a P depletion zone in rhizosphere) that leads to the reduction in phosphorus uptake and subsequently crop production (Hasan et al., 2016). This limitation in P availability could be overcome with the application of AM fungus. Prior reports have shown that different plants such as *H. coronarium* (Labidi et al., 2015), *Mentha crispa* (Urcoviche et al., 2015), *Nigella sativa* (Abadi et al., 2015), *Triticum aestivum* (Ortas et al., 2018), *Zea mays* (Garcés-Ruiz et al., 2017), *Malus domestica* (Gąstoł et al., 2016), *Borago officinalis* (Rahimi et al., 2017) showed mycorrhizal mediated increase in the acquisition of P. At the same time, the availability of a high concentration of P in soil elicits the mycorrhizal-associated plant to reduce the uptake of P by downregulating P transporters expressed in the root of symbionts (Bücking et al., 2012). Moreover, the high N and P applications negatively affected the N, P, K, Mn, and Zn uptake from soil by AM-associated root of lettuce (Azcón et al., 2003). The high level of P also reduces the fungal spore production and ramification of external hyphae of AM (Abadi et al., 2015). A field study conducted from 1990 to 1991 proved that at the early stages of maize, mycorrhization increased P uptake under reduced and conventional tillage (McGonigle and Miller, 1993).

The mycorrhizal mediated enhancement in the soil P content was attained by different strategies adopted by the fungal and plant partners of the symbiotic association. Extension of root surface was achieved by the development of root hairs and it helps the water and mineral absorption. Root hairs have an important role in macro and micromineral acquisition, especially in the mobilization of P; thus, crop plants with long and dense root hairs can establish even in a mineral-deficient condition (Zou et al., 2019). Mycorrhizae induced root structural transformations and associated enhancement in P uptake (Janeeshma and Puthur, 2019). A comparative study of the capability of P uptake and related root modifications was observed in 13 different plant species, showing mycorrhizal association (Wen et al., 2019). The plants selected for this study include *Pennisetum*

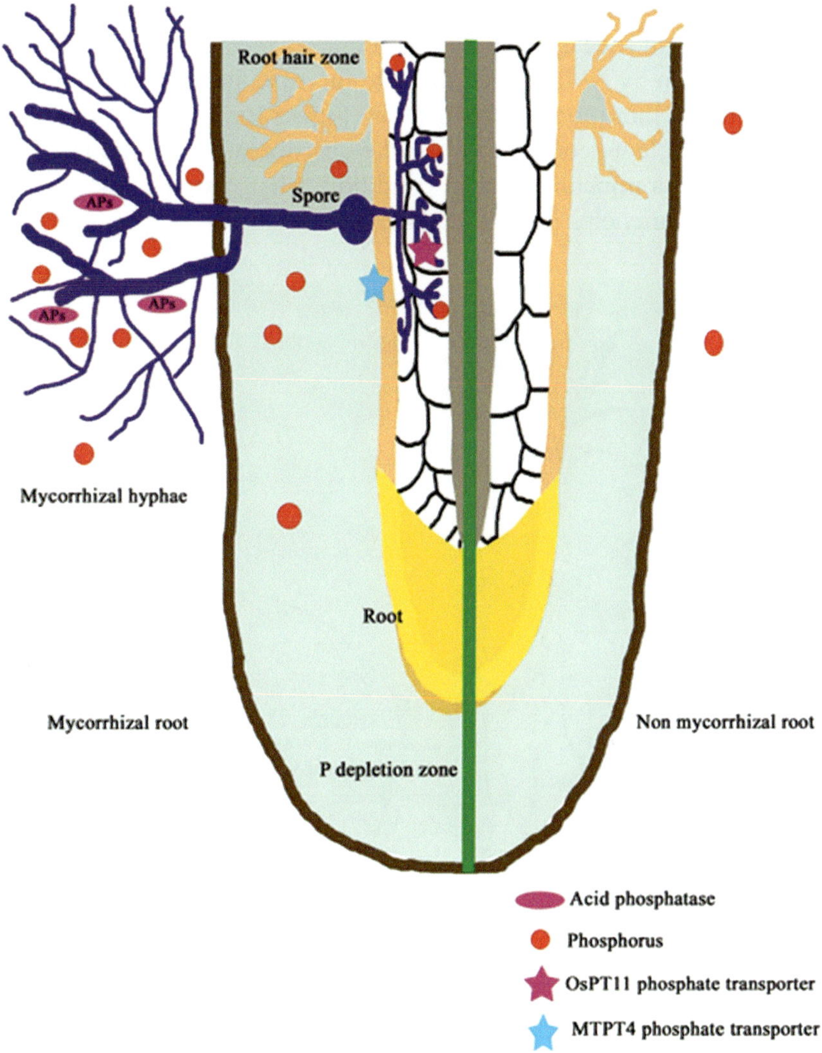

FIGURE 10.1 Factors influencing mycorrhizal mediated phosphorus acquisition.

americanum, T. aestivum, Hordeum vulgare, Sorghum bicolor, Z. mays, Medicago sativa, Glycine max, Vigna radiata, Pisum sativum, Cicer arietinum, Vigna faba, Allium fistulosum, Allium tuberosum and they exhibited prominent variation in root architecture. Members of Gramineae with thinner roots had higher root branching intensity and lower colonization by AMF and root exudates release. However, members of Fabaceae and species of *Allium* with thicker roots showed a high rate of colonization by AMF, which helps to increase the absorption surface of the root and thus more P-mobilization (Figure 10.2). This study clearly pointing that plants compensate for the absence of root hairs by AMF association in a mineral-deficient condition, where extensively growing fungal mycelium helps in the P uptake.

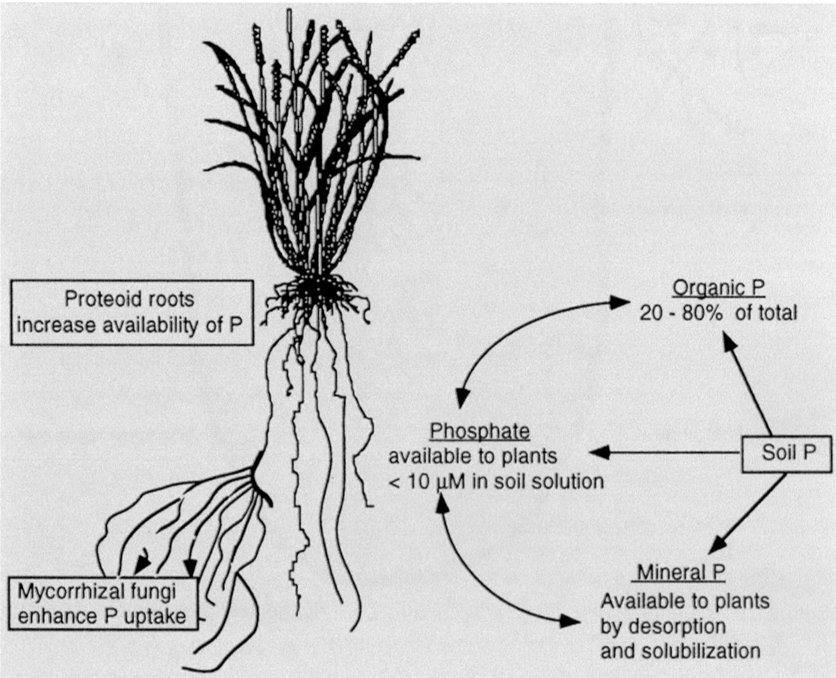

FIGURE 10.2 Plant root geometry and morphology are important for maximizing P uptake because root systems with a higher ratio of surface area to volume will more effectively explore a larger volume of soil (Lynch, 1995). For this reason, mycorrhizae are also important for plant P acquisition since fungal hyphae significantly increase the volume of soil that plant roots explore (Smith and Read, 1997). In certain plant species, root clusters (proteoid roots) are formed in response to P limitations. These specialized roots exude high amounts of organic acids (up to 23% of net photosynthesis), which acidify the soil and chelate metal ions around the roots, resulting in the mobilization of P and some micronutrients (Marschner, 1995).

Source: Reproduced with permission from Daniel et al. (1998). Copyright © 1998, Oxford University Press.

Bolan (1991) observed a faster movement of P into the mycorrhizal hyphae than the movement to root cells. He suggested that the fungal hyphae have a better affinity toward P and the threshold concentration of P was lower in the same as compared to root cells. When Cress et al. (1979) compared the kinetics of P absorption in mycorrhizal and nonmycorrhizal roots of tomato, mycorrhizal roots showed greater affinity toward $H_2PO_4^-$ at its lower concentrations. The prior reports of Sanders and Tinker (1973) also supported these findings, pointing toward an important observation that the diffusion resistance of P absorption by plant roots can compensate for the mycorrhizal hyphae.

The enhancement in the bioavailability of P in the rhizosphere depends on the ability of AM to elicit the overproduction of root exudates and phosphatase (Figure 10.2). The potential of extraradical hyphae of AM fungus to release acid phosphatase was scrutinized in *A. fistulosum* associated with *Rhizophagus clarus* (Sato et al., 2015). In this study, the activity-staining technique revealed that the solution collected from the mycorrhizae colonized soil showed acid phosphatase activity. Mycorrhization aided the establishment of plants in Cape Fynbos in South Africa, which is characterized by highly leached, sandy, mineral deficient, and acidic soils (Maseko and Dakora, 2013). They concluded that the release of root exudate from AM-associated plants improves P solubilization, which was identified as the major reason for enhancing P nutrition in mycorrhizal symbionts. When *R. irregularis* increased P uptake in *M. sativa,* the dual inoculation with both *R. irregularis* and *Pseudomonas alcaligenes* did not elevate the P content in plants (Zhang et al., 2014). Even though the dual inoculation showed high acid phosphatase activity, the availability of Phytate-P acts as the limiting factor for P uptake. However, the P transfer to plant roots could be increased by applying excess phytate. Sometimes, both acid phosphatase and alkaline phosphatase are involved in the mycorrhizal mediated P absorption, and it was proved by the study conducted in *Chrysanthemum indicum* associated with *G. mosseae* and *Acaulospora laevis* (Prasad et al., 2012). A similar result was observed in *Carica papaya*, when associated with *Glomus claroideum*. A study conducted in *Allium cepa* inoculated with *G. margarita*, or *G. etunicatum* suggested that hyphal exudate helps to solubilize inorganic P in the soil (Tawaraya et al., 2006) (Figure 10.3). It points toward the potential of mycorrhizal plants to utilize the inorganic P than nonmycorrhizal plants. Koide and Kabir (2000) extended this study to organic P utilization, an *in vitro* system of *G. intraradices* with Ri-TDNA (transfer DNA of root hair inducing plasmid) transformed carrot roots showed that the AMF hyphae hydrolyzed both 5-bromo-4-chloro-3-indolyl phosphate and phenolphthalein diphosphate (representing the organic compounds with phosphorus).

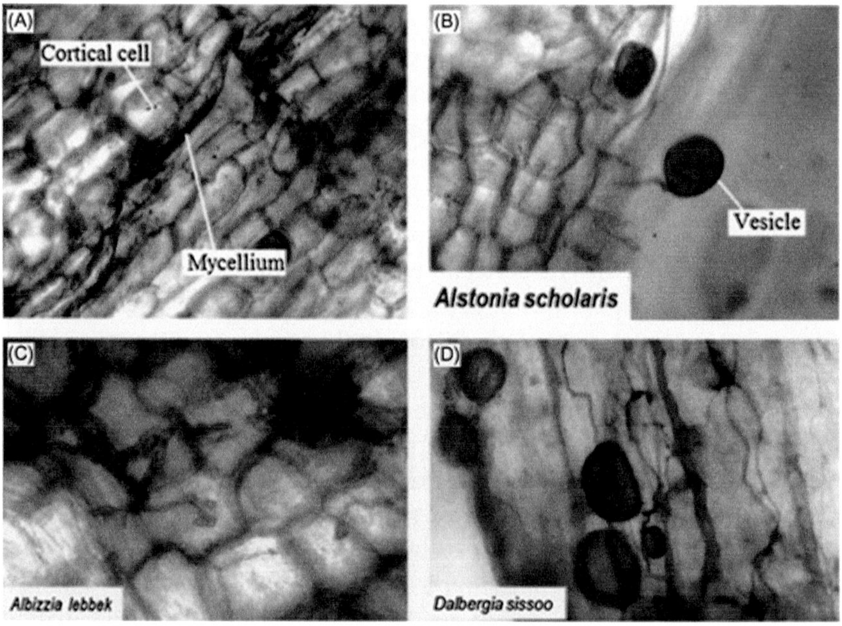

FIGURE 10.3 Mycorrhiza root infection observed under a compound microscope (OLYMPUS BX 60) – 430×. (A) VAM mycelium in the cortical cells of the root, (B) mycorrhiza infection in *Alstonia* roots showing vesicles, (C) mycorrhiza infection in *Albizia lebbeck* roots, and (D) mycorrhiza infection in *Dalbergia sissoo* roots.

Source: Reproduced with permission from Maiti and Dipita (2020). Copyright © Elsevier (2020).

Mycorrhizal hyphae have a great capacity to accumulate the absorbed P in the form of orthophosphate or soluble polyphosphate or polyphosphate granule. This potential allows the flow of P to hyphae. This high concentration of P was maintained in the hyphae with the help of the activity of polyphosphatase. So, as per the concentration gradient, the transfer of P to the host root occurs by the passive transport to the fungus-root interface and subsequent active transport into the root (Bolan, 1991).

P transporters have a crucial role in the mycorrhizal-mediated enhancement of mineral acquisition. AM induced upregulation of *Oryza sativa* phosphate transporter (OsPT11), and its accumulation was independent of P availability in the rhizosphere (Paszkowski et al., 2002). The transporters involved in the P transport between arbuscules and plant cytoplasm were investigated in *Medicago truncatula. A* novel P transporter (MTPT4) was discovered, and its expression was specified to arbuscule containing root cells (Harrison et al., 2002). RNA localization and gene expression analysis help to find the site of expression of StPT3 (P transporter gene in potato). It revealed that its action was specified to the mycorrhizal structures formed in the root cells (Rausch et al., 2001). It was clear that P absorption in plants depends on the surface area of absorption, root exudates, and P transporters, which are improved by AM association and hence P assimilation.

10.3 NITROGEN

When normal growth and development of plants are affected by N deficiency, the mycorrhizal association helps them to increase their uptake (Hodge and Fitter, 2010). AM association induced the activity of different enzymes (nitrate reductase, glutamine synthetase, and glutamine synthase) involved in the nitrogen metabolism of the host plant (Subramanian and Charest, 1999). Moreover, this symbiotic association activates nitrogen transporters in host roots (Guether et al., 2009). The potential of *G. intraradices* in N assimilation was analyzed with the help of ^{15}N and extraradical hyphae showed an accumulation of arginine with this labeled N (Jin et al., 2005). The common mycorrhizal networks (CMNs) between legumes and nonlegumes facilitate enhancement in N assimilation (He et al., 2003). The most crucial role in the N uptake mechanisms of AM was contributed by the tripartite symbiosis established between the host plant, AM fungi, and N-fixing bacteria. ^{15}N isotopic tracing method was used to analyze the role of tripartite symbiosis between *G. mosseae*, *Rhizobium* SH212, and *Glycine max* in nitrogen metabolism. They found that this dual inoculation enhanced the N uptake compared to the individual microbial treatment.

10.4 CONCLUSIONS

Mycorrhizae are abundant in natural environments and infect a wide range of tree species, forming mutualistic partnerships in which the fungi get sugars from the host plant to enhance the efficiency with which the host plant absorbs minerals from the soil. The nutritional condition of the host plant and the location where the mycelium is growing are important factors in determining the degree of mycelial development. When a nutrient such as phosphorus is deficient, mycorrhizal infection is more likely to occur, whereas when a nutrient such as nitrogen is deficient, infection is more likely to occur. There is also a relationship between the volume of root surface fungus and bacteria (which is closely related to biomass) and the amount of nitrogen in the shoots. On the contrary, it points toward the feature that establishing a microbial consortium in the rhizosphere always enhances the potential of mineral absorption by crop plants. However, in this context, it has to be remembered that soil microbial population density and diversity are crucially decreasing due to the application of the excess amount of agrochemicals.

KEYWORDS

- **fungal diseases**
- **mycorrhiza**
- **phosphate**
- **roots uptake**
- **root infections**

REFERENCES

Abadi, B. H. M.; Ganjali, H. R.; Mobasser, H. R. Effect of Mycorrhiza and Phosphorus Fertilizer on Some Characteristics of Black Cumin. *Biol. Forum* **2015,** *7*(1), 1115.

Al-Karaki, G. N. Growth of Mycorrhizal Tomato and Mineral Acquisition Under Salt Stress. *Mycorrhiza* **2000,** *10*(2), 51–54.

Azcón, R.; Ambrosano, E.; Charest, C. Nutrient Acquisition in Mycorrhizal Lettuce Plants Under Different Phosphorus and Nitrogen Concentration. *Plant Sci.* **2003,** *165*(5), 1137–1145.

Bolan, N. S. A Critical Review on the Role of Mycorrhizal Fungi in the Uptake of Phosphorus by Plants. *Plant Soil* **1991,** *134*(2), 189–207.

Bücking, H.; Liepold, E.; Ambilwade, P. The Role of the Mycorrhizal Symbiosis in Nutrient Uptake of Plants and the Regulatory Mechanisms Underlying these Transport Processes. *Plant Sci.* **2012,** *4*, 108–132.

Burghelea, C.; Zaharescu, D. G.; Dontsova, K.; Maier, R.; Huxman, T.; Chorover, J. Mineral Nutrient Mobilization by Plants from Rock: Influence of Rock Type and Arbuscular Mycorrhiza. *Biogeochemistry* **2015,** *124*(1/3), 187–203.

Clark, R. B.; Zeto, S. K. Mineral Acquisition by Arbuscular Mycorrhizal Plants. *J. Plant Nutr.* **2000,** *23*(7), 867–902.

Cress, W. A.; Throneberry, G. O.; Lindsey, D. L. Kinetics of Phosphorus Absorption by Mycorrhizal and Nonmycorrhizal Tomato Roots. *Plant Physiol.* **1979,** *64*(3), 484–487.

Daei, G.; Ardekani, M. R.; Rejali, F.; Teimuri, S.; Miransari, M. Alleviation of Salinity Stress on Wheat Yield, Yield Components, and Nutrient Uptake Using Arbuscular Mycorrhizal Fungi Under Field Conditions. *J. Plant Physiol.* **2009,** *166*(6), 617–625.

Daniel, P. S.; Robert J. R.; Ayling, S. M. Phosphorus Uptake by Plants: From Soil to Cell. *Plant Physiol.* **1998,** *116*(2), 447–453. https://doi.org/10.1104/pp.116.2.447.

Garcés-Ruiz, M.; Calonne-Salmon, M.; Plouznikoff, K.; Misson, C.; Navarrete-Mier, M.; Cranenbrouck, S.; Declerck, S. Dynamics of Short-Term Phosphorus Uptake by Intact Mycorrhizal and Non-Mycorrhizal Maize Plants Grown in a Circulatory Semi-Hydroponic Cultivation System. *Front. Plant Sci.* **2017,** *8*, 1471.

Gąstoł, M.; Domagała-Świątkiewicz, I.; Bijak, M. The Effect of Mycorrhizal Inoculation and Phosphorus Application on the Growth and Mineral Nutrient Status of Apple Seedlings. *J. Plant Nutr.* **2016,** *39*(2), 288–299.

Giri, B.; Mukerji, K. G. Mycorrhizal Inoculant Alleviates Salt Stress in *Sesbania aegyptiaca* and *Sesbania grandiflora* Under Field Conditions: Evidence for Reduced Sodium and Improved Magnesium Uptake. *Mycorrhiza* **2004,** *14*(5), 307–312.

Giri, B.; Kapoor, R.; Mukerji, K. G. Improved Tolerance of *Acacia nilotica* to Salt Stress by Arbuscular Mycorrhiza, *Glomus fasciculatum* may be Partly Related to Elevated K/Na Ratios in Root and Shoot Tissues. *Microb. Ecol.* **2007,** *54*(4), 753–760.

Guether, M.; Neuhäuser, B.; Balestrini, R.; Dynowski, M.; Ludewig, U.; Bonfante, P. A Mycorrhizal-Specific Ammonium Transporter from *Lotus japonicus* Acquires Nitrogen Released by Arbuscular Mycorrhizal Fungi. *Plant Physiol.* **2009,** *150*(1), 73–83.

Harrison, M. J.; Dewbre, G. R.; Liu, J. A Phosphate Transporter from *Medicago truncatula* Involved in the Acquisition of Phosphate Released by Arbuscular Mycorrhizal Fungi. *Plant Cell* **2002,** *14*(10), 2413–2429.

Hasan, M. M.; Hasan, M. M.; da Silva, J. A. T.; Li, X. Regulation of Phosphorus Uptake and Utilization: Transitioning from Current Knowledge to Practical Strategies. *Cell. Mol. Biol. Lett.* **2016,** *21*(1), 7.

He, X. H.; Critchley, C.; Bledsoe, C. Nitrogen Transfer Within and Between Plants Through Common Mycorrhizal Networks (CMNs). *Crit. Rev. Plant Sci.* **2003,** *22*(6), 531–567.

Hodge, A.; Fitter, A. H. Substantial Nitrogen Acquisition by Arbuscular Mycorrhizal Fungi from Organic Material has Implications for N Cycling. *Proc. Natl. Acad. Sci. USA* **2010,** *107*(31), 13754–13759.

Janeeshma, E.; Puthur, J. T. Direct and Indirect Influence of Arbuscular Mycorrhizae on Enhancing Metal Tolerance of Plants. *Arch. Microbiol.* **2019,** *202*(1), 1–16.

Janeeshma, E.; Puthur, J. T. Physiological and Metabolic Dynamism in Mycorrhizal and Non-Mycorrhizal *Oryza sativa* (var. Varsha) Subjected to Zn and Cd Toxicity: A Comparative Study. *Environ. Sci. Pollut. Res.* **2022,** *30*, 1–20.

Koide, R. T.; Kabir, Z. Extraradical Hyphae of the Mycorrhizal Fungus *Glomus intraradices* Can Hydrolyse Organic Phosphate. *New Phytol.* **2000,** *148*(3), 511–517.

Labidi, S.; Jeddi, F. B.; Tisserant, B.; Debiane, D.; Rezgui, S.; Grandmougin-Ferjani, A.; Sahraoui, A. L. H. Role of Arbuscular Mycorrhizal Symbiosis in Root Mineral Uptake Under $CaCO_3$ Stress. *Mycorrhiza* **2012,** *22*(5), 337–345.

Labidi, S.; Jeddi, F. B.; Tisserant, B.; Yousfi, M.; Sanaa, M.; Dalpé, Y.; Sahraoui, A. L. H. Field Application of Mycorrhizal Bio-Inoculants Affects the Mineral Uptake of a Forage Legume (*Hedysarum coronarium* L.) on a Highly Calcareous Soil. *Mycorrhiza* **2015,** *25*(4), 297–309.

Maiti, S. K.; Dipita, G. Plant–Soil Interactions as a Restoration Tool. In *Climate Change and Soil Interactions*, Maiti, S. K., Ed.; Amsterdam, Netherlands, pp 689–730. https://doi.org/10.1016/B978-0-12-818032-7.00024-2, 2020.

Marschner, H.; Dell, B. Nutrient Uptake in Mycorrhizal Symbiosis. *Plant Soil* **1994,** *159*(1), 89–102.

Maseko, S. T.; Dakora, F. D. Plant Enzymes, Root Exudates, Cluster Roots and Mycorrhizal Symbiosis are the Drivers of P Nutrition in Native Legumes Growing in P Deficient Soil of the Cape Fynbos in South Africa. *J. Agric. Sci. Technol.* **2013,** *3*(5A), 331.

Mathur, N.; Vyas, A. Influence of Arbuscular Mycorrhizae on Biomass Production, Nutrient Uptake and Physiological Changes in *Ziziphus mauritiana* Lam. Under Water Stress. *J. Arid Environ.* **2000,** *45*(3), 191–195.

McGonigle, T. P.; Miller, M. H. Mycorrhizal Development and Phosphorus Absorption in Maize Under Conventional and Reduced Tillage. *Soil Sci. Soc. Am. J.* **1993,** *57*(4), 1002–1006.

Mirshad, P. P.; Puthur, J. T. Arbuscular Mycorrhizal Association Enhances Drought Tolerance Potential of Promising Bioenergy Grass (*Saccharum arundinaceum* Retz.). *Environ. Monitor. Assess.* **2016,** *188*(7), 425.

Mirshad, P. P.; Chandran, S.; Puthur, J. T. Characteristics of Bioenergy Grasses Important for Enhanced NaCl Tolerance Potential. *Russ. J. Plant Physiol.* **2014,** *61*(5), 639–645.

Ortas, I.; Bykova, A. The Effect of mycorrhiza inoculation and phosphorus application on phosphorus efficiency of wheat plants. *Commun. Soil Sci. Plant Anal.* **2018,** *49*(10), 1199–1207.

Paszkowski, U.; Kroken, S.; Roux, C.; Briggs, S. P. Rice Phosphate Transporters Include an Evolutionarily Divergent Gene Specifically Activated in Arbuscular Mycorrhizal Symbiosis. *Proc. Natl. Acad. Sci. USA* **2002,** *99*(20), 13324–13329.

Prasad, K.; Aggarwal, A.; Yadav, K.; Tanwar, A. Impact of Different Levels of Superphosphate Using Arbuscular Mycorrhizal Fungi and *Pseudomonas fluorescens* on *Chrysanthemum indicum* L. *J. Soil Sci. Plant Nutr.* **2012,** *12*(3), 451–462.

Rahimi, A.; Jahanbin, S.; Salehi, A.; Farajee, H. Changes in Content of Chlorophyll, Carotenoids, Phosphorus and Relative Water Content of Medicinal Plant of Borage (*Borago officinails* L.) Under the Influence of Mycorrhizal Fungi and Water Stress. *J. Biol. Sci.* **2017,** *17*(1), 28–34.

Rausch, C.; Daram, P.; Brunner, S.; Jansa, J.; Laloi, M.; Leggewie, G.; Amrhein, N.; Bucher, M. A Phosphate Transporter Expressed in Arbuscule-Containing Cells in Potato. *Nature* **2001,** *414*(6862), 462.

Sanders, F. E.; Tinker, P. B. Phosphate Flow Into Mycorrhizal Roots. *Pestic. Sci.* **1973,** *4*(3), 385–395.

Santander, C.; Aroca, R.; Ruiz-Lozano, J. M.; Olave, J.; Cartes, P.; Borie, F.; Cornejo, P. Arbuscular Mycorrhiza Effects on Plant Performance Under Osmotic Stress. *Mycorrhiza* **2017,** *27*(7), 639–657.

Sarkar, S.; Dey, A.; Kumar, V.; Batiha, GE-S.; El-Esawi, M. A.; Tomczyk, M.; Ray, P. Fungal Endophyte: An Interactive Endosymbiont with the Capability of Modulating Host Physiology in Myriad Ways. *Front. Plant Sci.* **2021,** *12*, 701800. doi: 10.3389/fpls.2021.701800.

Sato, T.; Ezawa, T.; Cheng, W.; Tawaraya, K. Release of Acid Phosphatase from Extraradical Hyphae of Arbuscular Mycorrhizal Fungus *Rhizophagus clarus*. *Soil Sci. Plant Nutr.* **2015,** *61*(2), 269–274.

Subramanian, K.; Charest, C. Acquisition of N by External Hyphae of an Arbuscular Mycorrhizal Fungus and its Impact on Physiological Responses in Maize Under Drought-Stressed and Well-Watered Conditions. *Mycorrhiza* **1999,** *9*, 69–75.

Tawaraya, K.; Naito, M.; Wagatsuma, T. Solubilization of Insoluble Inorganic Phosphate by Hyphal Exudates of Arbuscular Mycorrhizal Fungi. *J. Plant Nutr.* **2006,** *29*(4), 657–665.

Urcoviche, R. C.; Gazim, Z. C.; Dragunski, D. C.; Barcellos, F. G.; Alberton, O. Plant Growth and Essential Oil Content of *Mentha crispa* Inoculated with Arbuscular Mycorrhizal Fungi Under Different Levels of Phosphorus. *Indus. Crops Prod.* **2015,** *67*, 103–107.

Weisany, W.; Raei, Y.; Salmasi, S. Z.; Sohrabi, Y.; Ghassemi-Golezani, K. Arbuscular Mycorrhizal Fungi Induced Changes in Rhizosphere, Essential Oil and Mineral Nutrients Uptake in Dill/Common Bean Intercropping System. *Ann. Appl. Biol.* **2016,** *169*(3), 384–397.

Wen, Z.; Li, H.; Shen, Q.; Tang, X.; Xiong, C.; Li, H.; Jiayin P.; Megan H. R.; Shen, J. Tradeoffs Among Root Morphology, Exudation and Mycorrhizal Symbioses for Phosphorus-Acquisition Strategies of 16 Crop Species. *New Phytol.* **2019,** *223*, 882–895.

Zhang, H.; Wei, S.; Hu, W.; Xiao, L.; Tang, M. Arbuscular Mycorrhizal Fungus *Rhizophagus irregularis* Increased Potassium Content and Expression of Genes Encoding Potassium Channels in *Lycium barbarum*. *Front. Plant Sci.* **2017,** *8*, 440.

Zhang, L.; Fan, J.; Ding, X.; He, X.; Zhang, F.; Feng, G. Hyphosphere Interactions Between an Arbuscular Mycorrhizal Fungus and a Phosphate Solubilizing Bacterium Promote Phytate Mineralization in Soil. *Soil Biol. Biochem.* **2014,** *74*, 177–183.

Zou, Y. N.; Zhang, D. J.; Liu, C. Y.; Wu, Q. S. Relationships Between Mycorrhizas and Root Hairs. *Pak. J. Bot.* **2019,** *51*(2), 727–733.

CHAPTER 11

EFFECT OF MYCORRHIZAL INOCULATION ON CITRUS SEEDLING GROWTH AND NUTRIENT UPTAKE

ARADHNA KUMARI[1*], SANTOSH KUMAR SINGH[2], ABDEL RAHMAN MOHAMMAD AL-TAWAHA[3], DEVARAJAN THANGADURAI[4], SIMRAN PANIGATTI[4], and JEYABALAN SANGEETHA[5]

[1]*Department of Plant Physiology, College of Agriculture, Jawaharlal Nehru Krishi Vishwa Vidyalaya, Ganj Basoda, Madhya Pradesh, India*

[2]*Department of Soil Science, Dr. Rajendra Prasad Central Agricultural University, Pusa, Samastipur, Bihar, India*

[3]*Department of Biological Sciences, Al Hussein bin Talal University, Maan, Jordan*

[4]*Department of Botany, Karnatak University, Dharwad, Karnataka, India*

[5]*Department of Environmental Science, Central University of Kerala, Kasaragod, Kerala, India*

**Corresponding author.*

ABSTRACT

When cultivating citrus plants, it is critical to understand their nutrient requirements throughout the growing season to safeguard optimal growth. Because fertilizers are limited and costly, as well as because the long-term environmental impact is uncertain, it is recommended that plant root zone modulation methods, such as mycorrhizal inoculation, be used to

Microbial Fertilizer Technology for Sustainable Crop Production. Jeyabalan Sangeetha, Abdel Rahman Mohammad Al-Tawaha, & Devarajan Thangadurai (Eds.)

DOI: 10.1201/9781003594826-11

promote growth. Because citrus plants rely on mycorrhizae colonization for survival, it is advantageous to inject seedlings with mycorrhizae at the initial stages of development. Agricultural techniques, such as soil and plant management, have a crucial influence on the establishment of indigenous mycorrhizae in field crops that are propagated by seed. Mycorrhizal inoculation, on the other hand, makes it easier to employ horticultural plants (plants that are planted in the field as seedlings) throughout large agricultural regions than it is to employ non-inoculated plants. Farmers might be told, based on significant studies and observations, that arbuscular mycorrhizal fungus-injected seedlings can be utilized in field settings to boost production and quality while decreasing costs. This is also an excellent choice for citrus crops. Increased yield, fruit quality, fertilizer use efficiency, and stress tolerance in citrus crops can all be improved by inoculating them with mycorrhizal fungi. Even though some mycorrhizal fungi are extremely effective at increasing the above-mentioned characteristics, most of the mycorrhizal fungus species are advantageous for these yield-determining criteria.

11.1 INTRODUCTION

Due to exhaustive cultivation, our soils are becoming deficient in soil nutrients. To get sustainable yield and optimum plant growth and development of any crop, it is essential to incorporate different means of nutrients in the soil. It is possible to either use fertilizers or biological supplements of nutrients. It is also well known in the recent era that due to excessive use of fertilizers, soil health is deteriorating up to such an extent that further maintaining soil health is becoming next to impossible, as well as these means of chemicals in the form of fertilizers are accumulating in the human body and causing many diseases, most prominently different types of cancers. So, exploring different compatible means of biological formulations for supplementing the soil with nutrients is pertinent. Many of these are exploited in different crops, such as *Rhizobium* in leguminous crops; *Azotobacter* in sugarcane; *Azolla* and blue-green algae in rice fields for enriching nitrogen; phosphorus solubilizing bacteria (PSB) and mycorrhizae for enriching phosphorus; and sulfur bacteria for enriching sulfur.

Genus *Citrus* belongs to the family Rutaceae and sub-family Aurantioedae. Citrus fruits are extremely important fruit crops in terms of international trade. Among all the commonly cultivated citrus fruit plants, oranges have the most citrus production credits, but besides this, grapefruits, pomelos, lemons, and limes also have significant importance. As per the data from the United Nations Food and Agriculture Organization, the overall combined production of all types of citrus fruits was 144 million tons in the world in the year 2019, in which about half the production share was of oranges followed by tangerines/mandarins, lemons/limes, and grapefruit (UNFAO, 2020; USDA, 2021). According to the United Nations Conference on Trade and Development (UNCTAD), the production of citrus began to increase during the first decade of the twenty-first century, which is primarily the result of the expansion of the area of cultivation, betterment in transport facilities and packaging, increasing incomes, and consumer preference for nutrient-dense foods (UNFAO, 2020). In 2021–2022, the production of oranges was about 50.8 million tons in the world, dominated in production by countries such as Brazil, China, the USA, the European Union, and Egypt as the largest producers (USDA, 2022). On the other hand, citrus productivity has started to decline in many countries in recent years. Over and inadequate fertilization are one of the reasons for declining citrus production. Providing nutrition input for fruit crops has become a very costly affair nowadays. So, to save input costs, it is necessary to use manures and fertilizers very judiciously to maintain the productivity of citrus orchards. The fertilization should be adjusted according to soil and leaf analysis. Leaf analysis is an indication of the quality of the effective absorption of nutrients and their distribution to the different plant parts. Using biological means of nutrients and supplements is a better choice for soil health with enhanced productivity. This chapter aims to explore the potential of mycorrhizal inoculation in citrus seedlings for better growth and nutrient uptake.

Citrus growers sometimes found limitations of better rootstock to fight against disease, and pests and acquire better nutrient uptake. So, it is imperative to get better rootstock to enhance the quality of citrus to reduce the susceptibility of the citrus growing zone to the onset of new diseases (Souza et al., 2010). To accomplish this, the seedlings generated should be innately healthy to ensure the availability of high-quality fruits to citrus cultivators (Koller, 2013).

Some fungi help plants, and their use should be encouraged. One such fungus is called mycorrhiza. Mycorrhizal fungi have a unique symbiotic relationship with citrus plants that is essential for citrus growth. It is critical for the setting up of symbiosis between plants and fungi because the plant offers energy to the fungus for its growth and maintenance through its photosynthates, while the fungus in return supplies nutrients and water to the plant to maintain their symbiosis (Wu et al., 2011; Ortas, 2012c; Jayme and Quigley, 2014). There are various forms of mycorrhizae fungi, the most common of which are endo and ectomycorrhiza. One of the most ubiquitous plant-root symbionts, arbuscular mycorrhizae (an endo-mycorrhizae), is present in over 90% of the plant species found on the earth's surface. The citrus plant is also included in it, which is associated with about 90% of all mycorrhizae fungi (Gadkar et al., 2001). Arbuscular mycorrhizae (AM) fungi form distinctive structures, such as arbuscules and vesicular in the root, to penetrate the living cells of plants without harming them. In arbuscular mycorrhizal fungi (AMF), structures responsible for the exchange of materials between plants and fungus are called arbuscules and vesicles, which accumulate glycogen and lipids. These structures serve as a repository for arbuscular mycorrhizal fungi, produced in the periods of low fungal and host activity, such as low temperature (Smith and Smith, 2011). It is reported by Ortas (2019) that citrus is a strongly mycorrhizal-dependent plant. Citrus trees have especially short roots and root hairs, which means they have less surface area for taking water and nutrients. It was noted that rootstocks with larger and surplus root hairs are less mycorrhizal dependent than plants with fewer or fewer root hairs (Wu et al., 2009). Citrus plants are reported as soil and environment-selective plants. It is primarily grown in light and well-drained soils. In this condition, the citrus plant gets more benefits if infected with mycorrhizae (Ortas, 2012c). Read and Fremont (1935) were the first group of scientists who realized the importance of mycorrhizae for the growth of citrus plants. It was reported that the host plant owns a factor that affects the interaction of mycorrhizal fungi and plants. Mycorrhiza in citrus roots helps to bring in extra water and nutrients that the roots can't manage on their own, for making a healthier tree. Therefore, it is particularly important to use a mycorrhizal inoculum for healthy and better seedling growth. Besides this, mycorrhiza can be very helpful in reducing the requirement of applied P fertilizers, as well as helping in maintaining the water status of citrus plants. A lack or uneven spread of fungus can lead to unhealthy and lackluster trees and fruit, which proves the positive effects of mycorrhizal fungi on the citrus plant. The soil's ability to contain mycorrhizal inoculum can influence plant root growth which is mostly dependent on the existence of compatible mycorrhizal partners (Ortas, 2019).

11.2 SPECIES DIVERSITY OF AMF IN CITRUS PLANTS/ORCHARD

Wu et al. (2009) reported that endo mycorrhizae (specifically arbuscular mycorrhiza; AM) have a vast host range and near about 150 species of arbuscular mycorrhizal fungi (AMF) colonize on 225,000 species of the host plant. Nemec et al. (1981) expressed that in citrus orchard soils, not only a single species of AMF is found, but there is a great diversity of AM fungi, and at a unit time, most of these species could colonize citrus. The most abundant mycorrhizae spore in the citrus root zone is the *Glomus* species (Davies and Albrigo, 1994). More than 80% of the gross AM fungal spores are of *Glomus occultum* and *Glomus* species (Bhattacharya, 1999; Fidelibus et al., 2000). Some other species found in the citrus plant rhizosphere are *Sclerocystis*, *Gigaspora*, and *Acaulospora* species (Nemec et al., 1981; Vinayak and Bagyaraj, 1990; de Souza et al., 2002). Sharma et al. (2009) separated and recognized 12 AMF species from the soil of citrus orchards maintained at nine different locations of India's north western Himalayan region (NWHR) and re-inoculated them. They reported a significant increase in growth parameters such as plant height, plant diameter, root length, and leaf area in seedlings inoculated with *G. fasciculatum* and *G. mosseae*. It has been analyzed that once a plant has been infected with mycorrhizae, it will carry out the mycorrhizal infection for an extended time. The diversity of arbuscular mycorrhizal fungi in the citrus rhizosphere is shown in Table 11.1.

11.3 BENEFITS OF MYCORRHIZA IN CITRUS

11.3.1 ENHANCE SEEDLING GROWTH OF CITRUS CROPS

Mycorrhizal inoculation enhances seedlings growth in citrus seedlings (Brasil, 2005; Johnson and Graham 2013; Back et al., 2016). According to Johnson and Graham (2013), the usage of arbuscular mycorrhizal fungi (AMF) is a tremendous assistance for citrus seedlings since it acts as an alternate approach to boost citrus seedlings by shortening the seedling growth span. This is particularly extremely helpful for greenhouse owners, to achieve a faster grafting stage, which can decrease costs and supply of seedlings to the market in a shorter span. Brazilian Ministry of Agriculture, Livestock and Supply, recommended that with the application

TABLE 11.1 Diversity of Arbuscular Mycorrhizal Fungi in the Citrus Rhizosphere.

Host plant	Genera of AMF	Species of AMF
Citrus reticulata Blanco	*Acaulospora*	*A. bireticulata*, *A. elegans*, *A. laevis*, *A. morrowiae*, *A. scrobiculata*, *A. spinosa*
	Gigaspora	*Gi. decipiens*, *Gi. gigantea*, *Gi. margarita*, *Gi. rosea*
	Glomus	*G. claroideum*, *G. constrictum*, *G. coremioides*, *G. etunicatum*, *G. fasciculatum*, *G. hoi*, *G. intraradices*, *G. invermium*, *G. macrocarpum*, *G. mosseae*, *G. versiforme*, *G. viscosum*, *G. sinuosum*
	Scutellospora	*S. aurigloba*, *S. coralloidea*, *S. nigra*, *S. pellucida*, *S. verrucosa*, *S. weresubiae*
Unknown citrus plants	*Glomus*	*G. constrictus*, *G. etunicatus*, *G. fasciculatus*, *G. macrocarpus*, *G. microcarpus*, *G. monosporus*, *G. mosseae*
	Sclerocystis	*Sc. sinuosa*
Newhall navel	*Acaulospora*	*A. koskei*
	Entrophospora	*E. schenckii*
	Gigaspora	*Gi. Albida*
	Glomus	*G. caledonium*, *G. claroideum*, *G. coronatum*, *G. geosporum*, *G. etunicatum*, *G. mosseae*, *G.* sp. 1, *G.* sp. 2
	Scutellospora	*Sc.* sp. 1
Nine citrus rootstocks	*Glomus*	*G. aggregatum*, *G. clarum*, *G. diaphanum*, *G. etunicatum*, *G. fasciculatum*, *G. intraradices*, *G. microaggregatum*, *G. monosporum*, *G. mosseae*
	Sclerocystis	*S. sinuosa*
C. unshiu Marc., *P. trifoliata* L. Raf.	*Acaulospora*	*A. brieticulata*, *A. paulinae*, *A. rehmii*, *A. spinosa*
	Entrophospora	*En. baltica*
	Glomus	*G. aggregatum*, *G. chimonobambusa*, *G. claroideum*, *G. diaphanum*, *G. etunicatum*, *G. geosporum*, *G. intraradices*, *G. tenebrosum*, *G. tortuosum*
	Pacispora	*P. robigina*
	Scutellospora	*Sc. nigra*

Source: Reproduced with slight modification and permission from Qiang-Sheng et al. (2013), Copyright © Elsevier (2013).

of AMF inoculation, seedlings can be produced within 18–22 months (Brasil, 2005). Ortas (2012c) also reported that citrus seedlings colonized with AMF have better quality. So, its demand as seedlings among growers is also remarkably high. Tong et al. (2006) reported that inoculation of several mycorrhizae to citrus plants improves fibrous root growth in it and root and shoot dry weight also increased in comparison to their controls without inoculation of mycorrhiza. Ortas (2019) found in his experiment that citrus seedlings are mycorrhizal dependent; hence, in the condition of infertile and low fertile soils (absence of mycorrhizal fungi), plant development was hindered and seedlings didn't develop appropriately.

As nutrient uptake is a function of root architecture and is profoundly reliant on the root surface area, it is critical to choose the rootstocks that form a reasonable root framework for a particular soil (Ortas, 2012c). The use of AMF in citrus crops and other horticultural crops also permits the creation of better, overwhelming, and uniform rootstocks, with a remarkable survival index and better initial functioning and establishment in the field (Nunes et al., 2008). AMF is mutualistically linked with plants, causing enhancements to their nutritional status and a decrease in bad impacts of biotic (pests and diseases) or abiotic (nutritional imbalance, drought, and flood) stresses, which leads to the robustness of citrus seedlings. Back et al. (2016) reported that the higher the number of AMF, the more significant would-be seedlings growth and vegetative growth of rootstock. They further explained through their results that it is due to a more significant symbiotic activity between fungus and plants. There was no significant vegetative growth of rootstocks without AMF inoculation (Back et al., 2016). They further have shown that different rootstocks behave differently for AMF inoculation for seedling's growth and development with respect to height, stem diameter, leaf number, leaf area, and root and shoot dry mass (Back et al., 2016). They took two rootstocks {citrange "Fepagro C37 Reck" (*P. trifoliata* (L.) Raf. × *C. sinensis* (L.) Osbeck.) and "Kumquat tree" (*Fortunella hindsii* (L.) Swingle)} and inoculated them by several AMF species (such as, *Scutellospora heterogama*, *Gigaspora margarita*, *Glomus etunicatum*, and *Acaulospora* sp.). They interestingly noted that in the presence of AMF, the "Fepagro C37 Reck" showed preferable vegetative advancement over that of the kumquat tree. It means seedling growth of citrus plants is generally a characteristic which is dependent on the genotype of the rootstock. Li et al. (2013) found in their experiment that arbuscular mycorrhizal fungi of the *Diversispora spurca* species had a better impact on the growth, root framework engineering, and chlorophyll content of the kumquat tree (*Fortunella margarita*). Thus, the kumquat tree can respond to mycorrhization effectively; however, the AMF species is significant for its constructive outcome. These two experiments clearly showed that citrus species are very much dependent for their seedling's growth and vigor on AMF inoculation (Back et al., 2016), and it is also highly dependent on genotypes of rootstocks (Back et al., 2016; Li et al., 2016) and species of AMF (Ortas, 2012c; Back et al., 2016). Ortas (2012c) also mentioned a remarkable point that mycorrhizal fungi have the potential to boost seedling survival rates, plant growth rates, and the number of blooms produced by plant species.

11.3.2 ENHANCE WATER STATUS FOR CITRUS PLANTS

Citrus is the second most important horticultural commodity in the world, after bananas, in terms of global trade volume. Plants such as citrus are extremely vulnerable to drought stress. It was demonstrated that drought stress has a significant negative impact on citrus development and fruit quality, particularly in the tropics (Levy et al., 1979). Despite this, it is primarily cultivated in areas that are subjected to a lot of water stress, such as the desert. It has been shown that AMF improves the water relations of citrus plants. Several researchers have shown that mycorrhizal inoculation is not only helpful for the mineral nutrition of citrus plants but also has a considerable positive impact on the uptake and usage of water (Levy et al., 1983; Wu and Xia, 2004). Based on their findings, we can conclude that mycorrhizal soil inoculation can also help citrus trees manage their water relations as well as their carbohydrate metabolism, according to Srivastava et al. (2002).

Better plant growth under moisture stress was achieved indirectly because of AM fungal colonization. This was accomplished by altering soil moisture retention through the influence of glomalin on water-stable aggregates in the soil (Ortas, 2012c). Experimental evidence has proved that mycorrhizal fungus hyphae improve the physiology of plants by increasing the intake and transfer of water to the plants on which they colonize. Wu and Xia (2004) observed an increase in plant growth parameters such as plant height, stem diameter, leaf area, shoot dry weight, root dry weight, and plant dry weight when arbuscular mycorrhizal fungi were inoculated into soils with water contents of 20%, 16%, and 12%, respectively. Almost all of the investigations conducted on the relationship between mycorrhizae and citrus indicated that mycorrhiza-infected plants have higher stomatal conductivity than uninfected plants.

11.3.3 RESISTANCE AGAINST STRESS CONDITIONS

In addition, mycorrhizal inoculation improves plant resistance to abiotic stimuli such as salt, dehydration, and high temperatures, among other things (Ortas, 2012c; Wu et al., 2013). AMF enhances the activity of soil enzymes, such as dehydrogenase, phosphatase, and urease, as well as the activity of soil phosphatase, as AMF colonization increases in number. Furthermore, these enzymes are particularly beneficial for plants when they are forced to adapt to adverse environments.

11.3.4 ENHANCED NUTRIENT UPTAKE OF CITRUS PLANTS

Mycorrhizae are believed to have a critical role in nutrient uptake, according to numerous studies (Ortas, 2012b; Ortas and Ustuner, 2014; Wu et al., 2017a). Mycorrhizae can produce plant growth hormones that can alter the branching pattern of a plant's root system. The differences in nutrient uptake between plant species are correlated with root growth and architecture of plants, as well as the interaction of plant roots with AM fungal colonization (Marschner, 1995). It was discovered that AM species responded differently to different citrus cultivars for nutrient intake. This was particularly true for the less mobile nutrients such as phosphorus, zinc, copper, and ammonium ions. Mycorrhizae were found to boost nitrogen uptake by host plants in soils in arid or semi-arid conditions (Ortas, 2012c). When phosphorus is limited in the soil, it has been discovered that AMF increases the uptake of phosphorus by plants, resulting in improved plant development. Under field conditions, it was projected that inoculation with AM fungus might result in a reduction of 80% of the recommended phosphate fertilizer requirement, and thus save the requirement of chemical fertilizers (Jakobsen, 1995). The AM inoculation can also help to maintain citrus yield and quality even when only a small amount of nutrients is applied. Some agrarian regions have soils that are deficient in phosphorus nutrition, and some areas have residual phosphorus accumulation. This is also noteworthy that the availability of P is extremely low under high soil pH and clay content situations and this availability is further constrained by the formation of calcium phosphate. Ortas (2012a) investigated the impact of mycorrhizal infection on the amounts of phosphorus and zinc in plants. Mycorrhiza-infected plants exhibited higher nutritional content than non-inoculated plants, particularly in terms of P and Zn, and this discovery was most clearly demonstrated under fumigated soil conditions, according to the researcher.

Supposedly, in Ca-P-rich soils, mycorrhizal infection, root growth, and root zone acidification brought about by NH_4^+ enrichment are the key factors determining the amount of P available in the rhizosphere (Ortas and Rowell, 2004). Since peasants are increasing their use of P fertilizers without considering the results of soil and plant analyses, a significant amount of P is accumulating in the soil. Moreover, it has been stated that a substantial application of P fertilizer might prevent the establishment of infections and the production of spores of AMF. Large inputs of soluble P (for instance, with the application of superphosphates) may reduce or eliminate the benefits of mycorrhizal fungi by suppressing the development and action of the vegetative mycelium in the root zone of the plant (Abbott and Robson, 1984).

In his research, Marschner (1995) discovered that, as a biological entity, AMF boosts the nutrient uptake efficiency of responsive plants from their natural sources, even in soils that are deficient in nutrients. Many researchers have observed that when methyl bromide was employed in citrus orchards in the United States to eradicate soil-borne organisms, a nutritional deficit occurred, which may have occurred as a result of the eradication of mycorrhizae (Kleinchmidt and Gerdemann, 1972). Since viable indigenous mycorrhizae were eliminated because of the harmful effects of soil fumigation, plants became stunted (Menge et al., 1978; Timmer and Leyden, 1978). Citrus output and quality can be maintained even with reduced nitrogen inputs if the plant is infected with AM (Ortas, 2012c). Citrus plants reap the greatest advantage from mycorrhizal fungi, exhibiting significant uptake of P, Zn (Ortas et al., 2016; Navarro and Morte, 2019), as well as Mg and Fe (Navarro and Morte, 2019). In an experiment by Ortas et al. (2015), citrus plants were grown inoculated with eight distinct mycorrhizal species under three different soils and growth conditions. The plants were then examined after 12 months. According to their findings, mycorrhizal species considerably raised the P (percentage) concentration of the plants in all three soils studied. Similarly, Wu and Zou

(2009b) revealed that a single AMF infection dramatically increased the P content of seedlings, leaves, and roots when compared to the same that did not have an AMF infection. Ortas et al. (2002a) investigated the influence of phosphorus and zinc levels on citrus development after mycorrhizal inoculation. The presence of exotic and/or indigenous native mycorrhizae in the soil has a significant contribution to the growth of citrus seedlings when there is a limited supply of P and Zn in the soil (Ortas et al., 2016). Ortas et al. (2016) also concluded that plants infected with mycorrhizae had greater Zn concentrations on average when compared to control plants. Inoculation with mycorrhizae may be used to successfully raise the P and Zn concentrations in citrus seedlings (Ortas et al., 2015).

As a result of his investigation, Ortas (2019b) discovered that the AM fungus is essential for the growth of citrus trees in terms of P and Zn nutrition. A factorial experiment was designed and started to investigate the effects of mycorrhizal fungi, P, and Zn dosages, as well as two soil types on the growth and nutrient uptake of the sour orange plant under greenhouse circumstances, and the results were promising. Ortas (2019b) discovered that mycorrhizal inoculation of citrus seedlings resulted in an increase in some essential growth metrics, including shoot height, shoot, and root dry matter, as well as root infection. Ortas (2019b) discovered that mycorrhizae-inoculated seedlings outperformed their control counterparts in terms of growth. Most significantly, Ortas (2019b) discovered that mycorrhizal inoculation resulted in a considerable increase in Zn and P uptake in plant tissue. *Rhizoglomus clarus* inoculated citrus plants showed an increase in P concentration of up to 83% with only a small amount of P application, as well as an increase in Zn concentration of up to 54%. Surprisingly, it was also shown that in the case of a high quantity of P application, the increase in P and Zn concentrations was inhibited (Ortas, 2019a). Antunes and Cardoso (1991) and Jifon et al. (2002) found that infection with an AM fungus (*G. intraradices*) reduced the growth of *Citrus aurantium* seedlings grown in soil with a high P supply, which is in consensus with previous findings. Antunes et al. (1991) infected their citrus plants with VAM and collected the plants six months after the inoculation to measure shoot dry weight, P and K uptake, root colonization, and the number of spores in 50 cm^3 of soil volume. They demonstrated through their studies that VAM fungus infection resulted in a considerable increase in dry matter yields as well as P and K contents, at both zero and 50 ppm soluble P levels, as well as at all rock phosphate levels.

However, when they used a higher concentration of phosphorus, they discovered contradictory outcomes. At 100 parts per million of soluble P, the improvement in growth parameters of VAM plants was comparable to that of non-VAM plants, and at 200 parts per million of soluble P, the development of the plant was much less than that of non-VAM plants. They demonstrated that greater P concentrations resulted in lower root colonization and spore development. They proposed a theory, stating that the growth depression observed in VAM-infected plants at higher P levels was most probably caused by the toxicity of P. It was also hypothesized that elevated leaf P and K concentrations may have interfered with the carbohydrate distribution and usage pattern of these symbiotic organisms. Several studies have found that mycorrhizal plants have larger Cu and B contents (Timmer and Leyden, 1980; Rocha et al., 1994; Marschner and Dell, 1994) than non-mycorrhizal plants (Timmer and Leyden, 1980; Marschner and Dell, 1994). When grown under a controlled greenhouse, the extraradical mycelium of AM fungi can provide 80% of P, 25–250% of N, 10% of K, 70% of Ca, 25–67% of Zn, and 60% of Cu (Marschner and Dell, 1994). It is extremely beneficial for plants because it allows them to receive nutrition at a low cost while also being environmentally friendly.

It was discovered by Wang et al. (2009) that arbuscular mycorrhizal fungus might influence the activation of mineral elements and increase the available iron concentrations in soil by altering the form of iron present in the soil. Using the trifoliate orange [*Poncirus trifoliata* (L.) Raf.] as a model, Wang and Xia (2009) conducted an experiment to determine the effect of the arbuscular mycorrhizal fungus (*Glomus versiforme*) on the growth and iron uptake of the plant at various pH levels of the nutrient solution. *G. versiforme* colonization resulted in considerable improvements in plant height, stem diameter, leaf number, and dry mass. According to their findings, the mycorrhizal fungus considerably increased the accumulation of chlorophyll, active iron, total iron, and root Fe chelate reductase activity when grown in the presence of nitrogen. Nutrient deficiency will occur in plants if the rootstocks' root systems are not able to efficiently absorb nutrients or if there is less mycorrhizal infection (in general, heavy fungicide application and phosphate fertilizer application kill mycorrhiza), and the plants will show signs of nutritional deficiency (Ortas, 2012c). It was discovered by Treeby (1992) that mycorrhizal fungus can boost the availability of iron to the host plant in acidic soil, but this was not the case in alkaline soil.

Menge et al. (1978) demonstrated that the concentrations of zinc, copper, and manganese were positively altered when the plant's roots were infected with mycorrhizal fungi. Using citrus tangerine leaves as a model, Youpensuk et al. (2008) revealed that AM fungus enhanced the amounts of P and Mg in the leaves. Plants injected with mycorrhizae showed a gradual improvement in the dry-weight ratio of their shoots and roots due to increased P and Zn availability (Ortas et al., 2002a). They also discovered that the total P and Zn content of the plant grew greater in mycorrhizae-treated plants when compared to non-treated plants when P and Zn were supplied externally to the plants. They discovered that the contribution of mycorrhizae to this was significantly higher when compared to the application of fertilizers for the same purpose.

The results of numerous studies on citrus seedlings inoculated with AMF inoculation have demonstrated a statistically significant increase in total dry weight, P and K contents of roots and leaves of the seedlings, as well as Ca, Mg, Fe, Cu, and Mn contents of roots and leaves of the seedlings when compared to non-AMF inoculated control citrus plants (Srivastava et al., 2002; Wu and Zou, 2009b; Wu and Zou, 2009c).

In general, tilling-less soils increase mycorrhizal activity, which in turn promotes nutrient uptake by plants and helps in the improvement of soil quality in the long run. It was reported that tillage decreased gross AMF colonization, spore density, hyphal length and density, and phosphatase activity in the soil (Wang et al., 2011).

11.3.5 ENHANCE SOIL FERTILITY

Mycorrhizal hyphae can improve soil aggregation and thereby boost soil fertility. In poor fertility and coarse-textured soil situations, mycorrhizae were found to be quite viable (Ortas, 2012c). In less fertile soil conditions or with slow-release fertilizer application, mycorrhizal colonization may aid the plant in nutrient dissolution in the soil. Graham and Syvertsen (1985) stated that under low-supplement soil conditions, mycorrhizae will generally flourish and benefit their host plants only when nutrient availability is low-to-moderate. AMF performs a significant function in the production and stability of soil aggregates, along with the physical fertility and quality of soil (Wright and Upadhyaya, 1998) by creating glomalin, which is assessed in the soil as a glomalin-related soil protein. The summary of the effect of AMF is given in Figure 11.1.

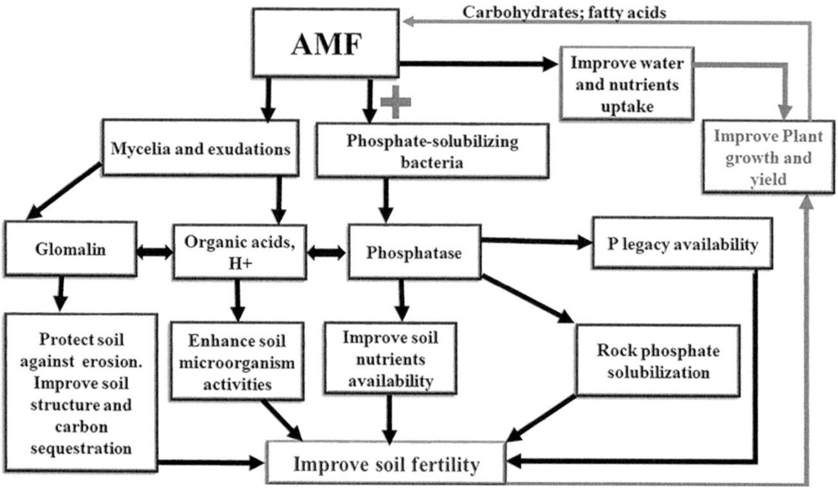

FIGURE 11.1 Effects of arbuscular mycorrhizal fungi on improving soil fertility.

Source: Adapted from Fall et al. (2022), https://creativecommons.org/licenses/by/4.0/

11.3.6 ENHANCE RESISTANCE TO SOIL DISEASES

Mycorrhizae not only aid in the absorption of nutrients and water by plants, but they also serve to strengthen the plant's tolerance against plant disease and harmful organisms (Ortas, 2012c). Graham et al. (1986) showed that citrus plants that have been inoculated with mycorrhizal fungi have increased tolerance to diseases. The release of antibiotics into the soil by mycorrhizae can also aid in the management of pests such as pathogenic fungi and nematodes, as well as the reduction of the risk of infection in humans. It is noteworthy that the occurrence of mycorrhizal fungi in the plant root framework recuperates the overall well-being of the plant which leads to a reduction in the likelihood of pest damage (Ortas, 2012c). Mycorrhizal inoculation can be used to protect host plants from soil-borne diseases, which can be harmful to them. The utilization of AM fungus boosted the resilience of plants to disease as well as infections in the soil and roots.

11.3.7 HIGH QUALITY OF FRUITS

AMF infection in citrus plants improves the photosynthetic efficiency of plants along with the overall health of citrus plants which leads to a better quality of fruits.

11.3.8 DECOMPOSITION OF ORGANIC MATTER

AMF was found to be engaged in the decomposition of organic matter through the formation of organic acids and other substances (MacFall, 1994) and the recycling of mineral nutrients in soils.

11.3.9 MAKE CITRUS PLANTS TOLERANT AGAINST VARIOUS STRESS

AMF hardens citrus plants against drought, heat, and salinity stress by enhancing plant water status. Wu and Zou (2009a) discovered that mycorrhizae inoculation increases root concentrations of K^+, Ca^{2+}, and Mg^{2+} at all salinity levels. They also stated that mycorrhizal inoculation appears to have the potential to improve citrus salt tolerance. The complex network of mechanisms mediated by AMF to alleviate drought stress symptoms in plants is shown in Figures 11.2 and 11.3.

11.4 MECHANISM OF NUTRIENT UPTAKE BY MYCORRHIZAE

To help plants absorb nutrients, mycorrhizal fungi collaborate with them by expanding the mycorrhizal hyphae network beyond the plant's root zone. Mycorrhizal inoculation changes the rooting pattern of plants, and experiments have revealed that the nutrient absorption capacity of inoculated roots is significantly greater than that of non-inoculated roots. AM fungi were found to be obligatory biotrophs, which means that they require green plants for carbohydrates accessible since they can't absorb carbohydrates from any source other than the plant cells themselves (Ortas and Rafique, 2017). In exchange for nutrients, mycorrhizal fungi receive approximately 20% of the photosynthesized carbon from the plant (Parniske, 2008). Furthermore, vitamin C stimulates nutrient absorption as

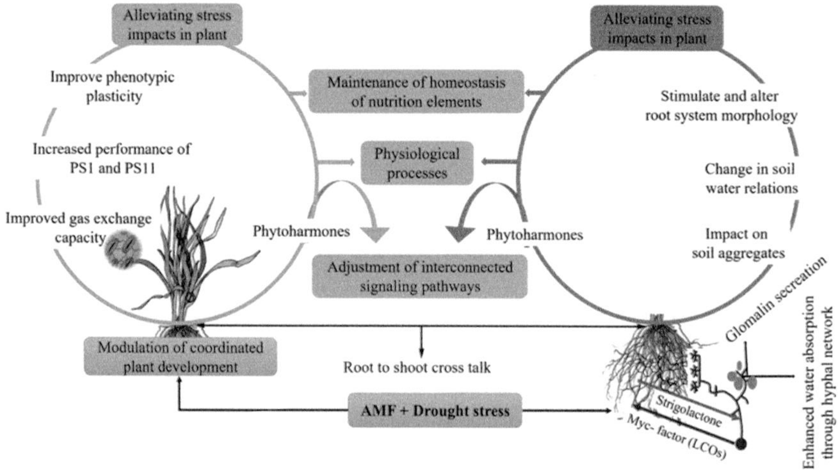

FIGURE 11.2 Complex network of mechanisms mediated by AMF to alleviate drought stress symptoms in plants. Through the establishment of mycorrhizal symbiosis and/or exudation of specific compounds such as strigolactones, plants adapt different strategies to alleviate deleterious effects of stress. This plant AMF association alters root physiology and plant growth by acting on different physiological, biochemical, and molecular mechanisms that essentially enhance water and nutrient uptake. Altered root physiology impacts root to shoot cross talk and contribute to the maintenance of homeostasis water, hormones, and ions in the plant. This alleviates the drought-induced growth impairment, and in turn, plants adapt their phenotype according to stress conditions by regulating the expression of stress-related genes and proteins. AMF symbiosis helps maintain the plant in water uptake, producing a large amount of external mycelium and increasing the effective root surface area by fungal hyphae or by increasing the lateral root formation. The exchange of water and nutrients between the two partners takes place through the arbuscules formed within the root cells of the host plant.

Source: Adapted from Ali et al. (2019), https://creativecommons.org/licenses/by/4.0/

well as translocation. Because the AM fungal hyphae have been stretched beyond the approachability of the roots, they can carry immobile nutrients (phosphorus, copper, zinc, and partially nitrate) in the soil that is not acquired by non-inoculated plants (Ortas, 2003; Marschner, 2012). Aside from that, AM fungi are responsible for the uptake and transfer of nutrients in barren soils (macro and micro). Because their hyphal diameter spans from 2 to 20 m and they have the capability of taking nutrients from an area of 25 cm^2 surrounding their roots for translocation, AM fungi are versatile and adaptable to a wide range of climatic circumstances (Jansa et al., 2003; Munkvold et al., 2004). It is not only that AM fungi improve nutrient transport into the plant, but they can also save their host plants from the toxicity of heavy metals by lowering the uptake of heavy metals in heavily polluted soils (Ortas and Rafique, 2017). Nutrients are taken up and transported fast by extraradical mycelium (ERM) to the intraradical mycelium (IRM), this is known as the MP (mycorrhizal uptake pathway). This process can take place up to several centimeters away from the plant cell, forcing the plant cell to respond through the interfacial apoplast (apical membrane). Ammonium (NH_4^+) transporters as well as orthophosphate (Pi) transporters are found in the perifungal membrane and are expressed in the AM fungus-infected roots (Bucher, 2007; Guether et al., 2009; Smith and Smith, 2011). The DP (direct uptake pathway) absorbs nutrients from the rhizosphere (immediate vicinity of roots) into the root epidermal and hair cells, while the MP (mycorrhizal

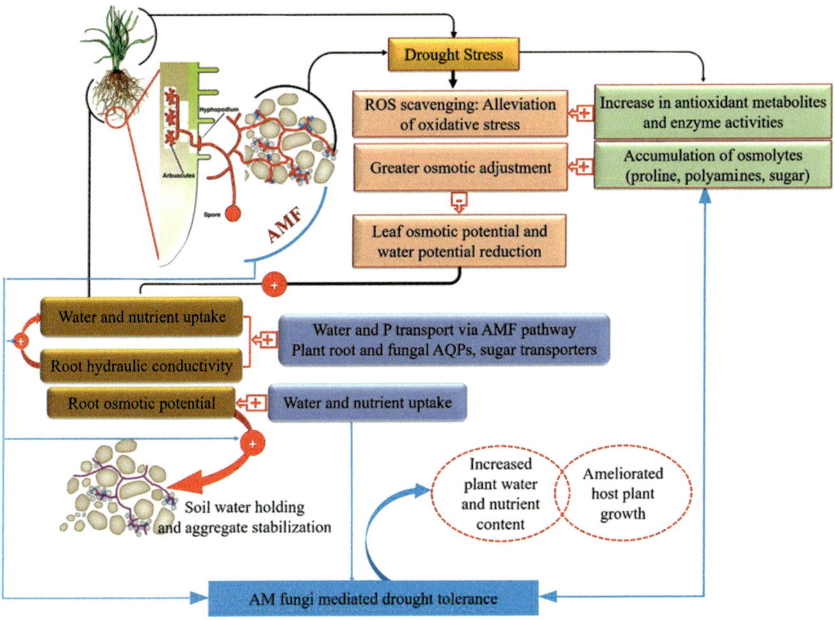

FIGURE 11.3 Role of AMF in drought tolerance. Arbuscular mycorrhizal fungi (AMF) symbiosis helps plants to maintain and regulate different processes in plants to cope with deleterious effects of drought stress, through either direct or indirect interaction, on plant growth performance. "+" and "–" symbols indicate an increase and a decrease in the production and accumulation of specific compounds.

Source: Adapted from Ali et al. (2019), https://creativecommons.org/licenses/by/4.0/

transport pathway) transports mycorrhizal hyphae that reach out far away from the rhizosphere can transport nutrients to cortical cells significantly through the MP (as shown in Figure 11.3), which is a highly regulated and rapid transit system of delivering nutrients.

Nutritional transport occurs through two pathways: the direct (plant) uptake pathway (DP) and the mycorrhizal uptake pathway (MUP) (MP). Alternatively, plants can absorb nutrients through transporters that are situated in the epidermis or root hairs (yellow colored), or through the MP, which includes uptake of nutrients by fungal transporters in the extraradical mycelium (red or green colored), the transport of nutrients through the hyphae from the ERM to the IRM (see mycorrhizal interface), and the uptake of nutrients from the mycorrhizal interface by mycorrhiza-inducible (orange colored). The colonization of a single host root by numerous fungal species, each of which differs in the efficiency with which they are capable to withdraw nutrients from the soil and pass these nutrients to their host (Bücking and Kafle, 2015), is shown by the presence of red and green fungal structures in Figure 11.4.

11.5 UNEVEN GROWTH OF CITRUS FRUIT AND MYCORRHIZAL FUNGI EFFECTS ON CITRUS

Mycorrhizal fungi grow in the soil and attach themselves to tree roots, where they thrive and spread. Unfortunately, a single mycorrhiza spore on the tree's roots is not enough to make a difference. The fungus must be directly attached to the root to get its benefits. Because of this, fungus growing on just one section of roots may result in uneven growth of citrus fruit, with the fruit on some branches bigger, healthier, and brighter (differing color) than on other branches of the same tree. If we notice uneven growth of citrus fruit, it may be caused by an uneven spread of mycorrhizal fungi on the roots. If this is the case, or if the citrus tree just seems to be failing in fruiting, the application of mycorrhizal fungi fertilizer to the soil may be beneficial. This fertilizer may be an inoculum, a small collection of spores that attach to the roots and grow into the beneficial fungus. Application of a lot of inoculums to many sites will result in their growth and spread, but the process is slow. Good coverage of mycorrhizal fungi will perk the plants in terms of their growth and nutrient uptake.

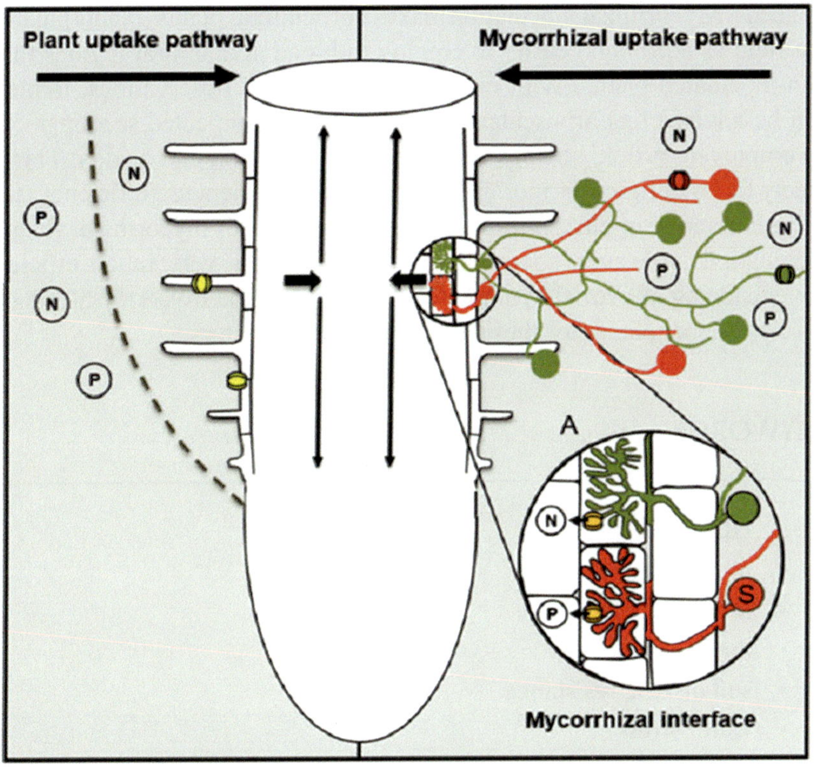

FIGURE 11.4 Plant uptake and mycorrhizal uptake pathway. Plants can take up nutrients by transporters that are situated in epidermis or root hairs (yellow colored) (called DP—direct uptake pathway) or via the mycorrhizal uptake pathway (MP) that comprises the uptake of nutrients by fungal transporters in the extra radical mycelium (red or green colored), the transport through the hyphae from the ERM to the IRM (mycorrhizal interface), and the uptake from the mycorrhizal interface by mycorrhiza inducible plant transporters in the periarbuscular membrane (orange colored). Indicated by the red and green fungal structures is the colonization of one host root by multiple fungal species that can differ in their efficiency with which they are capable to take up nutrients from the soil and transfer these nutrients to their host.

Source: Adopted from Bücking and Kafle (2015), https://creativecommons.org/licenses/by/4.0/

11.6 CONCLUSION

Soil and plant management practices have a significant role in the development of indigenous mycorrhizae in field crops that are propagated by seeds. Mycorrhizal inoculation makes horticultural plants (plants put into the field as seedlings) easier to employ in broad agricultural regions than non-inoculated plants. With extensive research and observations, farmers can be advised that arbuscular mycorrhizal fungus-injected seedlings can be employed in field settings for increased production and quality. This is nicely fit for citrus crops too. Yield, fruit quality, nutrient use efficiency, and stress tolerance in citrus crops can be enhanced with mycorrhizal fungus inoculation. However, some mycorrhizal fungus is very much efficient for enhancing above said parameters, but most of the mycorrhizal fungus species are beneficial for these yield determining factors.

KEYWORDS

- **citrus**
- **inoculation**
- **mycorrhizal fungus**
- **nutrient uptake**
- **soil disease resistance**
- **water stress**

REFERENCES

Abbott, L. K.; Robson, A. D. Colonization of the Root-System of Subterranean Clover by 3 Species of Vesicular Arbuscular Mycorrhizal Fungi. *New Phytol.* **1984,** *96* (2), 275–281.

Antunes, V.; Cardoso, E. J. B. N. Growth and Nutrient Status of Citrus Plants as Influenced by Mycorrhiza and Phosphorus Application. *Plant Soil* **1991,** *131*, 11–19.

Back, M. M.; Altmann, T.; de Souza, P. V. D. Influence of Arbuscular Mycorrhizal Fungi on the Vegetative Development of Citrus Rootstocks. *Pesq. Agropec. Trop. Goiânia.* **2016,** *46* (4), 407–412.

Bahadur, A.; Batool, A.; Nasir, F.; Jiang, S.; Mingsen, Q.; Zhang, Q.; Pan, J.; Liu, Y.; Feng, H. Mechanistic Insights into Arbuscular Mycorrhizal Fungi-Mediated Drought Stress Tolerance in Plants. *Int. J. Mol. Sci.* **2019,** *20* (17), 4199. doi: 10.3390/ijms20174199.

Bhattacharya, P. Use of Biofertilizers in Citrus. In *Citriculture*; Singh, S., Ed.; National Research Centre for Citrus: Nagpur, India, 1999; pp 194–204.

Brasil Ministério da Agricultura, Pecuária e Abastecimento. Instrução Normativa n° 24 de 16 de dezembro de 2005. Normas para produção, comercialização e utilização de mudas. Diário Oficial da União, Brasília, DF, n. 243, 20 dez. 2005. Seção 1., 2005

Bucher, M. Functional Biology of Plant Phosphate Uptake at Root and Mycorrhiza Interfaces. *New Phytol.* **2007,** *173*, 11–26.

Bücking, H.; Kafle, A. Role of Arbuscular Mycorrhizal Fungi in the Nitrogen Uptake of Plants: Current Knowledge and Research Gaps. *Agronomy* **2015,** *5*, 587–612.

Davies, F. S.; Albrigo, L. G. Citrus. CAB International: Wallingford, UK, 1994; p 254.

de Souza, P. V. D.; Schmitz, J. A. K.; de Freitas, R. S.; Carniel, E.; Carrenho, R. Identification and Quantification of Native Arbuscular Mycorrhizae Fungi of Citrus in the State of Rio Grande do Sul, Brazil. *Pesquisa Agropecuaria Brasileira* **2002,** *37* (4), 553–558.

Fall, A. F.; Nakabonge, G.; Sekandi, J.; Founoune-Mboup, H.; Apori, S. O.; Ndiaye, A.; Arfang, B.; Ngom, K. Roles of Arbuscular Mycorrhizal Fungi on Soil Fertility: Contribution in the Improvement of Physical, Chemical, and Biological Properties of the Soil. *Front. Fung. Biol.* **2022,** *3*, 723892. doi: 10.3389/ffunb.2022.723892

Fidelibus, M. W.; Martin, C. A.; Wright, G. C.; Stutz, J. C. Effect of Arbuscular Mycorrhizal (AM) Fungal Communities on Growth of 'Volkamer' Lemon in Continually Moist or Periodically Dry Soil. *Sci. Hortic.* **2000,** *84* (1-2), 127–140.

Gadkar, V.; David-Schwartz, R.; Kunik, T.; Kapulnik, Y. Arbuscular Mycorrhizal Fungal Colonization. Factors Involved in Host Recognition. *Plant Physiol.* **2001,** *127* (4), 1493–1499.

Graham, J. H.; Syvertsen, J. P. Host Determinants of Mycorrhizal Dependency of Citrus Rootstock Seedlings. *New Phytol.* **1985,** *101* (4), 667–676.

Graham, J. H.; Timmer, L. W.; Fardelmann, D. Toxicity of Fungicidal Copper in Soil to Citrus Seedlings and Vesicular-Arbuscular Mycorrhizal Fungi. *Phytopath.* **1986,** *76*, 66–70.

Guether, M.; Neuhauser, B.; Balestrini, R.; Dynowski, M.; Ludewig, U.; Bonfante, P. A Mycorrhizal-Specific Ammonium Transporter from *Lotus japonicus* Acquires Nitrogen Released by Arbuscular Mycorrhizal Fungi. *Plant Physiol.* **2009,** *150*, 73–83.

Jakobsen, I. Transport of Phosphorus and Carbon in VA Mycorrhizas. In *Mycorrhiza*; Varma, A., Hock, B., Eds.; Springer-Verlag: Berlin, 1995; pp 297–324.

Jansa, J.; Mozafar, A.; Frossard, E. Long-Distance Transport of P and Zn through the Hyphae of an Arbuscular Mycorrhizal Fungus in Symbiosis with Maize. *Agronomie* **2003,** *23*, 481–488.

Jayme, B.; Quigley, M. Influence of Arbuscular Mycorrhiza on Growth and Reproductive Response of Plants under Water Deficit: A Meta-Analysis. *Mycorrhiza* **2014,** *24* (2), 9–119.

Jifon, J. L.; Graham, J. H.; Drouillard, D. L.; Syvertsen, J. P. Growth Depression of Mycorrhizal Citrus Seedlings Grown at High Phosphorus Supply is Mitigated by Elevated CO_2. *New Phytol.* **2002,** *153* (1), 133–142.

Johnson, N. C.; Graham, J. H. The Continuum Concept Remains a Useful Framework for Studying Mycorrhizal Functioning. *Plant Soil* **2013,** *363* (1), 411–419.

Kleinchmidt, G. D.; Gerdemann, J. W. Stunting of Citrus Seedlings in Fumigated Nursery Soils Related to the Absence of Endomycorrhizae. *Phytopathology* **1972,** *62*, 1447–1453.

Koller, O. L. Citricultura Catarinense. Florianópolis: Epagri, 2013; p 319.

Levy, Y.; Shalhevet, J.; Bielorai, H. Effect of Irrigation Regime and Water Salinity on Grapefruit Quality. *J. Am. Soc. Hortic. Sci.* **1979,** *104* (3), 356–359.

Levy, Y.; Syvertsen, J. P.; Nemec, S. Effect of Drought Stress and Vesicular Arbuscular Mycorrhiza on Citrus Transpiration and Hydraulic Conductivity of Roots. *New Phytol.* **1983,** *93* (1), 61–66.

Li, X.; Zeng, R.; Liao, H., Improving Crop Nutrient Efficiency through Root Architecture Modifications. *J. Integr. Plant Biol.* **2016,** *58*, 193–202.

Li, Y.; Zou, Y. N.; Wu, Q. S. Effects of *Diversispora spurca* Inoculation on Growth, Root System Architecture and Chlorophyll Contents of Four Citrus Genotypes. *Int. J. Agric. Biol.* **2013,** *15* (2) 342–346.

MacFall, J. S. Effects of Ectomycorrhizae on Biogeochemistry and Soil Structure. In *Mycorrhizae and Plant Health*; Pfleger, F. L., Linderman, R. G., Eds.; APS Press: St. Paul, 1994; pp 213–237.

Marschner, H.; Dell, B. Nutrient-Uptake in Mycorrhizal Symbiosis. *Plant Soil* **1994,** *159* (1), 89–102.

Marschner, H. In *Mineral Nutrition of Higher Plants*; Academic Press: London, 1995; p 889.

Marschner, P. In *Marschner's Mineral Nutrition of Higher Plants*; Academic Press: London, 2012; p 672.

Menge, J. A.; Johnson, E. L. V.; Platt, R. G. Mycorrhizal Dependency of Several Citrus Cultivars under Three Nutrient Regimes. *New Phytol.* **1978,** *81* (3), 553–559.

Munkvold, L.; Kjøller, R.; Vestberg, M.; Rosendahl, S.; Jakobsen, I. High Functional Diversity within Species of Arbuscular Mycorrhizal Fungi. *New Phytol.* **2004,** *164*, 357–364.

Navarro, J. M.; Morte, A. Mycorrhizal Effectiveness in *Citrus macrophylla* at Low Phosphorus Fertilization. *J. Plant Physiol.* **2019,** *232*, 301–310.

Nemec, S.; Menge, J. A.; Platt, R. G.; Johnson, E. L. V. Vesicular-Arbuscular Mycorrhizal Fungi Associated with Citrus in Florida and California and Notes on their Distribution and Ecology. *Mycologia* **1981,** *73* (1), 112–127.

Nunes, J. L. S.; de Souza P. V. D.; Marodin G. A. B.; Fachinello J. C. Development Increase of 'Aldrighi' Peach Rootstocks by Indigenous Arbuscular Mycorrhizal Fungi. *Ciência e Agrotecnologia* **2008,** *32* (6), 1787–1793.

Ortas, I. Effect of Selected Mycorrhizal Inoculation on Phosphorus Sustainability in Sterile and Non-Sterile Soils in the Harran Plain in South Anatolia. *J. Plant Nutr.* **2003,** *26*, 1–17.

Ortas, I. Do Maize and Pepper Plants Depend on Mycorrhizae in Terms of Phosphorus and Zinc Uptake? *J. Plant Nutr.* **2012a,** *35*, 1639–1656. https://doi.org/10.1080/01904167.2012.698346.

Ortas, I. Mycorrhiza in Citrus: Growth and Nutrition. In *Advances in Citrus Nutrition*; Srivastava, A. K., Ed.; Springer-Verlag: The Netherlands, 2012b. https://doi.org/10.1007/978-94-007-41713_23.

Ortas, I. Mycorrhiza in Citrus: Growth and Nutrition. In *Advances in Citrus Nutrition*; Srivastava, A. K., Ed.; Springer: Amsterdam, 2012c; pp 333–351.

Ortas, I.; Akpinar, C.; Demirbas, A. Effect of Mycorrhizal Species on Growth and Nutrient Uptake by Seedlings of Citrus (*Citrus sinensis*) under Three Soil Growth Conditions. *Curr. Hortic.* **2015,** *3* (2), 61–64.

Ortas, I.; Akpinar, C.; Demirbas, A. Sour Orange (*Citrus aurantium* L.) Growth is Strongly Mycorrhizal Dependent in Terms of Phosphorus (P) Nutrition Rather than Zinc (Zn). *Commun. Soil Sci. Plant Anal.* **2016,** *47*, 2514–2527.

Ortas, I. Comparison of Indigenous and Selected Mycorrhiza in Terms of Growth Increases and Mycorrhizal Dependency of Sour Orange under Phosphorus and Zinc Deficient Soils. *Eur J Hortic. Sci.* **2019a,** *84* (4), 218–225.

Ortaş, İ. Effect of Mycorrhizal Inoculation on Citrus Seedling Growth and Nutrient Uptake. *Acta Hortic.* **2019b,** *1253*, 77–84.

Ortas, I.; Ortakci, D.; Kaya, Z. Various Mycorrhizal Fungi Propagated on Different Hosts have Different Effect on Citrus Growth and Nutrient Uptake. *Commun. Soil Sci. Plant Anal.* **2002a,** *35* (13-14), 1923–1944.

Ortaş, I.; Rafique, M. The Mechanisms of Nutrient Uptake by Arbuscular Mycorrhizae. In *Mycorrhiza-Nutrient Uptake, Biocontrol, Ecorestoration*; Varma, A., Prasad, R., Tuteja, N., Eds.; Springer Nature, 2017; pp 1–20.

Ortas, I.; Rowell, D. L. Effect of Ammonium and Nitrate on Indigenous Mycorrhizal Infection, Rhizosphere pH Change, and Phosphorus Uptake by Sorghum. *Commun. Soil Plant Anal.* **2004,** *33* (1-2), 259–272.

Ortas, I.; Ustuner, O. The Effects of Single Species, Dual Species and Indigenous Mycorrhiza Inoculation on Citrus Growth and Nutrient Uptake. *Eur. J. Soil Biol.* **2014,** *63*, 64–69.

Parniske, M. Arbuscular Mycorrhiza: The Mother of Plant Root Endosymbioses. *Nat. Rev. Microbiol.* **2008,** *6*, 763–775.

Qiang-Sheng, W.; Srivastava, A. K.; Ying-Ning, Z. AMF-Induced Tolerance to Drought Stress in Citrus: A Review. *Sci. Hortic.* **2013,** *164*, 77–87.

Read, H. S.; Fremont, T. Factors that Influence the Formation and Development of Mycorrhizal Associations in Citrus Roots. *Phytophatology* **1935,** *25*, 645–647.

Rocha, M. R. D.; Oliveira, E. D.; Correa, G. C. D.; Da-Rocha, M. R.; De-Oliveira, E. Effect of Phosphorus Doses and Mycorrhizal Fungi on Mineral Nutrition and Growth of 'Cleopatra' Mandarin (*Citrus reshni* Hort ex Tan.). *Pesquisa Agropecuaria Brasileira* **1994,** *29* (5), 725–731.

Sharma, S. D.; Kumar, P.; Singh, S. K.; Patel, V. B. Indigenous AM Fungi and *Azotobacter* Isolates, and their Screening from Citrus Seedlings at Different Levels of Inorganic Fertilizers Application. *Indian J. Hortic.* **2009,** *66* (2), 183–189.

Smith, S. E.; Smith, F. A. Roles of Arbuscular Mycorrhizas in Plant Nutrition and Growth: New Paradigms from Cellular to Ecosystem Scales. *Annu. Rev. Plant Biol.* **2011,** *62*, 227–250.

Souza, E. L. S.; Schwarz, S. F.; Oliveira, R. P. Porta-enxertos para os citros no Rio Grande do Sul. In *Indicações técnicas para a citricultura do Rio Grande do Sul*; Fepagro: Porto Alegre, 2010; pp 19–29.

Srivastava, A. K.; Singh, S.; Marathe, R. A. Organic Citrus: Soil Fertility and Plant Nutrition. *J. Sustain. Agric.* **2002,** *19* (3), 5–29.

Timmer, L. W.; Leyden, R. F. Stunting of Citrus Seedlings in Fumigated Soils in Texas and its Correction by Phosphorus Fertilization and Inoculation with Mycorrhizal Fungi. *J. Am. Soc. Hortic. Sci.* **1978,** *103* (4), 533–537.

Timmer, L. W.; Leyden, R. F. The Relationship of Mycorrhizal Infection to Phosphorus-Induced Copper Deficiency in Sour Orange Seedlings. *New Phytol.* **1980,** *85* (1), 15–23.

Tong, R.; Yang, X.; Li, D. Effects of Interspecies Difference of Arbuscular Mycorrhizal Fungi on *Citrus grandis* cv. Changshou Shatian Seedlings Vegetative Growth and Mineral Contents. *Yingyong Shengtai Xuebao* **2006,** *17* (7), 1229–1233.

Treeby, M. T. The Role of Mycorrhizal Fungi and Nonmycorrhizal Microorganisms in Iron Nutrition of Citrus. *Soil Biol. Biochem.* **1992,** *24* (9), 857–864.

UNFAO. Citrus Fruit, Fresh and Processed: Statistical Bulletin. UN Food and Agriculture Organization, 2020. https://www.fao.org/3/cb6492en/cb6492en.pdf

USDA. Citrus: World Markets and Trade. US Department of Agriculture, [Online] 2022. https://downloads.usda.library.cornell.edu/usdaesmis/files/w66343603/bv73d549r/1v53m4335/citrus.pdf

Vinayak, K.; Bagyaraj, D. J. Vesicular-Arbuscular Mycorrhizae Screened for Troyer Citrange. *Biol. Fertil. Soils* **1990,** *9* (4), 311–314.

Wang, P.; Ji-Hong, L.; Ren-Xue, X.; Qiang-Sheng, W.; Ming-Yuan, W.; Tao, D. Arbuscular Mycorrhizal Development, Glomalin-Related Soil Protein (GRSP) Content, and Rhizospheric Phosphatase Activitiy in Citrus Orchards under Different Types of Soil Management. *J. Plant Nutr. Soil Sci.* **2011,** *174* (1), 65–72.

Wang, M.; Xia, R. Effects of Arbuscular Mycorrhizal Fungi on Growth and Iron Uptake of *Poncirus trifoliata* under Different pH. *Weishengwu Xuebao* **2009,** *49* (10), 1374–1379.

Wright, S. F.; Upadhyaya, A. A Survey of Soils for Aggregate Stability and Glomalin, a Glycoprotein Produced by Hyphae of Arbuscular Mycorrhizal Fungi. *Plant Soil* **1998,** *198* (1), 97–107.

Wu, Q. S.; Levy, Y.; Zou, Y. N. Arbuscular Mycorrhizae and Water Relations in Citrus. *Tree Forestry Sci. Biotechnol.* **2009,** *3* (S1), 105–112.

Wu, Q. S.; Srivastava, A. K.; Zou, Y. N. AMF-Induced Tolerance to Drought Stress in Citrus: A Review. *Sci. Hortic.* **2013,** *164*, 77–87. https://doi.org/10.1016/j.scienta.2013.09.010.

Wu, Q. S.; Srivastava, A. K.; Zou, Y. N.; Malhotra, S. K. Mycorrhizas in Citrus: Beyond Soil Fertility and Plant Nutrition. *Indian J. Agric. Sci.* **2017,** *87*, 427–443.

Wu, Q. S.; Xia, R. X. Effects of Arbuscular Mycorrhizal Fungi on Plant Growth and Osmotic Adjustment Matter Content of Trifoliate Orange Seedling Under Water Stress. *Zhiwu Shengli yu Fenzi Shengwuxue Xuebao* **2004,** *30* (5), 583–588.

Wu, Q. S.; Zou, Y. N. Arbuscular Mycorrhizal Symbiosis Improves Growth and Root Nutrient Status of Citrus Subjected to Salt Stress. *Scienceasia* **2009a,** *35* (4), 388–391.

Wu, Q. S.; Zou, Y. N. Mycorrhizal Influence on Nutrient Uptake of Citrus Exposed to Drought Stress. *Philipp. Agric. Sci.* **2009b,** *92* (1), 33–38.

Wu, Q. S.; Zou, Y. N. The Effect of Dual Application of Arbuscular Mycorrhizal Fungi and Polyamines upon Growth and Nutrient Uptake on Trifoliate Orange (*Poncirus trifoliata*) Seedlings. *Not. Bot. Horti Agrobot. Cluj-Na.* **2009c,** *37* (2), 95–98.

Wu, Q. S.; Zou, Y. N.; Heb, X. H. Differences of Hyphal and Soil Phosphatase Activities in Drought-Stress Endomycorrhizal Trifoliate Orange (*Poncirus trifoliata*) Seedlings. *Sci. Hortic.* **2011,** *129* (1), 294–298.

Youpensuk, S.; Lordkaew, S.; Rerkasem, B. Arbuscular Mycorrhizal Fungi Associated with Tangerine (*Citrus reticulata*) in Chiang Mai Province, Northern Thailand, and their Effects on the Host Plant. *Sci. Asia* **2008,** *34*, 259–264.

CHAPTER 12

COMBINED INTERACTION BETWEEN *AZOSPIRILLUM* SPECIES AND ARBUSCULAR MYCORRHIZA (AM) FUNGI ON AGRICULTURAL CROPS

P. SARANRAJ[1*], NEETHU ASOKAN[2], ABDEL RAHMAN MOHAMMAD AL–TAWAHA[3], P. SIVASAKTHIVELAN[4], HIBA ALATRASH[5], SIDDHARTHA PATI[6], ABDEL RAZZAQ AL-TAWAHA[7], AMREENA SULTAN[8], R. Z. SAYYED[9], DEVARAJAN THANGADURAI[10], JEYABALAN SANGEETHA[11], MUHAMMAD AASIM[12], EKATERINA KOZUHAROVA[13], MUNIR TURK[14], TAHA A. AL-ISSA[14], and SAMIA KHANUM[15]

[1]*Department of Microbiology, Sacred Heart College (Autonomous), Tirupattur, Tamil Nadu, India*

[2]*Department of Life Sciences, Sri Sathya Sai University for Human Excellence, Kalaburagi, Karnataka, India*

[3]*Department of Biological Sciences, Al-Hussein Bin Talal University, Maan, Jordan*

[4]*Department of Agricultural Microbiology, Faculty of Agriculture, Annamalai University, Chidambaram, Tamil Nadu, India*

[5]*University of Helsinki, Helsinki, Finland*

[6]*NatNov Bioscience Private Limited, Balasore, Odisha, India*

[7]*National Agricultural Research Center (NARC), Baq'a 19381, Jordan*

[8]*Division of Vegetable Science, SKUAST-K, Shalimar, Srinagar, Jammu and Kashmir, India*

Microbial Fertilizer Technology for Sustainable Crop Production. Jeyabalan Sangeetha, Abdel Rahman Mohammad Al-Tawaha, & Devarajan Thangadurai (Eds.)

DOI: 10.1201/9781003594826-12

[9]*Department of Microbiology, PSGVP Mandal's Arts, Science and Commerce College, Shahada, Maharashtra, India*

[10]*Department of Botany, Karnatak University, Dharwad, Karnataka, India*

[11]*Department of Environmental Science, Central University of Kerala, Kasaragod, Kerala, India*

[12]*Department of Plant Protection, Faculty of Agricultural Sciences and Technologies, Sivas University of Science and Technology, Sivas, Turkey*

[13]*Medical University of Sofia, Faculty of Pharmacy, Department of Pharmacognosy, Sofia, Bulgaria*

[14]*Department of Plant Production, Faculty of Agriculture, Jordan University of Science and Technology, Irbid, Jordan*

[15]*Department of Botany, University of the Punjab, Lahore, Pakistan*

**Corresponding author.*

ABSTRACT

Microorganisms in the rhizosphere of plants play an important role in boosting soil fertility and, as a result, in improving the overall health of the plant. Plant and AMF association has evolved over at least 500 million years, making it one of the biological world's oldest symbiotic relationships. Arbuscular mycorrhizal (AM) fungi are soil fungi that can be found worldwide. Their significance in natural and semi-natural ecosystems is widely acknowledged, and their impact can be seen in increased plant productivity and diversity, as well as increased plant resistance to biotic and abiotic stresses. On the other hand, *Azospirillum* is a plant growth-promoting bacterium that can help many different types of plants grow and make more food. The most common theory about how it can help plants grow is that it can make phytohormones that help roots grow, plants absorb water and minerals, and in many cases, make plants more productive. It has been demonstrated that a synergistic relationship between AMF and *Azospirillum* spp. can promote plant development when compared to a single inoculation with either of them. The recommendation is that dual inoculation is preferred over inoculation with either *Azospirillum* or AMF alone since it allows for improved nutrition, pathogen resistance, and abiotic stress reduction, all of which are achieved through increased nutritional uptake and reduced abiotic stress.

12.1 INTRODUCTION

Agriculture depends on fertilizers produced at the expense of fossil fuels susceptible to fluctuations in oil markets. Plants require adequate phosphorus, nitrogen, and other beneficial nutrients for good yield and productivity (Grant et al., 2001). However, the low efficiency of nitrogen and phosphorous fertilizers, pollution, acidification of soils, and emission of greenhouse gases are some economic and environmental factors that have caused to increase in the use of inorganic fertilizers. However, sustainable crop production requires the use of an integrated nutrient management system involving organic and inorganic sources of plant nutrients.

With the accelerating population, there occurs a need for plenty of food crops to meet the augmenting requirements. Crops require several nutrients to obtain maximum potential yield. Most of the plants grow by absorbing nutrients from the soil and their ability to do this depends on the nature and characteristics of the soil. Soil texture and acidity were known to determine the extent to which the nutrients were available to the plants. The nutrients, which are prerequisites for plants, are found to occur naturally in the soil; nevertheless, sometimes these are substituted with lime or fertilizers (Aseri et al., 2018).

The microbes dwelling in the soil are found to interact with plants predominantly in the rhizosphere and mainly in the conversion of innumerable nutrients in the soil. Nevertheless, the nutrients are usually immobilized sometimes fixed, and applied when they are indirectly available to the plants. Microbes are known to play a fundamental part in enabling the accessibility of nutrients. Among the chief nutrients, nitrogen and phosphorus are the most applied nutrients for microbial transformation (Alexander, 1961).

Arbuscular mycorrhizal (AM) fungi are symbiotic associations of fungi with almost all taxonomic groups of plants with the list of uninfected species fewer compared to the infected ones. These symbiotic associations are highly advantageous to plants in several ways, such as augmenting the nutrients (especially phosphorus and nitrogen) accessibility and availability, improving water uptake, inducing disease resistance and high yield (Lekberg and Koids, 2005).

Arbuscular mycorrhiza (AM) is one of the most-known mycorrhiza and is commonly termed as a universal plant symbiosis. The significant characteristics are the lack of host specificity compared to further types known. Previous studies on arbuscular mycorrhizal fungi conducted in the last few decades envisaged their occurrence in a wide variety of hosts and different habitats and were observed to possess variability in quality and quantity (Jalaluddin, 2005). These symbiotic fungi are typically geographically ubiquitous and are commonly found in association with agricultural crops, shrubs, tropical tree species, and some temperate trees. However, there are reports of specific members lacking arbuscular mycorrhiza, namely, the mustard family, cabbages, cauliflowers, and canola (Wright et al., 2002). It has been observed that their nutritional requirements are not specific. The fungi family in these associations are usually the nonseptate Zygomycetes and Phycomycetes fungi. Arbuscular mycorrhizal fungi are commonly known are *Glomus*, *Gigaspora*, *Acaulospora*, *Entrophospora*, and *Scutellospora*, of which *Glomus* is the most common fungus (Sureshkumar Singh et al., 2003; James et al., 2008).

AM fungi are soil fungi that are found all over the world. Their importance in natural and semi-natural ecosystems is widely recognized, and their impact is manifested in increased plant productivity and diversity, as well as increased plant resistance to biotic and abiotic stresses. Arbuscular mycorrhizal fungi are found in soils all over the world (Smith and Rea, 2008). In today's world, these are widely used in a variety of sectors including agriculture, horticulture, and forestry programs. They are also used for environmental reclamation, crop yield enhancement, and crop health improvement, as well as to limit the use of harmful agrochemicals (Gianinazzi et al., 2002; Johansson et al., 2004).

Bacteria that are known to inhabit the rhizosphere and plant roots thereby enhance plant growth by different mechanisms and are referred to as plant growth-promoting bacteria (PGPR). With the collective global concern for food and environmental quality, the use of PGPR for reducing chemical inputs in agriculture is possibly beneficial and feasible. PGPR is applied on several crops to boost their growth, seed emergence, and crop yield. Some of the PGPRs have also been known to be commercialized for agricultural purposes (Dey et al., 2004; Herman et al., 2008). These organisms such as *Azospirillum*, *Azotobacter*, *Pseudomonas,* and *Bacillus* have been shown to take possession of the roots of several plants and thereby increase the plant height, flower number, fruit number, and total weight of plants such as the tomato (Minorsky, 2008). The significant benefit of PGPR is that it has effective antibacterial properties against several plant pathogens and pests. PGPR was also reported to facilitate biological control indirectly by inducing systemic resistance against innumerous plant diseases (Jeliyanon and Kloepper, 2002).

Azospirillum is a free-living, plant growth-promoting bacterium, capable of promoting growth and yield of numerous plant species. The prime theory concerning its growth promotion capacity lies in its ability to produce various phytohormones that improve root growth, induce adsorption of water and minerals that benefit the yield, and in many cases, harvest more productive plants (Dobbelaere et al., 2003). It has also been reported to have agronomic and ecological significance.

Azospirillum genus covers free living nitrogen-fixing rhizosphere bacteria that belong to α-subclass of roteobacteria. *Spirillum lipoferum* was the first species reported to be isolated by Beijerinck (1925). However, *Azospirillum* was seen to be forgotten for half a century and in 1970 regained the reputation of being the most-studied bacterium. Dobereiner and Day (1976) have previously reported that the nitrogen-fixing *Azospirillum* is a common root and soil inhabitant. Currently, about seven species are described under the genus, namely, *Azospirillum lipoferum*, *Azospirillum brasilense*, *Azospirillum amazonense*, *Azospirillum halopraeferens*, *Azospirillum irakense*, *Azospirillum largimobile*, and *Azospirillum doebereinerae*.

Nitrogen fixation is known as one of the significant microbial processes or activities that transforms atmospheric nitrogen into ammonia and ultimately to proteins and amino acids. It was also observed that the free-living nitrogen-fixing bacteria and the associative nitrogen-fixing bacteria inhabit the rhizosphere and fix atmospheric nitrogen. Among the most-researched bacteria, *Azospirillum* was found to have significant potential for commercial applications and most commonly as biofertilizers (Bashan and Holguin, 1997). Likewise, AM symbiosis is known to play an important role in plant mineral nutrition and health with several applications in sustainable agricultural systems (Ferrol et al., 2002). Arbuscular mycorrhizal fungi were found to have the ability to interact with a wide assortment of microorganisms residing in the rhizosphere. AM fungi of the genus *Glomus* are well distributed in both cultivated and non-cultivated soil (Mukerji and Kapoor, 1990).

In most of the developing and developed countries, *Azospirillum* is used alone or combined with PGPB and AM fungi for different crops and is of choice. This has also been known to be a possible agent to resolve environmental problems. There are few previous reports on significant progress in the genetic origin of *Azospirillum*, other PGPBs, and their benefits on plants (Bloemberg and Lugtenberg, 2001). AM fungi are of great value in promoting the growth and yield of tomatoes of all varieties (*Lycopersicon esculentum* L.). There was an increase in the absorption of minerals, particularly immobile nutrients, from the soil by the host when colonized by mycorrhizae. *Azospirillum* is one of the beneficial bacterial strains that play a crucial role in fixing nitrogen in the roots by harnessing the atmospheric nitrogen thereby improving the fertility of soil (Siddiqui and Mahmood, 2001).

12.2 ARBUSCULAR MYCORRHIZA (AM)

Mycorrhiza literally means "fungus root" and was first mentioned to be applied to fungus tree associations by the German Forest Pathologist A.B. Frank in the year 1885. It was documented that the majority of terrestrial plants form symbiotic associations with these fungi and this was known to be branded by the bi-directional movement of nutrients in which the carbon move to the fungus and inorganic nutrients move to the plant, providing a bond or link between the plant, root, and soil (Gerdemann, 2008).

Endomycorrhizal fungi are a set of mycorrhizal categories producing "arbuscules" and "vesicles," where arbuscules promote a bi-directional transfer of nutrients between the symbiotic plant and fungi. The trees have vesicles and are observed as storage organs that are usually formed at the tip of the fungal hyphae (Scannerini and Bonfante Fasolo, 2003). The vesicular arbuscular mycorrhiza (VAM) is so far known as the most common type of association among all types of mycorrhiza (Gerdemann, 2008). Arbuscular mycorrhizal fungi were stated to be previously known as vesicular arbuscular mycorrhiza or VAM (Morton, 1998). AM associations are formed by nonseptate Zygomycetes and Phycomycetes fungi. Some examples are *Glomus*, *Gigaspora*, *Acaulospora*, *Entrophospora*, and *Scutellospora* of which *Glomus* is the most common fungus (Remy et al., 1994).

AM fungi occur in almost all vegetation types and constitute an important component of the tropical soil microflora (Cardoso and Kuyper, 2006). They have been shown to increase the growth and yield of plants.

They are also reported to have positive effects on plants such as increased resistance to pathogens (Hampp et al., 1999), drought (Shi et al., 2002), or heavy metal stress (Blaudez et al., 2000). They have been identified to play roles in both nutrient mobilization and nutrient cycling. Their distribution and diversity in tropical ecosystems elsewhere appear to be receiving increased attention (Lovelock and Ewel, 2005). There were many reports on the association of AM fungi with several cultivated crops (Lakshman and Raghavendra, 1995). However, there are fewer studies on the influence of AM fungal status on fodder crop plants in India particularly in Tamil Nadu. AM fungi are known to play an important role in the growth and development of fodder crop plants, as they are observed to increase the uptake of diffusion-limited nutrients (Palipane and Bandara, 1985). Inoculation of AM fungi has already reported to improve the physiological conditions of fodder crops.

AM fungi occur as communities in soil and roots; they are likely to collectively contribute to nutrient uptake such as phosphorus and nitrogen (Jakobsen et al., 2001). Joner and Leyval (2001) have previously suggested that the use of AM fungal consortia that are adapted to metal-enriched soil should be considered for future investigation rather than individual AM fungal species. Further, it was observed that as AM fungi coexist and interact with plant growth-promoting rhizobacteria (PGPR) in soils, changes in microbial community structure may also affect the function of AM fungi, which necessitates experiments with unsterilized soils.

Arbuscular mycorrhizas are known to develop a constructive interaction with several plants, namely, the horticultural and forage field crops. There are reports mentioning the promotion of efficient nutrient absorption and increased plant growth and yield by fungi colonization. Arbuscular mycorrhiza's association with the soil has been stated to have a great role in monitoring soil fertility, soil erosion, and plant water stress. Arbuscular mycorrhiza fungi in crop production were observed to reduce fertilizer quantity when applied to farmland. AM fungi became significant for sustainable agriculture and the country's commercial development. Nevertheless, the progress on arbuscular mycorrhiza cultivation has not been very successful in the past several years, partly as the research became difficult and time-consuming. ARS researchers in their studies have reported optimal conditions for propagating the dual culture of root and fungus. It was also observed that pure arbuscular mycorrhiza fungi can be reproduced on gel plates (Gerdemann, 2008).

The arbuscular mycorrhiza fungi are well known for their ability to improve plant growth in low-phosphate soils by exploiting large areas of soil and actively transporting the phosphate back to the plants. Other benefits to the plants supplied by the arbuscular mycorrhiza colonization include increased absorption of nitrogen, potassium, magnesium, copper, zinc, boron, sulfur, molybdenum, and other elements that are transported back to the plant. Finally, it helps in retaining moisture around the root zone of plants (Srihari, 1997).

Most plants with numerous roots are found to possess arbuscular mycorrhiza, a type of fungi with a harmonious relationship with plant roots. The symbiotic association promotes the exchange of extra nutrients to plants by fungi from the soil, namely, zinc and phosphorus, in exchange for sugars or exudates for fungi. Fungi are an integral and normal component of the roots of about 80% of plants such as crops and trees. Arbuscular mycorrhiza is also known to help hold soil particles together. Around 150 species of fungi were observed to have favorites for different soil types and niches. In common, they are all capable of colonizing the roots of all susceptible plants, which was an important factor (Will and Sylvia, 1990).

Arbuscular mycorrhiza extends deep into the plant root systems and exploits the soil nutrients much more effectively than the plant alone. Several nutrients such as phosphorus and zinc transit very slowly through the soil solution; however, AM fungi in roots enable the easy absorption of these nutrients. A zone of nutrient depletion may occur near the root due to the removal of nutrients and slow down plant nutrient uptake which can be resolved by AM association (Tilak and Annapurna, 1993).

Applications of large quantities of inorganic fertilizer especially phosphorus reduce arbuscular mycorrhiza colonization and also reduce the fungal populations with the crop responsiveness. Exposure to high fertilizers wheat essentially loses its arbuscular mycorrhiza partner; nonetheless, peas, beans, and many pasture legumes may still persist and benefit from arbuscular mycorrhiza, however, to a lesser degree. Some fungicides leeched in soil are also found to reduce arbuscular mycorrhiza populations. Nevertheless, most herbicides do not seem to have a direct chemical effect on arbuscular mycorrhiza, but they kill the plants and block the food source for the arbuscular mycorrhiza fungi. There are previous reports on soil fumigants to eliminate all soil biota, including arbuscular mycorrhiza which is a challenge in horticulture, especially if the crop is particularly responsive to arbuscular mycorrhiza (Subba et al., 1985).

It has already been reported that retaining the stubble could return nutrients to the soil and the arbuscular mycorrhiza will take these directly to the plants. It was also noted that the burning of stubble kills arbuscular mycorrhiza, especially the hot burns. Previous research on peanut crops has reported a reduction of root length percentage from 72 to 16% when stubble was burned. This, however, translated to a reduction of arbuscular mycorrhiza colonized roots from 12 m per plant to 1.5 m per plant taking into consideration the differences in crop growth (Sureshkumar Singh et al., 2003).

Arbuscular mycorrhiza is well known to compete with other members of the soil biota for soil nutrients thereby increasing the competitive ability of their host plants. They increase the nodulation and nitrogen fixation in legumes by supplying phosphate. Arbuscular mycorrhiza is also reported to increase the tolerance of plants to diseases and pests by compensating for root damage and was observed to possess direct negative effects on the pathogens. Previous reports stated several soil biota grazing on arbuscular mycorrhiza hyphae and spores help to keep the fungi young and vigorous and release nutrients from the dead hyphae (Kung et al., 2008).

There are previous reports on arbuscular mycorrhizal fungi improve plant growth by augmented uptake of the nutrient phosphorus in soils with low fertility (Gerdemann, 2008) and also the uptake of nitrogen, calcium, zinc, copper, manganese, and iron from the soil. They prompt the capability to tolerate plant pathogenic microbes in addition to imparting moisture deficit. The uptake of phosphorous and other nutrients has been studied in several crops such as *Liriodendron tulipifera* L. to be efficient and enhanced in chili (Sreeramulu and Bagyaraj, 1986), tomato (Srihari, 1997), potatoes (Black and Tinker, 1977), cowpea, finger millet, and cotton (Bagyaraj and Manjunath, 1980) and uptake of nitrogen, calcium, zinc, copper, manganese, and iron by soybean (Ross and Harper, 1970), Nitrogen by soybean and maize (Kessal et al., 1985); nitrogen, calcium, zinc, copper, manganese, and iron by groundnuts (Krishna et al., 1982); and sulfur by red clover and maize (Gray and Gerdemann, 1973).

12.3 CONSEQUENCES OF ARBUSCULAR MYCORRHIZA (AM) ON AGRICULTURAL CROPS

AM fungi are known to form symbiotic associations that can enhance water and nutrient transport, particularly nitrogen and phosphorus minerals, thereby increasing the growth and yield of many plants (Carling and Brown, 1980). *Pseudomonas fluorescens* strains were reported to infect compatible species of cereals on infection of compatible species of cereals participating in the symbiotic association and root nodulation leading to more nitrogen fixation which increases the growth and yield (Dixon and Wheeler, 1986). Several reports have reported the effect of combined inoculation with arbuscular mycorrhizal fungi and *Pseudomonas fluorescens* to further increase the growth and yield of crops, namely, maize (Young et al., 1988). The multilateral association of symbiotic arbuscular mycorrhizal fungi, *Pseudomonas fluorescens*, and root nodule bacteria *Bradyrhizobium japonicum* were also reported to possess an augmenting effect on the yield and growth of maize.

Arbuscular mycorrhizal fungi are acknowledged to stimulate the phosphorous uptake of plants from the soil. Reports are showing phosphorous in soil inhibiting the arbuscular mycorrhizal growth. Above 50 ppm of PO_4 in the soil has been reported to inhibit severely arbuscular mycorrhizal colonization in previous studies. However, the application of insoluble phosphates, such as rock phosphate (Graham and Timmer, 1985), calcium phosphate, and bone powder was found effective in maintaining arbuscular mycorrhizal fungi. There are several studies stating that the long-term availability was an advantage of insoluble phosphate when compared to soluble phosphate.

Microbe–microbe interactions are mentioned to be crucial to understanding the dynamic processes and characteristics of rhizosphere formation and preservation. Mycorrhizal fungi and some other soil microorganisms are known to regulate mycorrhizal formation and functions. Conversely, mycorrhizas are seen to affect the establishment of rhizosphere populations. The microbial population especially the root-associated mycoflora in soil are observed to concentrate around plant roots stimulated by root exudates supplied by the plants (Curl and Truelove, 1986). The population dynamics of microorganisms were found to determine the competitive interactions at the root–soil interface by the limited carbon resources released by the plants. On the other hand, the added organic matters were reported to result in a flush of microorganisms (Johannson et al., 2003).

Sureshkumar Singh et al. (2003) studied the regeneration pattern and species diversity of AM fungi in shifting cultivated abandoned land (jhum fallow) and natural forest soils. The jhum fallow was known to contain a lower VAM fungal population and number of species than the natural forest. Around 44 arbuscular mycorrhizal species belonging to six genera, namely, *Acaulospora*, *Enterophospora*, *Gigaspora*, *Glomus*, *Sclerocystis,* and *Scutellospora* were already isolated from soils of jhum fallow and from around natural forest sites. There are particularly ten species of arbuscular mycorrhizal fungi that were found eliminated from the jhum fallow revealing that the shifting cultivation in the humid tropical soils causes reduction of arbuscular mycorrhizal fungal species.

Most plant species are capable of associating with fungi of a single family, Endogonaceae, to form an arbuscular mycorrhizal association (Gerdemann, 2008). This symbiotic association of the endomycorrhizal fungi and roots of plants is known to be beneficial with the exchange of nutrients between the symbionts. Mycorrhizal fungi usually profit from carbon substrates from plants and, in turn, the plants are provided with nutrients, especially phosphorous and nitrogen compounds, from the soil through the hyphal network of the fungi (Loyanachan, 2000; Ravarkar et al., 2000).

Ghosh et al. (2004) studied the variability in mycoflora population under low (25%), medium (50%), and waterlogged moisture conditions by treating rice with arbuscular mycorrhizal with *Azolla* combinations and individually with arbuscular mycorrhizal and *Azolla*. The variations were recorded as several CFU g^{-1} soil × 10^4 using rhizosphere soil. The fungal population was found to be high and well maintained both in medium moisture conditions and also when treated in combination with arbuscular mycorrhizal and *Azolla*. This condition was, however, found to be congenial for both mycorrhizal colonization and leaf phosphate accumulation in rice.

Solaiman and Hirata (2004) examined the effect of AM fungi inoculation on the growth and nutrient achievement of wetland rice in nurseries. The researchers grew seedlings on gamma-ray sterilized paddy soil in two different nurseries, one with a dry nursery and a wet nursery and other nurseries with or without arbuscular mycorrhizal fungi. It was observed that mycorrhizal seedlings had higher shoot biomass under both nursery conditions five weeks after sowing.

PGPR-inoculated plants on maturation produced higher biomass under field conditions and the grain yield was 14–21% higher than those un-inoculated. On the contrary, grain yield and shoot biomass were significantly uninfluenced by AM colonization under pot conditions. The plants that originated from the dry nursery were observed to have significantly high nitrogen, phosphorous, zinc, and copper concentrations of field-grown plants at harvest by pre-inoculation with arbuscular mycorrhizal over those left uninoculated.

Secilia and Bagyaraj (2004) reported the response of the wetland rice cultivar to inoculation with ten AM fungi to select the efficient mycorrhizal fungi. It was observed that *Glomus intraradices* and *Acaulospora* sp. caused an increase in grain yield and growth when sandy clay loam soil substrate, fertilized with nitrogen and potassium recommended levels (100 kg N ha^{-1} as ammonium sulphate and 50 kg K ha^{-1} as muriate of potash) and half the recommended level of P (25 kg ha^{-1} as super phosphate). These species were reported in previous studies to be efficient and suitable for inoculation into rice nurseries and were observed to possess increased grain yield and total biomass.

Kalimov et al. (2005) have conducted experiments to understand the special effects of phosphorous sources on the phosphatase activity of the rhizosphere and mycorrhizosphere of red clover inoculated with *Glomus mosseae* applied with a cultural system with three compartments. The experiment used plants harvested after eight weeks, and were measured for dry weight and phosphorous contents as well as soil phosphatase activity of root and hyphal compartment. It was concluded that acid phosphatase activity was higher than alkaline phosphatase activity and with organic phosphorus supplementation both activities were increased slightly after inoculation. However, the acid phosphatase activity was found to be higher than alkaline phosphatase in the mycorrhizosphere of all inoculated treatments. The dry weight, phosphorous content, and total phosphorous uptake of the plants were observed to increase significantly after inoculating mycorrhizal fungus. The report suggested that the amount of phosphorous uptake by hyphae accounted for 43.1% of the total phosphorous uptake of plants' KH_2PO_4 application, while the value was 60.8 % on the application of sodium phytate.

Kung et al. (2008) studied the influence of arbuscular mycorrhiza fungi inoculation on coppicing ability and drought resistance of *Senna spectabilis* in a screen house experiment. The result indicated the dependence of *Senna spectabilis* on mycorrhizal symbiosis. It was observed that the production of *Senna spectabilis* biomass increased by 26.9%, while under water-stressed conditions, the production augmented by 37% with watering. Supporting the results the analysis of variance revealed that the interaction between the mycorrhizal fungi and water stress was highly significant. Inoculating *Senna spectabilis* with arbuscular mycorrhizal was also reported to improve drought resistance. The reports of the studies have also mentioned that under drought conditions, inoculating *Senna spectabilis* increased total shoot length by 100%, the root collar diameter was increased by 74%, while the shoot dry weight by 35%, the root dry weight augmented by 39% and plant leaves number increased by 15%. It was observed that the inoculated plants had more water content in leaves than non-inoculated plants along with the inoculated *Senna spectabilis* plants took more days to show signs of drought stress. The study concluded that better growth responses of mycorrhizal plants were attributed to higher nutrient uptake and higher moisture absorption. It was also observed that the arbuscular mycorrhiza (AM) inoculation has a high potential in water-stressed environments in maintaining water relationships.

Pandey and Baink (2009) conducted some preliminary experiments with the medicinal plant *Aloe vera* to study the effectiveness of AM fungi. The *Aloe vera* plants grown co-inoculated with arbuscular mycorrhizal fungi showed an increase in plant growth (leaf, stem, and root dry biomass), root colonization, phosphorus, nitrogen, sugar, and barbaloin content over those grown un-inoculated. *Glomus mossae* inoculated with *Aloe vera* was found to be the finest arbuscular mycorrhizal association. The study investigated the effect of *Glomus mossae*, *Azotobacter* sp., rock phosphate, and soil type (% sand in garden soil v/v) on *Aloe vera* plant growth and barbaloin content (mg plant^{-1}). It was observed that the inoculation by both *Glomus mossae* and *Azotobacter* sp. augmented the yield and biomass of Barbaloin content plant^{-1}.

Rajasankar et al. (2013) isolated a pesticide-tolerant and phosphorus-solubilizing bacteria, *Pseudomonas* sp. SGRAJ09 from *Achillea clavennae* rhizosphere soil treated with pesticides to study on effects of PGPR. Rautaray et al. (2013) also isolated phosphate solubilization, IAA-producing *Pseudomonas fluorescens* and stated that PGPR strains enhanced growth of *Vigna mungo*. Reports also suggested that the amount of salt that affected land worldwide is estimated to be 900 million ha which is about 6% of the global total land mass.

12.4 *AZOSPIRILLUM* SPECIES

Azospirillum is a curve-shaped Gram-negative bacteria, with a size of about 1–5 μm in diameter, having peritrichous flagella for swarming motility and a polar flagellum for swimming and locomotion. The bacteria were observed to have reserved cell inclusion energy stored in the form of Poly β-hydroxy butyrate (PHB) granules for preventing the cell during the lack of energy (Okon et al., 1998). Dobereiner and Day (1976) stated *Azospirillum* sp. is an aerobic microorganism that grows under microaerophilic conditions. Surprisingly, it was also stated that the *Azospirillum* sp. is effective in fixing atmospheric nitrogen under microaerophilic conditions.

Azospirillum was acknowledged as PGPR as it was isolated from the rhizosphere of diverse grasses and cereals of tropical as well as temperate climates (Steenhoudt and Vanderleyden, 2000). The *Azospirillum* sp. has consistency and has proven as very promising PGPR since the 1970s and was the selective microflora for fixing the atmospheric nitrogen in the environment (Bashan et al., 2004). The global occurrence and distribution of *Azospirillum* from various global geographical regions were mentioned by Li and Hung (2017). They stated the occurrence of *Azospirillum* sp. in the rhizosphere and rhizoplane of crops which belong to the family Leguminosae, Cruciferceae, Gramineae, Solanaceae, and Tiliaceae. There are also reports by Aseri et al. (2018) on the occurrence of *Azospirillum* from continuous cultivation of the barely.

Carbon source is reflected to be the major nutrient requirement for the growth of bacteria and acts as a backbone for the bacteria cell. About 50% of the dry weight of bacteria is identified to be due to the carbon source. The salts present in the organic acids namely malate, succinate, lactate, and butyrate were preferably efficient and reasonable carbon and energy sources for promoting the growth of *Azospirillum* (Okon et al., 1998).

Chemotaxis is described as the process of movement of bacteria toward food (positive chemotaxis) and away from the toxic substance (negative chemotaxis). *Azospirillum* is known to display positive chemotaxis toward root exudates, organic acids, sugars, amino acids, and aromatic compounds (Lopez et al., 2004). After carbon sources, the most required nutrient requirement for microbial growth is nitrogen source. Ammonium or ammonia, nitrite, amino acids, nitrate, and molecular nitrogen were also reported to serve as nitrogen sources (Hartmann and Zimmer, 1994).

Azospirillum species were known to form cysts under unfavorable conditions such as desiccation (extreme dryness) and nutrient limitation (lack of essential nutrients). Tal et al. (1990) noticed that in conditions of stress and starvation, the build-up of abundant Poly-ß-hydroxybutyrate granules through encystations served as both carbon and energy sources for microorganisms. When compared to other free-living nitrogen-fixing bacteria, *Azospirillum* is stated to be highly efficient with nitrogenase properties and relatively better than other nitrogen-fixing microorganisms (Okon, 1985). Phytohormone production and nitrogen fixation processes are some of the potential attributes for the plant growth-promoting character of *Azospirillum* (Paul and Nair, 2020).

12.5 EFFECT OF *AZOSPIRILLUM* ON AGRICULTURAL CROP INTERACTION

Azospirillum is the most valuable bioinoculant in the PGPR group because it can colonize agricultural crop roots (Dobereiner and Day, 1976). Nitrogen fixation is previously identified to be the key function of the *Azospirillum* sp. The process of fixing atmospheric nitrogen was carried out only under anaerobic conditions because the major enzyme required for nitrogen fixation, nitrogenase was sensitive to oxygen which may cause the process to disturb in the presence of oxygen (Peng et al., 2020). Further studies on the *Azospirillum* bioinoculant have pinpointed the various beneficial effects on morphological and physiological changes in the inoculated roots that were found to improve the uptake of several minerals and also enhance water uptake (Karpagam et al., 2020). Some other physiological changes such as resistance against the abiotic stresses were also reported by the *Azospirillum* inoculated plants. Creus et al. (2020) inoculated the *Azospirillum* on wheat seedlings and observed the development of osmotic stress. The growth efficiency of wheat with *Azospirillum* inoculated and non-inoculated seedlings were compared. It was found that the bioinoculant inoculated seeds showed more fresh weight and yield of wheat when compared to non-inoculated seeds.

Previous reports mention that plants exposed to salt stress have a possibility of suffering from water deficiency. It was reported that about 10^8 cells of *Azospirillum brasilense* inoculated on root seedlings when exposed to mild and severe salt stress were found to significantly reverse the negative effects. Casanovas et al. (2003) performed field experiments with the bioinoculant *Azospirillum* on *Sorghum bicolor* and *Zea mays* and reported a significant increase in the growth, yield, and biochemical composition of crops along with efficient water and mineral uptake. Creus et al. (2020) stated in their research that the *Azospirillum* inoculated with the wheat seedlings and were observed to survive when exposed to the excess saline concentration of upto 320 mM NaCl for an average of three days. The inoculation technology with *Azospirillum* sp. extended to arid soil regions where the water scarcity was very high in order to protect the crops against drought conditions showed a significant increase in water content of the plant, relative water, water uptake potential, apoplastic water fraction, and lower cell wall modulus of elasticity values in *Azospirillum*-inoculated plants.

The beneficial effect of *Azospirillum* sp. in plants relies on the good colonization of roots and rhizosphere region according to previous studies. Root colonization is considered an important factor for creating the relationship of microorganisms with the plant, not only in infection caused by soil-borne phytopathogens (negative effect) but also for the beneficial association with the plant growth-supporting microorganisms (positive effect). The positive effect of the colonization process is the adhesion of the bacteria to the plant roots and this *Azospirillum*—root interaction is stated to be a two-step process. The first step is adsorption mediated by the help of bacterial proteins. The second step is anchoring involved by the bacterial polysaccharides. The root colonization of *Azospirillum* sp. was highly dependent on the active motility and chemotaxis toward the root exudates (Creus et al., 2020).

The distribution of *Azospirillum* isolates in the plant roots was studied in previous studies by using the technique that analyzes the GFP-protein and tag bacteria. Liu et al. (2010) confirmed in their research the colonizing pattern of *Azospirillum* sp. with plant roots. Some nitrogen-fixing strains of *Azospirillum lipoferum* and *Azospirillum brasilense* are known to be customary on the plant root surface but other strains are not capable of colonizing the root interior surface in the apoplast and intercellular spaces. This ability to colonize the root interior surface denoted its lower vulnerability to harsh conditions that are imposed by the soil environment which in turn supports the plant growth-promoting activities (Sturz and Nowak, 2020). The rhizobacteria which are establishing their relationship inside the plant roots are known as endophytes. These microorganisms are found to stimulate plant growth by producing plant growth-promoting hormones, enhance plant disease resistance against plant pathogens, and also improve the mobilization of nutrients in soil.

12.6 *AZOSPIRILLUM* AND AM FUNGI INTERACTION

Mycorrhizal fungi intermingle with several other microorganisms in the root, in the rhizosphere, mycorrhizosphere, and bulk soil. It was reported to be inhibitory or stimulatory and many are competitive while some are mutualistic. It is also known to adapt and change the physiology of plants including roots. Similarly, free-living nitrogen-fixing bacteria were identified to influence the behavior of mycorrhizal fungi indirectly by changing the host physiology (Sudhakar et al., 2019).

12.6.1 ENHANCEMENT OF *AZOSPIRILLUM* COLONIZATION BY MYCORRHIZAL FUNGI

Several reports are stating the association of nitrogen-fixing organisms with plants (Jain and Patriquin, 2015). Several sugars, amino acids, and growth regulators are exuded by plants that could fuel the rhizosphere microflora (Alexander, 1961). These microflora present in the rhizosphere interacted among themselves and other plant roots. It was reported that the synergistic interaction between the mycorrhizal fungi and bacteria was due to reasons such as hormonal effect, nutritional relations, and physical interactions. Barea and Azcon (1982) and Sterzelezyk and Pokojska (1984) in their studies have mentioned the production of phytohormones such as cytokinins and gibberellins by mycorrhizae.

Thimann (1974) stated that certain growth regulators affect the plant root growth rate. Alexander et al. (1989) cited that plant roots are infected by mycorrhizae which eventually changes the structure and physiology of the roots promoting a greater effect on rhizosphere microflora development.

There are several previous records signifying the importance of mycorrhiza fungi, nitrogen-fixing bacteria, and phosphate-solubilizing bacteria to influence plant growth and nutrition beneficially (Hajer et al., 2016). It is also noted that for the root systems to acquire nutrition one key determinant is the colonization of the roots with mycorrhiza (Mehnaz et al., 2017). There are reports on growth and nutrition enhancement in wheat varieties colonized by VAM fungi. It was stated that the colonization of the wheat varieties gradually increased the yield and the growth of the wheat (Nehi et al., 2019). Several studies have shown that VAM colonization or inoculation and biofertilization could enhance crop yield and nutrition compared to other noncolonized plants or crops. Bagyaraj and Menge (2008) reported the *Azotobacter* population in the root zone soil of mycorrhizal tomato plants. There are also reports on the other beneficial rhizosphere bacteria in plants. Wu et al. (2015) have reported the effects of Arbuscular mycorrhiza and *Azotobacter* sp., *Bacillus* sp., and phosphate solubilizing organisms as biofertilizers on plant yield, growth, and nutrition. It was observed that when used there was an enhancement in yield, nitrogen fixation, tolerance to stresses, and uptake of nutrition. Xia et al. (2015) have reported that *Glomus* species of mycorrhiza when inoculated in roots of plants benefitted their growth and nutrition.

Li and Hung (2017) in their experimental study isolated nitrogen-fixing organisms such as *Clostridium* sp. and *Azospirillum* sp. from Ectomycorrhizae signifying the existence of microbes within fungi. They also reported that these mycorrhizal fungi uptake phosphorus from soil and can be used up as ATP which is phosphate-containing energy source for nitrogenase activity during nitrogen fixation by *Azospirillum* sp.

Researchers studied on the effect of microbes on mycorrhizal association and found microbes could be isolated from ectomycorrhizal fungi mantle which were mostly Gram-negative bacteria. Meyer and Linderman (2018) experimented on sweet corn inoculated or colonized with mycorrhizae fungi and noticed an increase in specific microbe populations, namely, bacteria and actinomycetes in the rhizosphere and on rhizoplane beneficial for plants. Similarly, Klyuchnikov and Kozhevin (2019) observed and isolated the organism *Azospirillum brasilense* from potato roots colonized with the mycorrhiza *Glomus clarum*. Mycorrhization of the plant root also favored the formation of *Azospirillum lipoferum* on the rhizoplane according to the reports by Belimov et al. (2020).

12.6.2 ENHANCEMENT OF MYCORRHIZATION BY AZOSPIRILLUM

Several researchers have reported the influence of diazotrophic bacteria on the improved formation and development of AM fungi in the roots. Barea et al. (2005) proposed that the increased mycorrhizal colonization is due to the production of plant growth elements by the rhizosphere bacteria. *Azotobacter vinelandii* and *Azotobacter beijerinckii* were reported and identified to produce auxins, gibberellins, and cytokinins in culture. The production of plant growth substances by *Azospirillum brasilense* was also demonstrated *in vitro* by Kapulnik et al. (2005). It was hypothesized that these phytohormones cause morphological and physiological changes in the root system that could enhance plant growth.

Soil microbes and plant association are known to have synergistic interactions due to the formation of phytohormones. The formation of phytohormones helps in synergistic interactions between the soil micro-organisms that establish dual symbiosis with plants. There are previous studies on the growth and infection levels of endomycorrhizal fungi in tomatoes, alfalfa, and lavender treated with bacterial mix. There are reports of the application of cell-free supernatants or whole bacterial cultures of *Pseudomonas* and *Azotobacter* in crops to promote the shoot dry weight. The *Glomus* sp. root colonization was also reported to augment the pure plant hormones (Azcon and Barea, 2008).

Azospirillum brasilense pure cultures were detected to exhibit pectinolytic activities when grown in a medium with pectin. The pectinolytic activity of *Azospirillum* was found to soften the middle lamellae of the root cortex cells in a kind of sponge effect. Such a function of the pectinolytic enzyme isolated from *Pseudomonas putida* and *Azospirillum brasilense* was also previously reported to enhance and promote mycorrhizal root colonization (Umali-Garcia et al., 2010).

Pacovsky et al. (1985) have previously mentioned that the mycorrhizal inhabitation and vesicle formation in plant roots was supported by *Azospirillum* sp. and was examined to be due to greater availability of carbohydrates to the fungal endophyte. There are also several reports supporting the impact of *Azospirillum* sp. with mycorrhiza and elucidating the reason for the nitrogen dependence in the photosynthesis process. The co-inoculation of *Azospirillum* sp. with mycorrhiza was also reported by several researchers to enhance mycorrhization in maize and ryegrass, sorghum, palmarosa, and seots. Meyer and Linderman (2018) examined the augmentation of mycorrhization by VAM in clover plant roots that possessed growth-promoting rhizobacterium, *Pseudomonas putida*. Some research concepts have reported the acceleration of fungal spore and propagules germination during the pre-infection process of the mycorrhiza.

Soil bacteria and their culture filtrates were reported to play an important role in the *in vitro* development of *Glomus mosseae*. It was observed that the bacterial cell cultures and their metabolites *viz.* amino acids, plant hormones, and vitamins stimulated the hyphal growth at the plant roots during the pre-infection stages and also stimulated the new vesicles from germinated resting spores. The germination of *Glomus versiformae* spores with extensive hyphae along with bacteria, namely,

Pseudomonas and *Corynebacterium* were also reported in previous studies (Mayo et al., 1986).

Mycorrhiza formation has been reported to refine the nutritional value and fertility of the soil or the host. There are previous reports of isolation of *Azospirillum* sp. from VAM fungi (*Glomus fasciculatum*, *Glomus intraradices*, *Glomus scintillans*, *Glomus mosseae*, *Gigaspora gilmorei,* and *Endogone dusii*) surfaces that were pre-sterilized (Tilak et al., 1989). Studies have reported the enhanced nitrogenase activity of *Azospirillum* isolated from *Glomus* species.

The ectomycorrhizal fungi such as *Hebeloma crustuliniforme* and *Paxillus involutus* were reported to produce citric acid and malic acid which is a fungal carbon source and were also observed to metabolize even polyphenolic substances produced by the ectomycorrhizal (Duponnois and Garbaye, 1990). Tilak et al. (1990) revealed in his study that the cell-free extracts of the nitrogen-fixing bacteria, namely, *Azotobacter chroococcum*, *Azospirillum brasilense,* and *Azospirillum lipoferum* boosted the spore germination of *Glomus fasciculatum*.

Bacteria can upsurge the plant susceptibility to mycorrhizal colonization especially VAM. The bacteria were classified as mycorrhization helper bacteria (MHBs) and were reported by Duponnois (1992) to have mentioned the enzyme activities of endoglucanase, cellobiose hydrolase, pectate lyase and xylanase from the Douglas fir (*Laccaria laccata*). MHBs were known to soften the cell walls and middle lamella between the cells of the root cortex using producing specific enzymes to make fungal penetration easier. There are previous reports by Alarcon et al. (2002) stating the beneficial effects of phytostimulators such as *Azospirillum* on the activity level of AM fungi and also were stated to promote increased mycorrhization on root and their growth.

12.6.3 COMBINED INOCULATION OF AZOSPIRILLUM AND AM FUNGI FOR THE IMPROVEMENT IN THE NUTRITION AND PRODUCTIVITY OF THE HOST PLANT

There are noted reports on the multilateral relationship of the plant, the AM fungus, and *Azospirillum* as they have shown improved plant growth, yield, and productivity. Kumari and Balasubramanian (1993) in their study have mentioned the effect of the combined inoculation of AM fungi (*Glomus fasciculatum*, *Gigaspora margarita,* and *Acaulospora laevis*) with *Azospirillum brasilense* on growth, yield, vigor, and nutrient uptake of coffee seedlings grown in nurseries. There are also supportive reports of the benefits of dual inoculation of maize plants with *Azospirillum brasilense* and *Glomus mosseae*. They were observed to augment the development of mycorrhizal fungi to yield plants of a similar size, nitrogen content, and a higher phosphorous content (Barea et al., 1983).

Subba et al. (1985) reported that Pearl millet seed inoculation with *Azospirillum brasilense* with *Gigaspora margarita* or *Glomus fasciculatum* significantly gave higher shoot dry matter, root biomass, phosphorous content, and phosphorus uptake compared to inoculation with sole AM fungi. They have also studied on the dual inoculation effects of *Azospirillum brasilense* and various other AM fungi in barley. Dual inoculation treatment resulted in an increased in plant growth, alkaline phosphatases, root acid, and uptake of phosphorous, nitrogen, zinc, copper, and iron. It was observed that among the different AM fungi, soil inoculation with *Glomus mosseae* and *Glomus fasciculatum* produced considerably higher dry matter and grain yield than their corresponding controls in pot culture conditions.

Konde et al. (1988) in their study have mentioned that synchronized inoculation of either *Glomus* or *Gigaspora* along with *Azospirillum* promoted the fresh and dry weights and nitrogen and phosphorous uptake by shoots and bulbs of onion (*Allium cepa*) over consistent controls using sterile vertic ustropepts soil. Singh et al. (1990) have reported on the collaborative effects of *Glomus fasciculatum* and *Azospirillum brasilense* on prosperous yields of various genotypes of wheat. There were differences in AM root colonization and amplified grain yield with high P_2O_5 content in inoculated wheat plants grown under pot culture and field conditions.

Previous reports have mentioned high chlorophyll content, nitrate reductase activity (NRA), photosynthetic rate (Ps) and glutamine synthetase (GS) activity, and eventually an increased grain yield of wheat plants when co-inoculated with *Azospirillum brasilense* and *Glomus fasciculatum* (Panwar, 1991). Previous studies have reported the beneficial effect of *Glomus intraradices* and *Azospirillum lipoferum* on the growth of sorghum (Veeraswamy et al., 1992). There are reports of combined inoculation of Palmarosa (*Cymbopogon martinii*) with *Glomus aggregatum* and *Azospirillum brasilense* to increase yield, growth, and oil content over controls (Neelima and Janardhanan, 1996). Similarly, reports by Indi et al. (1990) in brinjal and studies by Sreeramulu et al. (1988) in maize also stated an improved yield of the plants when co-inoculated with *Glomus fasciculatum* and *Azospirillum* sp. (Figure 12.1).

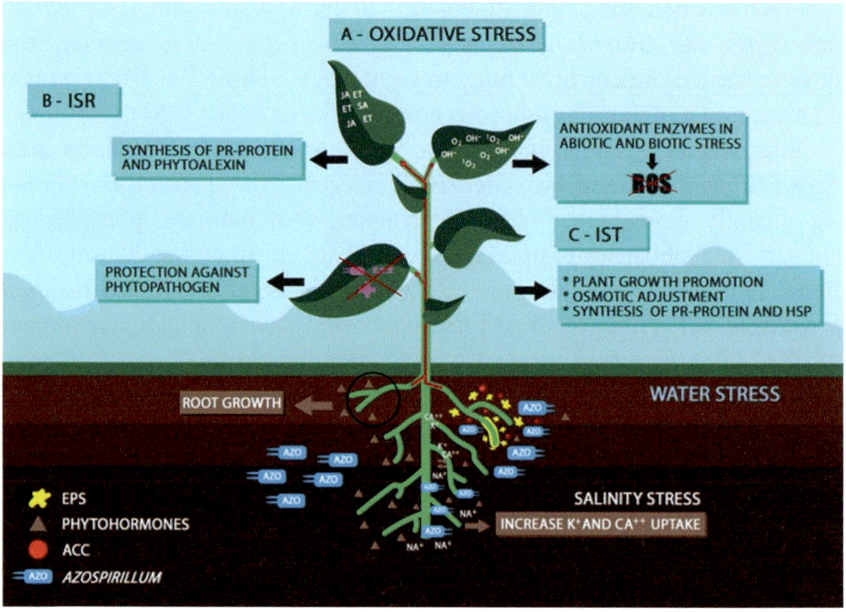

FIGURE 12.1 Mechanisms of tolerance of biotic and abiotic stresses induced by *Azospirillum* in plants. Tolerance to biotic stress includes induced systemic resistance (ISR), mediated by increased levels of phytohormones in the jasmonic acid (JA)/ethylene (ET) pathway independent of salicylic acid (SA), and systemic acquired resistance (SAR), a mechanism previously studied with phytopathogens controlled by intermediate levels of SA. Tolerance of abiotic stresses, named as induced systemic tolerance (IST), is mediated by antioxidants, osmotic adjustment, production of phytohormones, and defense strategies such as the expression of pathogenesis-related (PR) genes.

Source: Adapted from Josiane et al. (2018), https://creativecommons.org/licenses/by/4.0/

12.7 CONCLUSION AND RECOMMENDATIONS

Microorganisms in the rhizosphere of plants play a vital role in enhancing soil fertility and thereby improving plant health as a whole. Out of various groups of microorganisms that colonize the rhizosphere, AMFs are unique because they are partly present inside the root and partly outside the root, influencing other rhizosphere microbes and also plant growth.

Plant and AMF association has evolved over at least 500 million years and this association is one of the most ancient symbiotic relationships in the biological world. This hypothesis is also supported by recent molecular studies carried out on liverwort which is the most ancient plant. AM fungi and PGPR activity by microbes form a key component of the soil microbial population. Arbuscular Mycorrhizal fungi form symbiotic associations with most of the cultivated crop plants and they help plants in supplying required nutrition and protecting them against biotic and abiotic stresses. Many species of *Azospirillum* occurring in soil are also known to promote plant growth through nitrogen fixation, phytohormone production and protection against biotic and abiotic stresses. Synergistic interaction between AMF and *Azospirillum* spp. in promoting plant growth compared to single inoculation with either of them has been reported. Thus, dual inoculation improves nutrient uptake, plant pathogen resistance, and abiotic stress relief compared to AMF or *Azospirillum* alone.

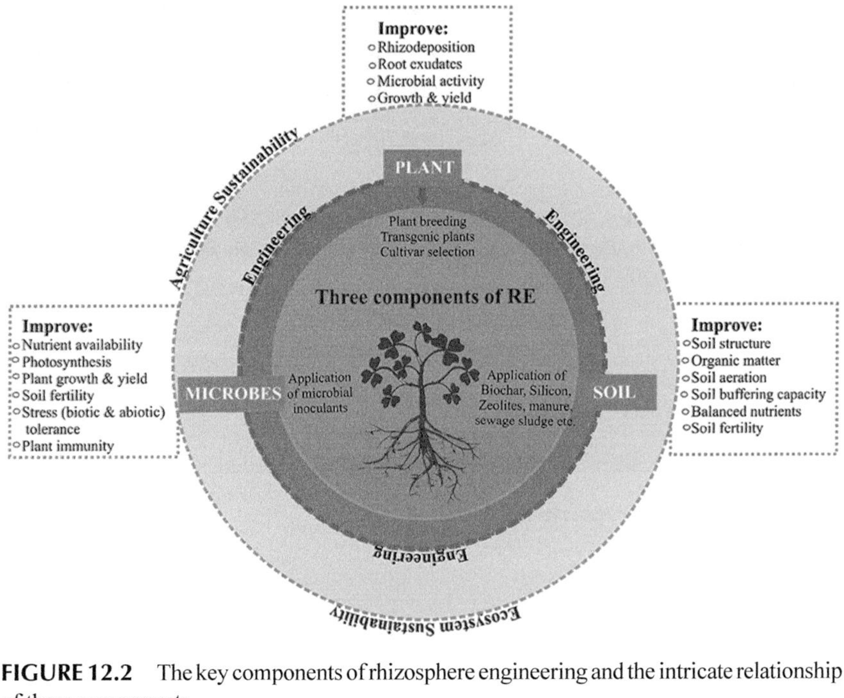

FIGURE 12.2 The key components of rhizosphere engineering and the intricate relationship of these components.

Source: Adapted from Sughra et al. (2021), Copyright © 2021 Hakim, Naqqash, Nawaz, Laraib, Siddique, Zia, Mirza and Imran. https://creativecommons.org/licenses/by/4.0/

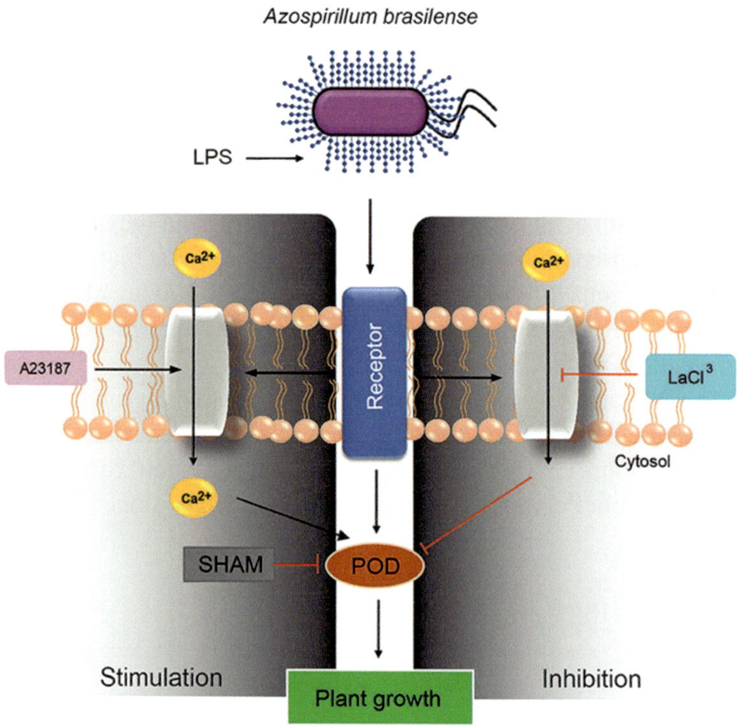

FIGURE 12.3 The relationship among LPS, calcium mobilization, and POD activity, to regulate plant growth. A23187, ionophore A23187; $LaCl_3$, lanthanum chloride; LPS, lipopolysaccharides; POD, peroxidase; SHAM, salicylhydroxamic acid.

Souce: Adapted from Hernaández-Esquivel et al. (2020), Creative Commons Attribution 4.0 International.

KEYWORDS

- **arbuscular mycorrhizal**
- ***Azospirillum***
- **nutritional uptake**
- **phytohormones**
- **rhizosphere**
- **soil fertility**

REFERENCES

Alarcon, A.; Davis, F. T.; Egilla, N. J.; Fox, T. C.; Estrada, L. A. A.; Ferrera, C. R. Short-Term Effects of *Glomus claroideum* and *Azospirillum brasilense* on Growth and Root Acid Phosphatase Activity of *Carcia papaya* L. under Phosphorus Stress. *Rev. Latinoam Microbiol.* **2002,** *44*, 31–37.

Alexander, T.; Toth, R.; Meier, R.; Webber, H. C. Dynamics of Arbuscle Development and Degeneration in Onion, Bean and Tomato with Reference to Vesicular Arbuscular Mycorrhizae in Grasses. *Canadian J. Bot.* **1989,** *67*, 2505–2513.

Aseri, G. K.; Jain, N.; Panwar, J.; Rao, A. W.; Meghwal, P. R. Biofertilizers Improve Plant Growth, Fruit Yield, Nutrition, Metabolism and Rhizosphere Enzyme Activities of Pomegranate (*Punica granatum* L.) in Indian Thar Desert. *Sci. Hortic.* **2018,** *117*, 130–135.

Azcon, A.; Barea, J. M. Effect of Soil Microorganisms on Formation of Vesicular Arbuscular Mycorrhizas. *Trans. Brit. Mycol. Soc.* **2008,** *84*, 536–537.

Bagyaraj, D. J.; Manjunath, A. Response of Crop Plants to Mycorrhizal Inoculation in an Unsterile Indian Soil. *New Phytol.* **1980,** *52*, 33–36.

Bagyaraj, D. J.; Menge, J. A. Interaction Between a VA Mycorrhiza and *Azotobacter* and their Effects on Rhizosphere Microflora and Plant Growth. *New Phytol.* **2008,** *80*, 567–573.

Barea, J. M.; Azcon, A. Production of Plant Growth Regulating Substances by the Vesicular Arbuscular Mycorrhizal Fungus *Glomus mosseae. Appl. Environ. Microbiol.* **1982,** *43*, 810–813.

Barea, J. M.; Azcon, R.; Hayman, D. S. Possible Synergistic Interactions between *Endogone* and Phosphate Solubilizing Bacteria in Low Phosphate Soil. In *Endomycorrhiza*; Sanders, et al., Ed.; Academic Press: New York, 2005; pp 409 – 417.

Barea, J. M.; Bonis, A. F.; Olivares, A. Interactions between *Azospirillum* and VA Mycorrhiza and their Effects on Growth and Nutrition of Maize and Rye Grass. *Soil Biol. Biochem.* **1983,** *15*, 705–709.

Bashan, Y.; Holguin, G. *Azospirillum*-Plant Relationships: Environmental and Physiological Advances (1990-1996). *Can. J. Microbiol.* **1997,** *43*, 103–121.

Bashan, Y.; Holguin, G.; de-Bashan, L. E. *Azospirillum*-Plant Relationships: Physiological, Molecular, Agricultural and Environmental Advances (1997–2003). *Can. J. Microbiol.* **2004,** *50*, 521–577.

Beijerinck, M. Uberein *Spirillium,* Watches Freien Stickst off Bidenkann. Central Baktparasit Infert II Abstract, 1925; *63*, pp 353–359.

Belimov, A. A.; Serebrennikova, N. V.; Stepanok V. V. Interaction of Associative Bacteria and an Endomycorrhizal Fungus with Barley upon Dual Inoculation. *Microbiology* **2020,** *68*, 104–108.

Black, R. L. B.; Tinker, P. B. Interaction between Effects of Vesicular Arbuscular Mycorrhiza and Fertilizer Phosphorus on Yields of Potatoes in the Field. *Nature* **1977,** *267*, 510–511.

Blaudez, D.; Jacob, C.; Turnau, K.; Colpaert, J. V.; Ahonen Jonnarth, U.; Finlay, R.; Botton, B.; Chalot, M. Differential Responses of Ectomycorrhizal Fungi to Heavy Metals *in vitro. Mycol. Res.* **2000,** *104*, 1366–1371.

Bloemberg G. V.; Lugtenberg B. J. J. Molecular Basis of Plant Growth Promotion and Biocontrol by Rhizobacteria. *Curr. Opin. Plant Biol.* **2001,** *4*, 343–350.
Carling, D. E.; Brown, M. F. Relative Effect of Vesicular Arbuscular Mycorrhizal Fungi on the Growth and Yield of Soybean. *Soil Sci. Soc. J.* **1980,** *44*, 528.
Casanovas, E. M.; Barassi, C. A.; Andrade, F. H.; Sueldo, R. J. *Azospirillum* Inoculated Maize Plant Responses to Irrigation Restraints Imposed during Flowering. *Cereal Res. Commun.* **2003,** *31*, 395-402.
Creus, C.; Graziano, M.; Casanovas, E.; Pereyra, M.; Simontacchi, M.; Puntarulo, S.; Barassi, C.; Lamattina, L. Nitric Oxide is Involved in the *Azospirillum brasilense*-Induced Lateral Root Formation in Tomato. *Planta* **2020,** *221*, 297–303.
Curl, E. A.; Truelove, B. In *The Rhizosphere*; Springer-Verlag: Berlin, 1986.
Dey, R.; Pal, K. K.; Bhatt, D. M.; Chauhan, S. M. Growth Promotion and Yield Enhancement of Peanut (*Arachis hypogaea* L.) by Application of Plant Growth Promoting Rhizobacteria. *Microbiol. Res.* **2004,** *159*, 371–394.
Dixon, R. O. D.; Wheeler, C. T. Nitrogen Fixation in Plants. Chapman and Hall, NY, 1986.
Dobbelaere, S.; Vanderleyden, J.; Okon, Y. Plant Growth Promoting Effects of Diazotrophs in the Rhizosphere. *Crit. Rev. Plant Sci.* **2003,** *22*, 107–149.
Dobereiner, J.; Day, J. M. First International on Nitrogen Fixation. In *Proceedings of International Symposium on Nitrogen Fixation*; Newton, W. E., Nyman, C. J., Ed.; Washington, State University Press: Pullman, WA, 1976.
Duponnois, R. Les bacteries auxiliaires de la mycorhization da Douglas (*Pseudotsuga menziesii* (Mirb) Franco) par, *Laccaria laccata* Souche S238. These de 'l' Universite de Nancy I, France, 1992.
Duponnois, R.; Garbaye, J. Some Mechanisms Involved in Growth Stimulation of Ectomycorrhizal Fungi by Bacteria. *Canad. J. Bot.* **1990,** *68*, 2148–2152.
Ferrol, N.; Barea, J. M.; Azcon, A. C. Mechanism of Nutrient Transport Across Interfaces in Arbuscular Mycorrhizas. *Plant Soil* **2002,** *244*, 231–237.
Gerdemann, J. W. The Endogonaceace in the Pacific Northwest. *Mycologia* **2008,** *5*, 1076.
Ghosh, N.; Chakrabarti, D.; Kushari, P.; Chatterjee, N. C. Integrated Effect of VAM and *Azolla* on Fluctuations in Fungal Population, Mycorrhization and Leaf Phosphates in Rice under Different Moisture Conditions. *Asian J. Exp. Sci.* **2004,** *18*, 27–35.
Gianinazzi, S.; Schuepp, H.; Barea, J. M.; Haselwandter, K. Mycorrhizal Technology in Agriculture: From Genes to Bioproducts Growth Promoting Rhizobacteria and Selected Mutants. *Canad. J. Microbiol.* **2002,** *47*, 404–411.
Graham, B.; Timmer, D. J. Selection of Efficient Vesicular-Arbuscular Mycorrhizal Fungi for Wetland Rice (*Oryza sativa* L.). *Biol. Fertility Soil* **1985,** *13*, 108–111.
Grant, C. A.; Flaten, D. N.; Tomasiewicz, D. J.; Sheppard, S. C. The Importance of Early Season Phosphorous Nutrition. *Canad. J. Plant Sci.* **2001,** *81*, 211–224.
Gray, L. E.; Gerdemann, J. W. Uptake of Sulphur-35 by Vesicular-Arbuscular Mycorrhizae. *Plant Soil* **1973,** *39*, 687–389.
Hajer, A. S.; Malibari, A. A.; Al-Zahrani, H. S.; Almaghrabi, O. A. Responses of Three Tomato Cultivars to Sea Water Salinity. *Afr. J. Biotechnol.* **2016,** *5*, 855–861.
Hampp, R.; Wiese, J.; Mikolajewski, S.; Nehls, U. Biochemical and Molecular Aspects of C/N Interaction in Ectomycorrhizal Plants: An Update. *Plant Soil* **1999,** *215*, 103–113.
Hartmann, A.; Zimmer, Z. Physiology of *Azospirillum*. In *Azospirillum/Plant Association*; Okon, Y., Ed.; CRC Press: Boca Raton, 1994; pp 15–39.

Herman, M. A. B.; Nault, B. A.; Smart, C. D. Effects of Plant Growth Promoting Rhizobacteria on Bell Pepper Production and Green Peach Aphid Infestations in New York. *Crop Prot.* **2008,** *27*, 996–1002.

Hernaández-Esquivel, A. A.; Castro-Mercado, E.; Valencia-Cantero, E.; Alexandre, G.; García-Pineda, E. Application of *Azospirillum brasilense* Lipopolysaccharides to Promote Early Wheat Plant Growth and Analysis of Related Biochemical Responses. *Front. Sustain. Food Sci.* **2020,** *4*, 579976. https://doi.org/10.3389/fsufs.2020.579976.

Indi, D. V.; Konde, B. K.; Wani, P. V.; Kale, P. N. Yield and Nutrient Uptake by Brinjal as Influenced by *Azospirillum brasilense* and *Glomus fasciculatum* Inoculation under Graded Phosphorus Levels. In *Current Trends in Mycorrhizal Research*; Jalali, B. L., Chand, H., Eds.; Proceedings of the National Conference on Mycorrhiza, Haryana Agricultural University: Hissar, India, 1990; pp 157–159.

Jain, D. K.; Patriquin, D. G. Root Hair Deformation, Bacterial Attachment, and Plant Growth in Wheat-*Azospirillum* Associations. *Appl. Environ. Microbiol.* **2015,** *48*, 1208–1213.

Jakobsen, I.; Gazey, C.; Abbott, L. K. Phosphate Transport by Communities of Arbuscular Mycorrhizal Fungi in Intact Soil Cores. *New Phytol.* **2001,** *149*, 95–103.

Jalaluddin, M. Effect of Inoculation with VAM-Fungi and *Bradyrhizobium* on Growth and Yield of Soybean in Sindh. *Pakistan J. Bot.,* **2005,** *37*, 169–173.

Jeliyanon, K.; Kloepper, J. W. Mixtures of Plant Growth Promoting Rhizobacteria for Induction of Systemic Resistance Against Multiple Plant Diseases. *Biol. Control* **2002,** *24*, 285–291.

Johansson, P. M.; Johnsson, L.; Gerhardson, B. Suppression of Wheat Seedling Diseases caused by *Fusarium culmorum* and *Microdochium nivale* using Bacterial Seed Treatment. *Plant Pathol.* **2004,** *52*, 219–227.

Joner, E. J.; Leyval, C. Time-Course of Heavy Metal Uptake in Maize and Clover as Affected by Root Density and Different Mycorrhizal Inoculation Regimes. *Biofertilizers Soils* **2001,** *33*, 351–357.

Josiane, F.; Paula, C.; Mariangela, H. *Azospirillum*: Benefits that go Far Beyond Biological Nitrogen Fixation. *AMB Express* **2018,** *8*, 73. doi: 10.1186/s13568-018-0608-1.

Kamiov, C. D.; Balachandar, S.; Sundaram, S. P. Characterization of 1-amino Cyclopropane–1-Carboxylate Deaminase Producing *Methylobacteria* from Phyllosphere of Rice and their Role in Ethylene Regulation. *World J. Microbiol. Biotechnol.* **2005,** *25*, 1403–1411.

Kapulnik, Y.; Okon, Y.; Henis, Y. Changes in Root Morphology of Wheat Caused by *Azospirillum* Inoculation. *Canad. J. Microbiol.* **2005,** *31*, 881–887.

Karpagam, R.; Kannan, M.; Natarajan, S.; Srinivasan, K. Studies on the Efficacy of Foliar Feeding of Water Soluble Fertilizers on Growth Parameters and Yield of Brinjal Hybrid COBH.1. *South Indian Hortic.* **2020,** *52*, 139–142.

Kessal, C. V.; Singleton, P. W.; Hoben, H. J. Enhanced N-Transfer from a Soybean to Maize by Vesicular Arbuscular Mycorrhizal (VAM) Fungi. *Plant Physiol.* **1985,** *79*, 562–563.

Klyuchnikov, A. A.; Kozhevin, P. A. Dynamics of *Pseudomonas fluorescens* and *Azospirillum brasilense* Populations during the Formation of the Vesicular Arbuscular Mycorrhiza. *Microbiology* **2019,** *59*, 449–452.

Konde, B. K.; Tamble, A. D.; Ruikar, S. K. Yield of Nitrogen and Phosphorus Uptake by Onion as Influenced by Inoculation of VAM Fungi and *Azospirillum brasilense.* In *Mycorrhiza for Green Asia, Proceedings of the First Asian Conference on Mycorrhiza*;

Mahadevan A., Raman, N., Natarajan, K., Eds.; University of Madras, Chennai, India, 1988; pp 222–224.

Krishna, K. R.; Bagyaraj, D. J.; Rai, P. V. Response of Groundnut to VAM Mycorrhizal Inoculation in Black Clayey Soil. *Indian J. Microbiol.* **1982,** *22*, 206–208.

Kumari, S. M. P.; Balasubramanian, A. Effect of Combined Inoculation of VAM and *Azospirillum* on the Growth and Nutrient Uptake by Coffee Seedlings. *Indian Coffee* **1993,** *57*, 5–11.

Kung, J.; Rodel, D.; Lasco, U.; Lorettu, U.; Dela, U.; Cruz, E.; Reynaldo, C. D.; Husain, T. Effect of Vesicular Arbuscular Mycorrhiza (VAM) Fungi Inoculation on Coppicing Ability and Drought Resistance of *Senna spectabilis. Pakistan J. Bot.* **2008,** *40*, 2217–2228.

Lakshman, H. C.; Raghavendra, S. Effect of Leaf Litter and Soil in Different Proportions on Growth and Biomass Production in Six Timber Tree Species Colonized with Vesicular-Arbuscular Mycorrhizae. In *Mycorrhizae: Biofertilizers for the Future*; Adholeya, A., Singh, S., Eds.; Tata Research Institute: New Delhi, 1995; pp 445–448.

Lekberg, B.; Koids, D. Effect of Arbuscular Mycorrhiza Fungi Inoculation on Coppicing Ability and Drought Resistance of *Senna spectabilis. Pakistan J. Bot.* **2005,** *40*, 2217–2228.

Li, C. Y.; Hung, L. L. Nitrogen-Fixing (Acetylene-Reducing) Bacteria Associated with Ectomycorrhizae of Douglas-fir. *Plant Soil* **2017,** *98*, 425–428.

Liu, J.; Luo, J.; Ye, H.; Sun, Y.; Lu, Z.; Zeng, X. Medium Optimization and Structural Characterization of Exopolysaccharides from Endophytic Bacterium *Paenibacillus polymyxa* EJS-3. *Carbohydr. Polym.* **2010,** *79*, 206–213.

Lopez, B. C.; Alcaraz, C. F.; Carlos, M. Involvement of Sugars in the Response of Pepper Plants to Salinity: Effect of Calcium Application. *Asian J. Plant Sci.* **2004,** *3*, 455–462.

Lovelock, C. E.; Ewel, J. J. Links between Tree Species, Symbiotic Fungal Diversity and Ecosystem Functioning in Simplified Tropical Ecosystems. *New Phytol.* **2005,** *167*, 219–228.

Mayo, K.; Davis, R. E.; Motta, J. Stimulation of Germination of Spores of *Glomus versiforme* by Spore Associated Bacteria. *Mycologia* **1986,** *78*, 426–431.

Mehnaz, S.; Weselowski, B.; Lazarovits, G. *Azospirillum canadense* sp. nov., A Nitrogen-Fixing Bacterium Isolated from Corn Rhizosphere. *Int. J. Syst. Evolut. Microbiol.* **2007,** *57*, 620–624.

Meyer, J. R.; Linderman, R. G. Response of Subterranean Clover to Dual Inoculation with Vesicular Arbuscular Mycorrhizal Fungi and a Plant Growth Promoting Bacteria, *Pseudomonas putida. Soil Biol. Biochem.* **2018,** *18*, 185–190.

Minorsky, P. V. On the Inside. *Plant Physiol.* **2008,** *146*, 323–324.

Morton, J. B. Taxonomy of VA Mycorrhizal Fungi: Classification, Nomenclature and Identification. *Mycotaxon* **1998,** *32*, 267–324.

Mukerji, K. G.; Kapoor, A. Taxonomy of VAM Fungi Special Reference to Indian Taxa. In *Perspectives in Mycological Research II*; Agarwal, G. P., Ed., Today and Tomorrow's Printers and Publishers: New Delhi, 1990; pp 7–16.

Neelima, R.; Janardhanan, K. K. Response of Dual Inoculation with VAM and *Azospirillum* on the Yield and Oil Content of Palmarosa (*Cymbopogon martinii* var. *motia*). *Microbiol. Res.* **1996,** *151*, 325–328.

Nehi, D. B.; McGee, P. A.; Torrisi, V.; Patinson, G. S. Patterns of Arbuscular Mycorrhiza Down the Profile of a Heavy Textural Soil do not Reflect Associated Colonization Potential. *New Phytol.* **2019,** *142*, 495–503.

Okon, Y. *Azospirillum* as a Potential Inoculant for Agriculture. *Trends Biotechnol.* **1985,** *3*, 223–228.

Okon, Y.; Bloemberg, G. V.; Lugtenberg, B. J. J. Biotechnology of Biofertilization and Phytostimulation. In *Agricultural Biotechnology*; Altman, A., Ed.; Marcel Dekker: New York, 1998.

Pacovsky, R. S.; Fuller, G.; Paul, E. A. Influence of Soil on the Interactions between Endorhizosphere and *Azospirillum* in Sorghum. *Soil Biol. Biochem.* **1985,** *17*, 525–531.

Palipane, J. B.; Bandara, J. M. R. S. Influence of Vesicular-Arbuscular Mycorrhizal Fungi on the Growth of Coffee and Cacao Seedlings. *Sri Lanka J. Agric. Sci.* **1985,** *22*, 73–79.

Pandey, T.; Baink, C. Effect of *Rhizobium* and VAM-Inoculation on Nodulation, Symbiotic Nitrogen Fixation and Soybean Yield in Tropical Fields. *Biol. Fertility Soil* **2009,** *6*, 165–169.

Panwar, J. D. S. Effect of VAM and *Azospirillum brasilense* on Photosynthesis, Nitrogen Metabolism and Grain Yield in Wheat. *Indian J. Plant Physiol.* **1991,** *34*, 357–361.

Patriquin, D. G.; Dobereiner, J.; Jain, D. K. Sites and Processes of Association between Diazotrophs and Grasses. *Can. J. Microbiol.* **1983,** *29*, 900–915.

Paul, D.; Nair, S. Stress Adaptations in a Plant Growth Promoting Rhizobacterium (PGPR) with Increasing Salinity in the Coastal Agricultural Soils. *J. Basic Microbiol.* **2020,** *48*, 378–384.

Peng, G.; Wang, H.; Zhang, G.; Hou, W.; Liu, Y.; Wang, E. T.; Tan, Z. *Azospirillum melinis* sp. nov., A Group of Diazotrophs Isolated from Tropical Molasses Grass. *Int. J. Syst. Evolut. Microbiol.* **2020,** *56*, 1263–1271.

Rajasankar, R.; Gayathry, G. M.; Sathiavelu, A.; Ramalingam, C.; Saravanan, V. S. Pesticide Tolerant and Phosphorus Solubilizing *Pseudomonas* sp. Strain SGRAJ09 Isolated from Pesticides Treated *Achillea clavennae* Rhizosphere Soil. *Ecotoxicology* **2013,** *22*, 707–717.

Rautaray, S. K.; Ghosh, B. C.; Mittra, B. N. Effect of Fly Ash, Organic Wastes and Chemical Fertilizers on Yield, Nutrient Uptake, Heavy Metal Content and Residual Fertility in a Rice-Mustard Cropping Sequence under Acid Lateritic Soils. *Bioresour. Technol.* **2013,** *90*, 275–283.

Ravarkar, K. P.; Juma, N. G.; Mc Gill, W. B. In *Vesicular Arbuscular Mycorrhiza Hyphal Bridges and N-Transfer from Legume to Non-Legume Plant*, Extended Summary International Conference on Managing Natural Resources for Sustainable Agriculture Production in the 21st Century, New Delhi, India, 2000, pp 657–658.

Remy, N.; Chakrabarti, D.; Kushari, D. P.; Chatterjee, N. C. Integrated Effect of VAM and *Azolla* on Fluctuations in Fungal Population, Mycorrhization and Leaf Phosphates in Rice Under Different Moisture Conditions. *Asian J. Exp. Sci.* **1994,** *18*, 27–35.

Ross, J. P.; Harper, J. A. Effect of Endogone Mycorrhizae on Soybean Yield. *Phytopathology* **1970,** *60*, 1552–1556.

Scannerini, S.; Bonfante Fasolo, P. Comparative Ultrastructural Analysis of Mycorrhizal Associations. *Can. J. Bot.* **2003,** *61*, 917–943.

Secilia, L.; Bagyaraj, D. J. Selection of Efficient Vesicular-Arbuscular Mycorrhizal Fungi for Wetland Rice (*Oryza sativa* L.). *Biol. Fertility Soil* **2004,** *13*, 108–111.

Shi, L. B.; Guttenberger, M.; Kottke, I.; Hampp, R. The Effect of Drought on Mycorrhizas of Beech (*Fagus sylvatica* L.): Changes in Community Structure, and the Content of Carbohydrates and Nitrogen Storage Bodies of the Fungi. *Mycorrhiza* **2002,** *12*, 303–311.

Siddiqui, Z. A.; Mahmood, I. Effects of Rhizobacteria and Root Symbionts on the Reproduction of *Meloidogyne javanica* and Growth of Chickpea. *Bioresour. Technol.* **2001,** *79*, 41–45.

Singh, C. S.; Pal, K. K.; Tilak, B. R.; Saxena, A. K. Suppression of Maize Root Disease Caused by *Macrophomina phaseolina*, *Fusarium moniliforme* and *Fusarium graminacearum* by Plant Growth Promoting Rhizobacteria. *Microb. Res.* **1990,** *156*, 209–223.

Smith, S. E.; Read, D. J. In *Mycorrhizal Symbiosis*, 3rd Ed.; Academic: London, 2008.

Solaiman, M. Z.; Hirata, H. Effect of Arbuscular Mycorrhizal Fungi Inoculation of Rice Seedlings at the Nursery Stage upon Performance in the Paddy Field and Greenhouse. *Plant Soil* **2004,** *19*, 11–12.

Sreeramulu, K. R.; Bagyaraj, D. J. Field Response of Chilli to VAM Mycorrhiza on Black Clayey Soil. *Plant Soil* **1986,** *93*, 299–302.

Sreeramulu, K. R.; Dyala, D. D.; Reddy, T. K.; Patil, R. H. Interaction Effects of VA Mycorrhiza and *Azospirillum* on Growth and Uptake of Nutrients in Maize. *Indian J. Microbiol.* **1988,** *28*, 247–250.

Srihari, P. C. Performance of Arbuscular Mycorrhizal Fungi in Chilli (*Capsicum annuum* L.) and Tomato (*Lycopersicon esculentum* Mill.). Ph.D. Thesis, University of Agricultural Sciences, Dharwad, 1997.

Steenhoudt, O.; Van der Leyden, J. *Azospirillum* a Free-Living Nitrogen Fixing Bacterium Closely Associated with Grasses: Genetic, Biochemical and Ecological Aspects. *FEMS Microbiol. Rev.* **2000,** *24*, 487–506.

Sterzelezyk, E.; Pokojska, B. A. Production of Auxins and Gibberelin-Like Substances by Mycorrhizal Fungi, Bacteria and Actinomyces, Isolated from Soil and the Mycorrhizosphere of Pine (*Pinus silvestris* L.). *Plant Soil* **1984,** *81*, 185–194.

Sturz, A. V.; Nowak, J. Endophytic Communities of Rhizobacteria and the Strategies Required to Create Yield Enhancing Associations with Crops. *Appl. Soil Ecol.* **2000,** *15*, 183–190.

Subba, R. N. S.; Tilak, K. V. B. R.; Singh, C. S. Synergistic Effect of Vesicular-Arbuscular Mycorrhizas and *Azospirillum brasilense* on the Growth of Barley in Pots. *Soil Biol. Biochem.* **1985,** *17*, 119–121.

Sudhakar, P.; Gangwar, S. K.; Satpathy, B.; Sahu, P. K.; Ghosh, J. K.; Saratchandra, B. Evaluation of Some Nitrogen Fixing Bacteria for Control of Foliar Diseases of Mulberry (*Morus alba*). *Indian J. Sericult.* **2019,** *39*, 9–11.

Sughra, H.; Tahir, N.; Muhammad, S. N.; Iqra, L.; Muhammad, J. S.; Rabisa, Z.; Muhammad, S. M.; Asma, I. Rhizosphere Engineering with Plant Growth Promoting Microorganisms for Agriculture and Ecological Sustainability. *Front. Sustain. Food Sci.* **2021,** *5*, 617157. https://doi.org/10.3389/fsufs.2021.617157.

Sureshkumar Singh, S. S.; Tiwar, C.; Dkhar, M. S. Species Diversity of Vesicular Arbuscular Mycorrhizal (VAM) Fungi in Jhum Fallow and Natural Forest Soils of Arunachal Pradesh, North Eastern India. *Trop. Ecol.* **2003,** *44*, 207–215.

Tal, J. J.; Krieg, N. R.; Dobereiner, J. A Taxonomic Study of the *Spirillum lipoferum* Group with Description of a New Genus, *Azospirillum* gen. nov. and Two Species, *Azospirillum lipoferum* (Beijerinck) comb. nov. and *Azospirillum brasilense* nov. *Canad. J. Microbiol.* **1990,** *24*, 967–980.

Thimann, K. V. Fifty Years of Plant Hormone Research. *Plant Physiol.* **1974,** *54*, 450–453.

Tilak, K. V.; Annapurna, K. Role of VAM in the Improvement of Crop Production and Plant Nutrition. In *Nitrogen*; Abrol, Y. P., Ed.; Indian National Science Academy: New Delhi, 1993; pp 155–164.

Tilak, K. V.; Dwivedi, A.; Li, C. Y. Enhancement of Spore Germination of *Glomus fasciculatum* by Bacterial Cell Free Extracts. *Indian J. Exp. Biol.* **1990,** *28*, 373–375.

Tilak, K. V.; Li, C. Y.; Iwan, H. O. Occurrence of Nitrogen-Fixing *Azospirillum* in Vesicular Arbuscular Mycorrhizal Fungi. *Plant Soil* **1989,** *116*, 286–288.

Umali-Garcia, M.; Hubbell, D. H.; Gaskins, M. H.; Dazzo, F. B. Association of *Azospirillum* with Grass Roots. *Appl. Environ. Microbiol.* **2010,** *39*, 219–226.

Veeraswamy, J.; Padmavathi, T.; Venkateswarlu, K. Interaction Effects of *Glomus intraradices* and *Azospirillum lipoferum* on Sorghum. *Indian J. Microbiol.* **1992,** *32*, 305–308.

Will, M. E.; Sylvia, D. M. Interaction of Rhizosphere Bacteria, Fertilizer and Vesicular Arbuscular Mycorrhizal Fungi with Sea Oats. *Appl. Environ. Microbiol.* **1990,** *56*, 2073–2079.

Wright, S. F.; Nichols, K. Glomalin: A Hiding Place for a Third World's Stored Soil Carbon. *Agric. Res.* **2002,** *50*, 4–9.

Wu, S. C.; Caob, Z. H.; Lib, Z. G.; Cheunga, K. C.; Wong, M. H. Effects of Biofertilizer Containing N-Fixer, P and K Solubilizers and AM Fungi on Maize Growth: A Greenhouse Trial. *Geoderma* **2015,** *125*, 155–166.

Xia, Y.; Embley, T. M.; O'Donnell, A. G. Phylogenetic Analysis of *Azospirillum* by Direct Sequencing of PCR-Amplified 16S rDNA. *Syst. Appl. Microbiol.* **2015,** *17*, 197–201.

Young, C.; Juang, T. C.; Chao, C. C. Effect of *Rhizobium* and VAM-Inoculation on Nodulation, Symbiotic Nitrogen Fixation and Soybean Yield in Tropical Fields. *Biol. Fertility Soil* **1988,** 6165–6169.

CHAPTER 13

NITROGEN FIXATION IN RICE-BASED FARMING SYSTEM AND IMPLICATION FOR SUSTAINABLE INTENSIFICATION IN DEVELOPING COUNTRIES

ABDEL RAHMAN MOHAMMAD AL-TAWAHA[1*], GIRMA BIRRU[2], ALLA ALEKSANYAN[3], ABHIJIT DEY[4], SIDDHARTHA PATI[5], ABDEL RAZZAQ AL-TAWAHA[6], PUTERI E. MEGAT WAHAB[7], AMREENA SULTAN[8], R. Z. SAYYED[9], DEVARAJAN THANGADURAI[10], JEYABALAN SANGEETHA[11], VEDAVYAS SHIVANAND CHAVAN[10], MUHAMMAD AASIM[12], EKATERINA KOZUHAROVA[13], MUNIR TURK[14], TAHA A. AL-ISSA[14], DEVENDRA KUMAR PANDEY[15], VISHNU D. RAJPUT[16], ABEER JUBILY[17], and IFTIKHAR ALI[18]

[1]*Department of Biological Sciences, Al Hussein bin Talal University, Maan, Jordan*

[2]*Agroecosystem Management Research Unit, USDA-ARS, NE, USA*

[3]*Institute of Botany aft. A.L. Takhtajyan NAS RA/Department of Geobotany and Plant Eco-Physiology, Yerevan, Armenia*

[4]*Department of Life Sciences, Presidency University, Kolkata, India*

[5]*NatNov Bioscience Private limited, Balasore, Odisha, India*

[6]*National Agricultural Research Center (NARC), Baq'a 19381, Jordan*

[7]*Department of Crop Science, Faculty of Agriculture, University Putra Malaysia, Serdang, Selangor, Malaysia*

[8]*Division of Vegetable Science, SKUAST-K, Shalimar, Srinagar, Jammu and Kashmir, India*

[9]*Department of Microbiology, PSGVP Mandal's Arts, Science and Commerce College, Shahada, Maharashtra, India*

Microbial Fertilizer Technology for Sustainable Crop Production. Jeyabalan Sangeetha, Abdel Rahman Mohammad Al-Tawaha, & Devarajan Thangadurai (Eds.)

DOI: 10.1201/9781003594826-13

[10]Department of Botany, Karnatak University, Dharwad, Karnataka, India

[11]Department of Environmental Science, Central University of Kerala, Kasaragod, Kerala, India

[12]Department of Plant Protection, Faculty of Agricultural Sciences and Technologies, Sivas University of Science and Technology, Sivas, Turkey

[13]Medical University of Sofia, Faculty of Pharmacy, Department of Pharmacognosy, Sofia, Bulgaria

[14]Department of Plant Production, Faculty of Agriculture, Jordan University of Science and Technology, Irbid, Jordan

[15]Department of Biotechnology, School of Bioengineering and Biosciences, Lovely Professional University, Phagwara, Punjab, India

[16]Academy of Biology and Biotechnology, Southern Federal University, Rostov-on-Don, Russia

[17]Centre for Agricultural Research, General Commission for Scientific Agricultural Research, Syria

[18]Center for Plant Sciences and Biodiversity, University of Swat, Charbagh, Pakistan

**Corresponding author.*

ABSTRACT

Rice is the most important crop on the planet, providing a staple meal for more than 2.4 billion people. It occupies the greatest area of plantation in the Asia-Pacific region. About 90% of the world's rice is consumed in the Asian region. Scientists have predicted that rice consumption in the Asian region would increase faster than population growth. The availability of nitrogen is most frequently related to bacteria such as *Herbaspirillum*. Inorganic nitrogen fertilizers being environmentally and economically costly, attention to the use of alternative sources of nitrogen fertilizer has been renewed to give confidence for sustainable agriculture. Biological nitrogen fixation has been demonstrated to have many agronomic, economic, and environmental benefits for rice-based farming systems. This chapter discusses traditional and new aspects of biological nitrogen fixation of rice-based farming systems with examples primarily chosen from developing countries.

13.1 INTRODUCTION

Rice (*Oryza sativa*) is the second-most significant crop in the world, and around 80% of it is grown in irrigated wetland conditions, making it one of the most productive crops on the planet (Roger and Watanabe, 1986). It occupies the greatest area of plantation in the Asia-Pacific region. About 90% of the world's rice is consumed in the Asian region. Hossain (1995) reported that rice consumption in the Asian region would increase faster than population growth. He also expected that by the year 2025, consumption of rice would increase by more than 50%. Similarly, Britto and Kronzucker (2004) reported that by the year 2034, rice production will have to increase to more than 60%. In spite of the growing concern about the harmful consequences of chemical use on the environment and the environment's ability to provide food, high inputs, particularly fertilizers, pesticides, and herbicides, have resulted in improved yields, as highlighted by experts in a number of countries (Turk et al., 2003a, 2003b, 2003c; Nikus et al., 2004a, 2004b; Tawaha and Turk, 2004; Turk and Tawaha, 2004; Hayyawi et al., 2018; Amanullah et al,. 2019; Alu'datt et al., 2019; Hani et al., 2019). Soil fertility refers to the soil's ability to support plant growth while also optimizing crop output. Organic and inorganic fertilizers applied to the soil can help to improve the soil's fertility. Nuclear techniques provide information that can be used to improve soil fertility and crop output while also reducing environmental impact (Al-Tawaha et al., 2005; Lee et al., 2005; Sulpanjan et al., 2005; Tawaha et al., 2005; Assaf et al., 2006; Turk et al., 2006; Al-Kiyyam et al., 2008; Al-Ajlouni et al., 2009; Al-Tawaha et al., 2010a, 2010b, 2018a, 2018b). On the other hand, sustainability in an organic farming system is based on managing organic matter to maintain soil health and provide nutrients to developing plants, whereas sustainability in an industrial farming system is based on managing organic matter to maintain soil health and provide nutrients to developing plants (Turk and Tawaha, 2001; Tawaha and Turk, 2002a, 2002b; Turk and Tawaha, 2002a, 2002b; Tawaha et al., 2003; Abebe et al., 2005; Al-Tawaha et al., 2005, 2010a, 2010b, 2017; Elser et al., 2007; Al-Kiyyam et al., 2008; Hameed et al., 2008; Ananthi et al., 2017; Abu Obaid et al., 2018). Biological nitrogen fixation is a process carried out only by prokaryotic microorganisms including root nodule bacteria (*Rhizobium*, *Bradyrhizobium*, *Sinorhizobium*, and *Azorhizobium*), actinomycetes (*Frankia*), and cyanobacteria (*Anabaena* and *Nostoc*) (Young, 1992; Young and Haukka, 1996). Rice has the ability to fix the atmospheric nitrogen in the soil through their relationship with bacteria such as *Herbaspirillum* (Yoshida and Ancajas, 1971; Dommergues et al., 1973; James and Olivares, 1998; Reinhold-Hurek, 1998; Biswas et al., 2001; Tan et al., 2001; Peng et al., 2002). Biological nitrogen fixations in a cropping system are thought to have several beneficial effects including saving energy, cost of production, environment, labor, and soil, and alleviating hunger. Most of the earlier research on the prospect of biological nitrogen fixation in a rice-based farming system was conducted in developing countries. Hoflich et al. (1995) found that *Rhizobium leguminosarum* bv. *Trifolii* develops natural, intimate associations with rice root. They also reported that this association could increase yield under experimental field conditions. Many factors such as rice varieties, cultural conditions, climatic factors, and inoculant strains are the major factors affecting the association between rice and rhizobia (Biswas et al., 2000). Research conducted at the International Rice Research Institute (IRRI), in the Philippines, found that rice suffers from a disparity of its N incest and is supplied as fertilizer, resulting in 60% loss of applied N fertilizer (IRRI, 1999). It is now generally decided that increasing the ability of the rice system to fix its N, would be one of the most significant contributions biotechnology could make to agriculture. The advantages of nitrogen fixation for rice-based farming systems are numerous. The use of biological nitrogen fixation in agriculture has been shown to have numerous agronomic, economic, and environmental benefits.

13.2 AGRONOMIC AND ENVIRONMENTAL BENEFITS

It is possible to improve the rate of seeding emergence, radical elongation, height and dry matter, plumule length, cumulative leaf and root areas, grain yield and straw yield, single-leaf net photosynthetic, nutrient uptake, and soil properties in rice-cropping systems by incorporating biological nitrogen fixation into the system (Tien et al., 1979; Hussain et al., 1987; Islam and Whitton, 1992; Watababe and Liu, 1992; Banu et al., 1993; Tuzun and Kloepper, 1994; Hoflich et al., 1995; Islam and Hesham, 1995; Chabot et al., 1996; Noel et al., 1996; Handlesman and Staab, 1996; Heshem, 1997; Nautiyal, 1997; Yanni et al., 1997; Heshem, 1998). A study found that the use of biological nitrogen fixation can enhance the environment by lowering the level of ground water pollution caused by nitrate and decreasing greenhouse gas emissions (Watanabe, 1992; Wani et al., 1995; Vance, 1997; Nautiyal, 1997; Van Kammen, 1997).

13.3 SOME OF THE USEFUL MICROORGANISMS FOR NITROGEN FIXATION IN RICE CROPPING SYSTEMS

13.3.1 AZOLLA

Azolla is an aquatic fern that is native to Asia, Africa, and the Americas. It lives naturally on the water surface of rice fields (Gevrek, 2000). *Azolla* has been successfully used as feed for pigs, ducks, and fish (Khan, 1983) and can also be used in salads and sandwiches because it has a high protein content of 23–37% (Wagner, 1997). *Azolla* has been of interest to botanists and agriculturists for years because it's considered a nitrogen source in the rice-based farming system (Talley, 1977; Berja, 1983; Khan, 1983; Watanbe, 1984; Watanabe 1985; Gevrek, 2000). Azolla has been used as a green manure for wetland rice-based systems in central to southern China and northern Vietnam for centuries (Watanabe and Liu, 1992). At present, Azolla is used in more than a million hectares of wetland rice-based systems in central to southern China (Gevrek, 2000). Azolla are thought to have several beneficial effects in wetland rice-based systems including producing nitrogen fertilizer for increasing crop yields, maintaining soil fertility, and suppressing the growth of aquatic weeds (Watanbe, 1985; Gevrek, 1999; Gevrek, 2000). Azolla can have a symbiotic relationship with the nitrogen-fixating, blue-green algae *Anabaena azollae*. Further, the subtle fern gives nutrients and a defensive leaf cavity for the anabaena, which in turn provides nitrogen for the fern (Gevrek, 1999; Gevrek, 2000). Azolla can fix 4 kg N/ha/day, it can provide a potential nitrogen source for rice (Ladha et al., 1993). Farmers, however, generally have little economic inducement to select Azolla over inorganic N since additional costs such as labor, irrigation, and seed/inoculum, are involved (Ladha et al., 1997). Azolla is extremely sensitive to stress conditions, including low temperature, moisture deficiency, and P deficiency (Peters et al., 1980; Subundhi and Watanabe, 1981; Vu et al., 1986). A considerable international effort will be needed to be directed into areas such as the selection of Azolla with improved characteristics such as stress and pest resistance.

13.3.2 CYANOBACTERIA

Cyanobacteria are organisms that have been known to fix a large amount of N and maintain the fertility of rice soils and postulated that rice fields have been known to provide ideal conditions for the growth of cyanobacteria (Islam and Whitton, 1992; Watababe and Liu, 1992; Banu et al., 1993; Heshem et al., 1995; Islam and Heshem, 1995; Heshem, 1997; Heshem, 1998). Takenaka et al. (1998) reported that cyanobacteria such as *Nostoc* and *Spirulina* have been used as a potential food because they contain protein and vitamins. On the other hand, Burja et al. (2001) reported that cyanobacteria are thought to have several beneficial effects including anti-HIV, and antimicrobial activities. They also reported that cyanobacteria such as *Anabaena* and *Nostoc* spp. have the ability to produce over 20 natural products having medicinal activity. Many studies have demonstrated the importance of cyanobacteria for rice cultivation. Hamdi (1982) reported that cyanobacteria are being used as biofertilizers in Burma, China, the Philippines, Thailand, Vietnam, India, Bangladesh, and Egypt. This may be related to the fact that cyanobacteria can contribute to decreasing the cost of chemical fertilizers as well as decreasing soil and water pollution caused by the use of chemical fertilizers (Islam and Whitton, 1992; Watababe and Liu, 1992; Banu et al., 1993; Heshem et al., 1995, 1997, 1998; Islam and Heshem, 1995). In developing countries, such as Burma, China, the Philippines, Bangladesh, and Egypt, in the total absence of chemical fertilizer, cyanobacteria can fix 25–30 kg N ha^{-1} $crop^{-1}$ (Watababe and Liu, 1992; Banu et al., 1993; Heshem, 1998). In a study conducted by Heshem (2001) in Bangladesh to study the problems and prospects of cyanobacterial biofertilizer for rice cultivation, he concluded that cyanobacteria increased rice grain yield in acid, saline, and red soil than those in calcareous soil and neutral soil. He also reported that cyanobacteria can support the rice plants by 30% N in saline, acid, and red soil. Further studies demonstrated that cyanobacteria increased rice grain yield under low soil fertility (Valiente et al., 2000). Heshem (2001) found that straw yields are influenced by cyanobacterial biofertilizers, and he stated that the highest straw yield was observed in acid soil in treatment with cyanobacterial biofertilizers. In saline soil, cyanobacterial biofertilizers caused a marked increase in N uptake in rice (Heshem et al., 1995; Heshem, 1997, 1998). Heshem (2001) found that cyanobacterial biofertilizers have significant effects on soil properties, and concluded that soil pH was reduced in calcareous and increased in acid, saline, and red soils by the application of biofertilizers. He also found that electrical conductivity (EC) was reduced significantly in acid, saline, red, and calcareous soils. Similarly, Uddin (1996) in Bangladesh found that cyanobacteria have the ability to reduce the electrical conductivity, as well as the pH of the soil. Heshem (2001) reported that the indigenous highest cyanobacterial populations were found in neutral and calcareous soils as compared with acid, saline, and red soils. In a study conducted by Banu et al. (1993), to study the growth and nitrogen fixation of cyanobacteria isolated from saline rice soils of Bangladesh, and concluded that the highest N_2 fixing capacity of cyanobacteria had been found in saline soil. Further studies demonstrated that cyanobacteria have the ability to improve the fertility status, chiefly of N, P, and S (Heshem, 2001).

13.3.3 RHIZOBIUM

Direct growth endorsement of no legumes by bacterial inoculation has been reported (Hoflich et al., 1995; Chabot et al., 1996; Noel et al., 1996; Yanni et al., 1997). Recently, Biswas et al. (2000) studied the association between rice and rhizobia using Rhizobium strains (*Rhizobium leguminosarum* Bv. *Trifolii* E11 and *Rhizobium* sp. IRBG74), and concluded that these bacteria can improve the rate of seedling emergence, radical, elongation, height and dry matter, plumule, length, cumulative leaf and root areas, and grain yield and straw yield. Similarly, Chaintreuil et al. (2000) reported a 20% increase in the shoot growth and grain yield of wild rice (*Oryza breviligulata*) grown in a greenhouse in response to inoculation with one *Bradyrhizobium* strain. In greenhouse experiments conducted by Peng et al. (2002) at the International Rice Research Institute (IRRI), Los Banos, Philippines, to demonstrate the effect of Rhizobial inoculation on grain yield of rice, they found that Rhizobial inoculation could increase grain yield by 16%. They also reported that the increase in grain yield was due to an increase in biomass production rather than harvest index. They also reported that Rhizobial inoculation could increase sink size by increasing spikes number of grains per spike. On the other hand, Yanni et al. (1997) found that Rhizobial inoculation increased rice grain yield by 45% in a field inoculation experiment. Recently Biswas et al. (2000) reported that inoculated rhizobia influences the yield status of inoculated plants by increasing root growth. Concerning the effect of Rhizobial inoculation on the photosynthesis of rice, Peng et al. (2002) reported that Rhizobial inoculation increased single-leaf net photosynthetic rates compared with the uninoculated control. They also reported that Rhizobial inoculation had limited impact on stomatal conductance, specific leaf weight, and leaf N content. Regarding the effect of Rhizobial inoculation on nutrient uptake of rice, Biswas et al. (2000) found that compared to an unrelated control, the Rhizobial inoculation increased uptake of N, P, K, and Fe and postulated that this increase might be related to morphological changes in rice roots, especially increased root number thickness and length. Similarly, Barber (1985) reported that higher Fe and K uptake is related to thicker roots. Tiller capacity is important in determining the productivity of rice plants. It has previously been reported that Rhizobial inoculation increased tiller production and postulated that this increase might be related to the change in hormonal balance (Biswas et al., 2000). Lebuhn and Hartman (1993) found that higher auxin concentrations in rhizosphere soils harboring rhizobia. Further studies demonstrated that Rhizobial inoculation could increase the levels of Indole-3-acetic acid (Biswas et al., 2000). Others have demonstrated that Rhizobial inoculation may induce the growth-promoting effects, including phytohormone production, fungal growth inhibition, N_2 fixation, antibiosis against phytopathogens, and induction of systemic disease resistance (Tien et al., 1979; Hussain et al., 1987; Urquiaga et al., 1992; Tuzun and Kloepper, 1994; Chabot et al., 1996; Handlesman and Staab, 1996; Nautiyal, 1997), thus improving the plant ability to take up elements from soil. Biswas et al. (2000) reported that inoculation of rice seedling (*Oryza sativa* L.); with *Rhizobium leguminosarum* Bv. *Trifolii* can improve growth possibly through mechanisms that include changes in growth physiology and foot morphology rather than biological nitrogen fixation. In the past, several workers have studied the association between rice and rhizobia, and reported several physiological responses in the host including such as the ability of rhizobia to attach to rice root system, elect the root hair deformation, and form nodule-like structures or form thick short lateral roots (Bender et al., 1990; Terouchi and Syono, 1990; Jing et al., 1992; Cocking et al., 1993; De Bruijin et al., 1995). Toenniessen (1991) found that rice has the genes essential to form nodules but does not produce the chemical signals. In a study conducted by Elbeltagy et al. (2001) to focus on the question of whether the bacteria has the ability to fix nitrogen in the rice plants, they found that *Herbaspirillum* sp. which can be isolated from wild rice species has the ability to fix nitrogen in *Oryza* species. They also reported that nitrogen-fixing endophytes are higher in stems than roots, and postulated that rice stems may provide a better niche. On the other hand, Cocking et al. (1994) stated that

Azorhizobium caulinodans ORS571 which can be isolated from stem nodules of the tropical legume *Sesbania rostrata* has the ability to active nitrogen fixation of rice. They also reported that we need a small number of legume genes to be transferred to rice to create the essential signals for effectual symbiotic nitrogen-fixing interaction with soil rhizobia. The most important factors affecting nitrogen fixation in rice-based farming systems. There are a variety of factors that influence nitrogen fixation in the rice farming system, including rice genotypes and pests as well as soil acidity and salinity, drought, and soil fertility, as well as pesticides and flooding. Inoculant strains are one of the most important factors that influence nitrogen fixation in the rice farming system (Watababe and Liu, 1992; Islam and Whitton, 1992; Banu et al., 1993; Islam and Heshem, 1995; Heshem et al., 1995; Heshem, 1997, 1998; Biswas et al., 2000).

13.4 INOCULANT STRAINS AND RICE VARIETIES

In general, the effects of bacterial strains on nitrogen fixation are opposing; studies demonstrated that some bacterial strains have the ability to fix nitrogen in rice; others have little effect (Cocking et al., 1994; Yanni et al., 1997; Prayitno et al., 1999; Elbeltagy et al., 2001). Recently, Eibeltagy et al. (2001) found that *Herbaspirillum* sp., which can be isolated from wild rice species, has the ability to fix nitrogen in *Oryza* species. Cocking et al. (1994) observed similar results and found that *Azorhizobium caulinodans* ORS571 which can be isolated from stem nodules of the tropical legume *Sesbania rostrata* has the ability to activate nitrogen fixation of rice. On the other hand, Biswas et al. (2000) reported that inoculation of rice seedlings (*Oryza sativa* L.), with *Rhizobium leguminosarum* Bv. *Trifolii* can improve growth possibly through mechanisms that include changes in growth physiology and root morphology their than biological nitrogen fixation.

Earlier studies showed positive growth responses of rice to certain Rhizobial strains (Yanni et al., 1997; Prayitno et al., 1999). The greenhouse experiments performed by Biswas et al. (2000) at the International Rice Research Institute demonstrated the effects of Rhizobial inoculation in the growth of lowland rice, and concluded that grain yield was higher when inoculated with strain E11, IRBG74, or IRBG271, and straw yield were increased when inoculated with strain E11 or IRBG74. In study conducted by Peng et al. (2002) to determine whether inoculation with rhizobia could influence leaf photosynthesis of rice plants under greenhouse conditions, and concluded that the photosynthetic rate increased by 14% following inoculation with strain IRBG74 compared with untreated control while the other two strains including E12 and IRBG271 increased photosynthetic rates by 9–12%, respectively. Concerning the response of rice cultivars to bacterial inoculation, Prayitno et al. (1999) reported significant differences between rice cultivars in their ability to respond to bacterial inoculation. Similarly, Shrestha and Ladha (1996) found genotypic variation in the promotion of rice nitrogen fixation. Further studies demonstrated variations between rice cultivars in their ability to respond to bacterial inoculation (Biswas et al., 2000). In general, the effects of fertilization on nitrogen fixation are contradictory; some research found that N_2 fixation in rice fields may be controlled by single elements may (Howarth et al., 1988; Smith, 1990). It has been reported that nitrogen fixation is inhibited by N fertilization (Howarth et al., 1988), and stimulated by P fertilization (Smith, 1990). It has previously been reported that a higher grain yield when rice plants were inoculated with rhizobia in the presence of higher N fertility (Yanni et al., 1997). Biswas et al. (2000) observed similar results and found that positive growth responses of rice to Rhizobial inoculation were improved rather than dormant at higher levels of added N fertilizer. In general, the effects of soil salinity on nitrogen fixation are contradictory and appear to be related to the salinity levels. Some researchers found that the highest N_2-fixing capacity of cyanobacteria were found in saline soil (Banu et al., 1993; Quesada and Fernandez-Valiente, 1996), while Pinckney et al. (1995) found that higher salinities inhibited cyanobacterial nitrogenase activity. Heshem (2001) found that the highest indigenous cyanobacterial populations were found in neutral and calcareous soils as compared with acid, saline, and red soils (Figure 13.1).

13.5 KEY FACTORS AFFECTING NITROGEN FIXATION IN RICE PRODUCTION SYSTEMS

Light is one of the most important factors affecting nitrogen fixations in rice plants. Many studies have demonstrated the importance of light in nitrogen fixations (Alimango and Yoshida, 1977; Rother et al., 1988; Rother and Whitton, 1989, Roger et al., 1991). It has been reported that nitrogenase activity is activated by light (Babout et al., 1987). In a study conducted at

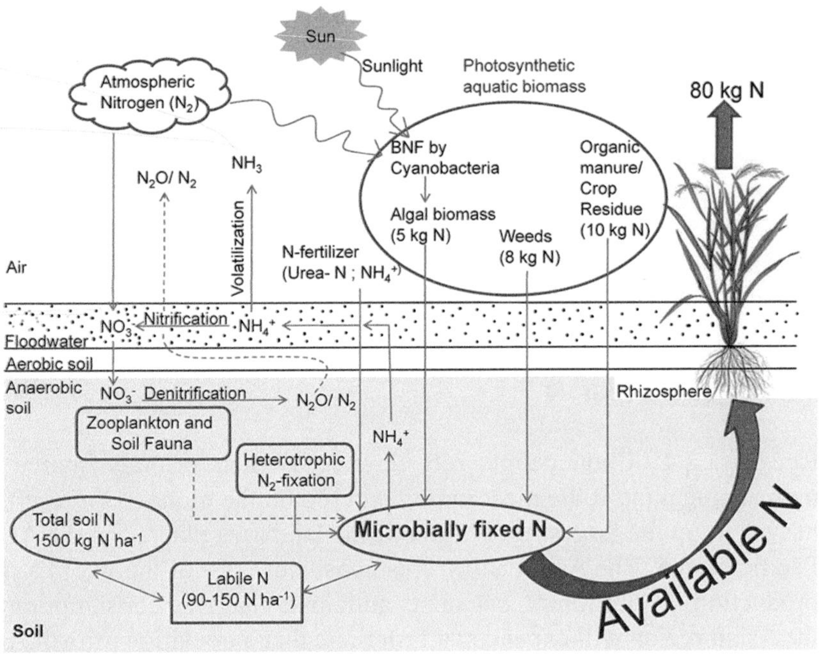

FIGURE 13.1 Nitrogen cycle and nitrogen contributions in rice field ecosystem.

Source: Reproduced with permission from Kaushik et al. (2018), Copyright © 2018, Springer Nature.

the International Rice Research Institute, Roger and Kulasooriya (1980) examined the long-term effects of light regime on nitrogenase activity and concluded that daily and seasonal variation in nitrogenase activity has been interpreted as a purpose of light variation. Quesada (1993) found in the Valencian rice field that some of cyanobacteria can be activated by highlight intensity, while some authors reported that highlight intensities could inhibit nitrogenase activity by inhibiting photosynthesis. In a study conducted by Quesada et al. (1998) assessed how light regime modulates N_2 fixation in Spanish rice fields and concluded that nitrogenase activity was higher under natural light than under experimental light intensity. Agricultural botanists and agriculturists have been speculating about the prospects of nitrogen fixation in rice-based farming systems for many years. Researchers have followed several approaches such as (1) the transfer to and expression of nitrogen fixation genes in rice itself (Dixon et al., 1997), (2) the detection of endophytic interactions of N-fixing organisms with nonlegumes, and (3) improvements in the presently available nitrogen-fixing systems.

On the other hand, many factors inhibit nitrogen fixation in rice-based farming systems such as acid and alkaline soils, soil salinity, and element deficiencies, it is now generally decided that finding cultivar-strain combinations resistant to these stresses would be one of the most significant contributions biotechnology could make to agriculture. A lot of efforts are required to increase understanding of plant genes involved in nodule-like structure development and nitrogen fixation.

13.6 CONCLUSION

Roughly, 2.4 billion people rely on rice as their primary source of nutrition, making it the most important crop on the globe. As a result, it has taken up the largest proportion of the land area planted in the Asia-Pacific region. The Asia-Pacific area consumes 90% of the world's rice production. In the future, scientists anticipate that rice consumption in the Asian region will expand at a faster rate than population growth. The presence of bacteria such as *Herbaspirillum* is most usually associated with the availability of nitrogen. Because inorganic nitrogen fertilizers are both environmentally and economically harmful, renewed emphasis has been paid to the use of alternative sources of nitrogen fertilizer in order to boost trust in the future of sustainable agriculture. Numerous agronomic, economic, and environmental benefits have been established for rice-based agricultural systems that utilize biological nitrogen-fixation techniques.

KEYWORDS

- **bacteria inoculation**
- **farming system**
- **nitrogen fixation**
- **rice**
- **sustainable agriculture**

REFERENCES

Abebe, G.; Hattar, B.; Al-Tawah, A. M. Nutrient Availability as Affected by Manure Application to Cowpea (*Vigna unguiculata* L. Walp.) on Calacarious Soils. *J. Agric. Soc. Sci.* **2005,** *1* (1), 1–6.

Abera, T.; Feyisa, D.; Yusuf, H.; Nikus, O.; Al-Tawaha, A. R. Grain Yield of Maize as Affected by Biogass Slurry and NP Fertilizer Rate at Bako, Western Oromiya, Ethopia. *Biosci. Res.* 2005, *2*, 31–38.

Al-Ajlouni, M. M.; Al-Ghzawi, A. L. A.; Al-Tawaha, A. R. Crop Rotation and Fertilization Effect on Barley Yield Grown in Arid Conditions. *J. Food Agric. Environ.* **2010,** *88* (3), 869–872.

Al-Juthery, H. W.; Habeeb, K. H.; Jawad, F.; Altaee, K.; AL-Taey, D. K. A.; Al-Tawaha, A. R. M. Effect of Foliar Application of Different Sources of Nano-Fertilizers on Growth and Yield of Wheat. *Biosci. Res.* **2010,** *15* (4), 3988–3997.

Alimango, B. V.; Yoshida, T. In Situ Acetylene-Ethylene Assay of Biological Nitrogen Fixation in Lowland Rice Soils. *Plant Soil* **1977,** *47*, 239–244.

Al-Kiyyam, M. A.; Turk, M., Al-Mahmoud, M.; Al-Tawaha, A. R. Effect of Plant Density and Nitrogen Rate on Herbage Yields of Marjoram under Mediterranean Conditions. *J. Am.-Eurasian J. Agric. Environ. Sci.* **2008,** *3*, 153–158

Al-Tawaha, A. R. M.; Nidal, O. Use of Sorghum and Maize Allelopathic Properties to Inhibit Germination and Growth of Wild Barley (*Hordeum spontaneum*). *Not. Bot. Horti Agrobo. Cluj-Na.* **2010,** *38* (3), 124–127.

Al-Tawaha, A. M.; Turk, M.; Hameed, K.; Assaf, T.; Tahhan, R.; Al-Jamali, A. F. Effects of Application of Olive Mill by-Products on Lentil Yield and their Symbiosis with Mycorrhizal Fungi under Arid Conditions. *Crop Res.* **2010a,** *40* (3), 66–75.

Al-Tawaha, A. R.; McNeil, D. L.; Yadav, S. S.; Turk, M.; Ajlouni, M.; Abu-Darwish, M. S.; Al-Ghzawi, A. L. A.; Al-udatt, M.; Aladaileh, S. Integrated Legume Crops Production and Management Technology. In *Climate Change and Management of Cool Season Grain Legume Crops*; Yadav, S. S., David, L., McNeil, R. R., Sharanagouda, A.; Eds.; Springer: Dordrecht, 2010b; pp 325–349.

Al-Tawaha, A. R.; Al-Karaki, G.; Al-Tawaha, A. R.; Sirajuddin, S. N.; Makhadmeh, I.; Wahab, P. E. M.; Youssef, R. A.; Al Sultan, W.; Massadeh, A. Effect of Water Flow Rate on Quantity and Quality of Lettuce (*Lactuca sativa* L.) in Nutrient Film Technique (NFT) under Hydroponics Conditions. *Bulg. J. Agric. Sci.* **2018a,** *245*, 791–798.

Al-Tawaha, A. R.; Al-Tawaha, A. R.; Aludatt, M.; Al-Ghzawi, A. L.; Wedyan, M.; Al-Obaidy, S. D. A.; Al-Ramamneh, E. A. D. Effects of Soil Type and Rainwater Harvesting Treatments in the Growth, Productivity and Morphological Trains of Barley Plants Cultivated in Semi-Arid Environment. *Aust. J. Crop Sci.* **2018b,** *12* (6), 975–979.

Al-Tawaha, A. R.; Turk, M. A.; Abu-Zaitoon, Y. M.; Aladaileh, S. H.; Al-Rawashdeh, I. M.; Alnaimat, S.; Al-Tawaha, A. R. M.; Alu'datt, M. H.; Wedyan, M. Plants Adaptation to Drought Environment. *Bulg. J. Agric. Sci.* **2017,** *23* (3), 381–388.

Alu'datt, M. H.; Rababah, T.; Alhamad, M. N.; Al-Tawaha, A.; Al-Tawaha, A. R.; Gammoh, S.; Ereifej, K. I.; Al-Karaki, G.; Hamasha, H. R.; Tranchant, C. C.; Kubow, S. Herbal Yield, Nutritive Composition, Phenolic Contents and Antioxidant Activity of Purslane

(*Portulaca oleracea* L.) Grown in Different Soilless Media in a Closed System. *Ind. Crops Prod.* **2019,** *141*, 111746.

Al Zoubi, O. M.; Al-Tawaha, A. R. Allelopathic Effect of Beetroot (*Beta vulgaris* L.) on Germination and Growth *Zea mays* and *Vigna umbellata*. *Int. J. Bot. Stud.* **2019,** *4* (4), 47–51.

Ananthi, T.; Amanullah, M. M.; Al-Tawaha, A. R. M. S. A Review on Maize-Legume Intercropping for Enhancing the Productivity and Soil Fertility for Sustainable Agriculture in India. *Adv. Environ. Biol.* **2017,** *11* (5), 49–63.

Banu, N. A.; Hashem, M. A.; Islam, M. R. Growth and Nitrogen Fixation of Blue-Green Algae (cyanobacteria) Isolated form Saline Rice Soils of Bangladesh. *Bangladesh J. Soil Sci.* **1993,** *24*, 63–72.

Barber, S. A. Potassium Availability at the Soil-Root Interface and Factors Influencing Potassium Uptake. In *Potassium in Agriculture*; Munson, R. D., Eds.; ASA, CSSA, and SSSA: Madison, WI, 1985; pp 309–324.

Bebout, B. M.; Paerl, H. W.; Crocker, K. M.; Prufert, L. E. Diel Interaction of Oxygenic Photosynthesis and N_2 Fixation in a Marine Mat Community. *Appl. Environ. Microbiol.* **1987,** *53*, 2353–2362.

Bender, G. L.; Preston, L.; Barnar, D.; Rolfe, B. G. Formation of Nodule-Like Structures on the Roots of Non-Legumes Rices and Wheat. In *Nitrogen Fixation: Achievements and Objectives*; Gresshoff, P. M., Roth, L. E., Stacey, G., Newton, W. E., Eds.; Chapman and Hall: London, 1990; p 825.

Berja, N. S. Growth of Four Species of Azolla as Affected by Temperature. *Aquatic Bot.* **1983,** *15*, 175–185.

Biswas, J. C.; Ladha, J. K.; Dazzo, F. B. Rhizibia Inoculation Improves Nutrient Uptake and Growth of Lowland Rice. *Soil Sci. Soc. Am. J.* **2000,** *64*, 1644–1650.

Britto, D. T.; Kronzucker, H. J. Bioengineering Nitrogen Acquisition in Rice: Can Novel Initiatives in Rice Genomics and Physiology Contribute to Global Food Security. *BioEssays* **2004,** *26*, 683–692.

Burja, A. M.; Banaigs, B.; Abou-Mansour, E.; Burgess, J. G.; Wright, P. C. Marine Cyanobacteria a Prolific Source of Natural Products. *Tetrahedron* **2001,** *57*, 9347–9377.

Chabot, R.; Antoun, H.; Cescas, M. C. Growth Promotion of Maize and Lettuce by Phosphate-Solubilizing *Rhizobium leguminosarum* Biovar *phaseoli*. *Plant Soil* **1996,** *184*, 311–321.

Chaintreuil, C.; Giraud, E.; Prin, Y.; Lorquin, J.; Ba, A.; Gillis, M.; de Lajudie, P.; Dreyfus, B. Photosynthetic *Bradyrhizobium* are Natural Endophytes of the African Wild Rice *Oryza breviligulata*. *App. Environ. Microbiol.* **2000,** *66*, 5437–5447.

Cocking, E. C.; Srivastava, J. S.; Cook, J. M.; Kothari, S. L.; Davey, M. R. Studies on Nodulation of Maize, Wheat, Rice and Oilseed Rape: Interactions of Rhizobia with Emerging Lateral Roots. In *Biological Nitrogen Fixation - Novel Associations with Non-Legume Crops*; Yanfu, N., Kennedy, I. R., Tingwei, C., Eds.; Qingdao Ocean University Press: Qingdao, pp 53–58.

Cocking, E. C.; Webster, G.; Batchelor, C. A.; Davey, M. R. Nodulation of Non-Legumes Crops. A New Look. *Agro-Food-Ind. Hi- Tech.* **1994,** *1*, 21–24.

De Bruijn, F. J.; Jing, Y.; Dazzo, F. B. Potential and Pitfalls of Trying to Extend Symbiotic Interactions of Nitrogen-Fixing Organisms to Presently Non-Nodulated Plants, Such as Rice. *Plant Soil* **1995,** *172*, 207–219.

Dixon, R.; Cheng, Q.; Shen, G. F.; Day, A.; Dowson-Day, M. Nif Gene Transfer and Expression in Chloroplast: Prospects and Problems. *Plant Soil* **1997,** *194*, 193–203.

Dommergues, Y.; Balandreau, J.; Rinaudo, G.; Weinhard, P. Non-Symbiotic Nitrogen Fixation in the Rhizosphere of Rice, Maize and Different Tropical Grasses. *Soil Biol. Biochem.* **1973,** *5*, 83–89.

Elbeltagy, A.; Nishioka, K.; Sato, T.; Suzuki, H.; Ye, B.; Haada, T.; Isawa, T.; Mitsui, H.; Minamisawa, K. K. Endophytic Colonization and in Planta Nitrogen Fixation by a *Herbaspirillum* sp. Isolated from Wild Rice Species. *Appl. Environ. Microbiol.* **2001,** *67*, 5285–5293.

Gallon, J. R. Reconciling the Incompatible: N_2 Fixation and O_2. *New Phytol.* **1992,** *122*, 571–609.

Gevrek, M. N. The Adaptability of *Azolla mexicana* in the Ege Region of Turkey. *Int. Rice Res. Notes (IRRN)* **1999,** *24*, 31.

Gevrek, M. N. A Study on Azolla Nitrogen Source in Rice Farming System. *Turk. J. Agric. For.* **2000,** *24*, 165–172.

Gyaneshwar, P.; James, E. K.; Mathan, N.; Reddy, P. M.; Reinhold-Hurek, B.; Ladha, J. Endophytic Colonization of Rice by a Diazotrophic Strain of *Serratia marcescens*. *J. Bact.* **2001,** *183*, 2634–2645.

Hamdi, Y. A. Application of Nitrogen Fixing Systems in Soil Management. FAO Soils Bulletin 49, Rome, Italy, 1982.

Handlesman, J.; Staab, E. Biocontrol of Soilborne Plant Pathogens. *Plant Cell* **1996,** *8*, 1855–1869.

Heshem, M. A. Role of Blue-Green Algae in Improving Soil Fertility and Reclaiming Salinity of Soil. Annual Report of BAURES Financed Research Project. BAU: Mymensingh, India, 1997.

Heshem, M. A. Ecophysiological Studies of Cyanobacteria in Paddy Soils Bangladesh. In *Development in Plant and Soil Sciences. Proceedings of the 7th International Symposium on N_2 fixation with Non-Legumes*; Malik, K. A., Mirza, M. S., Ladha, J. K., Eds.; Kluwer Academic Publishers: Dordrecht, The Netherlands, 1998; pp 333–344.

Heshem, M. A. Problem and Prospects of Cyanobacterial Biofertilizer for Rice Cultivation. *Aust. J. Plant Physiol.* **2001,** *28*, 881–888.

Heshem, M. A.; Islam, M. R.; Banu, N. A. Nitrogen Fixation and Salinity Reductions by Blue-Green Algae. *Bangladesh J. Agric. Sci.* **1995,** *22*, 251–256.

Hoflich, G.; Wiehe, W.; Buchholz, C. H. Rhizosphere Colonization of Different Crops with Growth Promoting *Pseudomonas* and *Rhizobium* Bacteria. *Microbial. Res.* **1995,** *150*, 139–147.

Hossain, M. Sustaining Food Security for Fragile Environments in Asia: Achievements, Challenges, and Implications for Rice Research. In *Fragile lives in Fragile Ecosystems*; IRRI: Los Banos, Philippines, 1995; pp 3–23.

Howarth, R. W.; Marino, R.; Cole, J. J. Nitrogen Fixation in Freshwater, Estuarine, and Marine Ecosystems. 2. Biogeochemical Controls. *Limnol Oceanogr.* **1988,** *33*, 688–701.

Hussain, A.; Arshad, M.; Hussain, A.; Hussain, F. Response of Maize (*Zea mays*) to *Azotobacter* Inoculation under Fertilized and Unfertilised Conditions. *Biol. Fertil. Soils* **1987,** *4*, 73–77.

Islam M., R.; Hashem, M. A. Role of Blue-Green Algae in Improving Soil Environment. *Bangladesh J. Training Dev.* **1995,** *8*, 112–118.

Islam, M. R.; Whitton, B. A. Cell Composition and Nitrogen Fixation by Cyanobacterium (Blue-Green Alge) *Calothrix* D764. *Microbios* **1992,** *69*, 77–88.

James, E. K.; Olivares, F. L. Infection and Colonization of Sugarcane and Other Graminaceous Plants by Endophytic Diazotrophs. *Crit. Rev. Plant Sci.* **1998,** *17*, 77–119.

Jeanjean, R.; Matthijs, H. C. P.; Onana, B.; Havaux, M.; Joset, F. Exposure of the *Cyanobacterium synechocystis* PCC6803 to Salt Stress Induces Concerted Changes in Respiration and Photosynthesis. *Plant Cell Physiol.* **1993,** *24*, 1073–1079.

Jing, Y.; Li, G.; Shan, X. Development of Nodule-Like Structure on Rice Roots in Nodulations and Nitrogen Fixation in Rice. International Rice Research Institute: Manila, 1992; pp 123–126.

Kaushik, M. S.; Kumar, A.; Abraham, G.; et al. Field Evaluations of Agrochemical Toxicity to Cyanobacteria in Rice Field Ecosystem: A Review. *J. Appl. Phycol.* **2018,** *31*, 471–489. https://doi.org/10.1007/s10811-018-1559-2.

Khan, M. M. A Primer on *Azolla* Production and Utilization in Agriculture. UPLB, PCARRD and SEARCA, Los Banos, 1983.

Ladha, J. K., De Brujin, F. J., Malik, K. A. Introduction: Assessing Opportunities for Nitrogen Fixation in Rice - A Frontier Project. *Plant Soil* **1997,** *194*, 1–10.

Ladha, J. K.; Tirol-Padre, A.; Reddy, C. K.; Ventura, W. Prospects and Problems of Biological Nitrogen Fixation in Rice Production: A Critical Assessment. In *New Horizons in Nitrogen Fixation*; Palacios, R., Mora, J., Newton, W. E., Eds.; Kluwer Academic Publishers: The Netherlands, 1993; pp 677–682.

Lebuhn, M.; Hartmann. Method for the Determination of Indole-3-Acetic Acid and Related Compounds for L-Tryptophan Catabolism in Soils. *J. Chromatogr.* **1993,** *629*, 255–266.

Nautiyal, C. S. Rhizosphere Competence of *Pseudomonas* sp. NBR 19926 and *Rhizobium* sp. NBR 19513 Involved in the Suppression of Chickpea (*Cicer arietinum* L.) Pathogenic Fungi. *FEMS Microb. Ecol.* **1997,** *23*, 145–158.

Nikus, O.; Al-Tawaha, A. M.; Turk, M. A. Effect of Manure Supplemented with Phosphate Fertilizer on the Fodder Yield and Quality of Two Sorghum Cultivars (*Sorghum bicolor* L.). *Biosci. Res.* **2004a,** *1*, 1–7.

Nikus, O.; Al-Tawaha, A. M.; Turk, M. A. Yield Response of Sorghum (*Sorghum bicolor* L.) to Manufacture Supplemented with Phosphate Fertilizer under Semi-Arid Mediterranean Conditions. *Int. J. Agric. Biol.* **2004b,** *6*, 889–893.

Noel, T. C.; Sheng, C.; Yost, C. K.; Pharis, R. P.; Hynes. M. F. *Rhizobium leguminosarum* as Plant Growth-Promoting Rhizobacterium: Direct Growth Promotion of Canola and Lettuce. *Can. J. Microbiol.* **1996,** *42*, 279–283.

Peng, S.; Biswas, J. C.; Ladha, J. K.; Gyaneshwar, P.; Chen, Y. Influence of Rhizobial Inoculation on Photosynthesis and Grain Yield of Rice. *Agron. J.* **2002,** *94*, 925-929.

Peters, G. A.; Toia, R. E.; Evans, W. R.; Crist, D. K.; Mayne, B. C.; Poole, R. E. Characterization and Comparisons of Five N_2-fixing *Azolla-Anabaena* Associations. I Optimisations of Growth Conditions for Biomass Increase and N Content in a Controlled Environment. Pl. *Cell Env.* **1980,** *3*, 261–267.

Pinckney, J.; Paer, H. W.; Bebout, B. M. Salinity Control of Benthic Microbial Mat Community Production in a Bahamian Hypersaline Lagoon. *J. Exp. Mar. Biol. Ecol.* **1995,** *187*, 223–237.

Prasanna, R.; Kumar, V.; Kumar. S.; Kumar, Y. A.; Tripathi, U.; Kumar, S. A.; Jain, M. C.; Gupta, P.; Singh, P. K.; Sethumathan, N. Methane Production in Rice Soil is Inhibited by Cyanobacteria. *Microbiol. Res.* **2002,** *157*, 1–6.

Prayitno, J.; Stefaniak, J.; Mclver, J.; Weinman, J. J.; Dazzo, F. B.; Ladha, J. K.; Baraquio, W.; Yanni, Y. G.; Rolfe, B. G. Interactions of Rice Seedlings with Bacteria Isolated from Rice Roots. *Austr. J. Plant. Physiol.* **1999,** *26*, 521–535.

Quesada, A.; Fernandez-Valiente, E. Relationship between Abundance of N_2-Fixing Cyanobacteria and Environmental Features of Spanish Rice Fields. *Micro. Ecol.* **1996,** *32*, 59–71.

Quesada, A.; Nieva, M.; Leganes, F.; Ucha, A.; Martin, M.; Prosperi, C.; Fernandez-Valiente, E. Acclimation of Cyanobacterial Communities in Rice Fields and Response of Nitrogenase Activity to Light Regime. *Micro. Ecol.* **1998,** *35*, 147–155.

Reinhold-Hurek, B.; Hurek, T. Life in Grasses: Diazotrophic Endophytes. *Trends Microbiol.* **1998,** *6*, 139–144.

Roger, P. A.; Jimenez, R.; Sartiago-Ardales, S. Methods for Studying Blue Green Algae in Rice Fields: Distributional Ecology Sampling Strategies, and Estimation of Abundance. IRRI Res. Paper Ser. 150; IRRI, Los Banos, Philippines, 1991.

Roger, P. A.; Kulasooriya, S. A. Blue-Green Algae and Rice. The International Rice Research Institute, Los Banos, Philippines, 1980.

Roger, P. A.; Watanabe, I. Technologies for Utilizing Biological Nitrogen Fixation in Wetland Rice: Potentialities, Current Usage and Limiting Factor. *Fert. Res.* **1986,** *9*, 39–77.

Rother, J. A.; Aziz, A.; Karim, N. H.; Whitton, B. A. Ecology of Deepwater Rice-Fields in Bangladesh. 4. Nitrogen Fixation by Blue-Green Algal Communities. *Hydrobiologia* **1988,** *169*, 43–56.

Rother, J. A.; Whitton, B. A. Nitrogenase Activity of Blue-Green Algae on Seasonally Flooded Soils in Bangladesh. *Plant Soil* **1989,** *113*, 47–52.

Shrestha, R. K.; Ladha, J. K. Genotypic Variation in Promotion of Rice Nitrogen Fixation as Determined by 15N Dilution. *Soil Sci. Am. J.* **1996,** *60*, 1815–1821.

Smith, V. H. Nitrogen, Phosphorus, and Nitrogen Fixation in Lacustrine and Estuarine Ecosystems. *Limnol Oceanog* **1990,** *35*, 1852–1859.

Subudhi, B. P.; Watanabe, I. Differential Phosphorus Requirements of *Azolla* Species and Strains in Phosphorus-Limited Continuous Culture. *Soil Sci. Plant Nut.* **1981,** *27*, 237–247.

Takenaka, H.; Yamaguchi, Y.; Sakaki, S.; Wataraj, K.; Tanaka, N.; Horim, S. H.; Tsuchida, M.; Yamada, A.; Nishimori, T.; Morinaga, T. Safety Evaluation of *Nostoc flagelliforme* (Nostocales. Cyanobacteria) as a Potential Food. *Food Chem. Toxicol.* **1998,** *36*, 1073–1077.

Talley, S. N.; Talley, B. J.; Rains, R. W. In *Nitrogen Fixation by Azolla in Rice Field*; Plenum Press: New York, 1977; pp 259–281.

Tan, Z.; Hurek, T.; Vinuesa, P.; Muller, P.; Ladha, J. K.; Reinhold-Hurek, P. Specific Detection of *Bradyrhizobium* and *Rhizobium* Strains Colonizing Rice (*Oryza sativa*) Roots by 16S-23S Ribosomal DNA Intergenic Spacer-Targeted PCR. *Appl. Environ. Microbiol.* **2001a,** *67*, 3655–3664.

Tan, Z.; Hurek, T.; Vinuesa, P.; Muller, P.; Ladha, J. K. Endophytic Colonization of Rice by a Diazotrophic Strain of *Serratia marcescens*. *J. Bacteriol.* **2001b,** *183*, 2634–2645.

Tawaha, A. M.; Singh, V. P.; Turk, M. A. A Review on Growth, Yield Components and Yield of Barley as Influenced by Genotypes, Herbicides and Fertilizer Application. *Res. Crop* **2003,** *4* (1), 1–9.

Tawaha, A. M.; Turk, M. A. Lentil (*Lens culinaris* Medic.) Productivity as Influenced by Rate and Method of Phosphate Placement in a Mediterranean Environment. *Acta Agronomica Hungarica*, **2002a,** *50* (2), 197–201.

Tawaha, A. M.; Turk, M. A. Field Pea Seeding Management for Semi-Arid Mediterranean Conditions. *J. Agron. Crop Sci.* **2004,** *190*, 86–92.

Tawaha, A. R. M.; Turk, M. A.; Lee, K. D.; Zheng, W. J.; Ababneh, M.; Abebe, G.; Musallam, I. W. Impact of Fertilizer and Herbicide Application on Performance of Ten Barley Genotypes Grown in Northeastern Part of Jordan. *Int. J. Agric. Biol.* **2005,** *7* (2), 162–166.

Terouchi, N.; Syono, K. *Rhizobium* Attachment and Curling in Asparagus, Rice and Oat Plants. *Plant Cell Physiol.* **1990,** *31*, 119–127.

Tien, T.; Gaskins, M. H.; Hubbell, D. H. Plant Growth Substances Produced by *Azospirililum brasilense* and their Effect on Growth of Pearl Millet (*Pennisetum americanum* L.). *Appl. Environ. Microbiol.* **1979,** *37*, 1016–1024.

Toenniessen, G. H. Potentially Useful Genes for Rice Genetic Engineering. In *Rice Biotechnology in Agriculture*; Khush, G. S., Toenniessen, G. H., Eds.; International Rice Research Institute-CAB International: Wallingford, Oxon, UK, 1991; pp 253–280.

Turk, M. A.; Tawaha, A. M. Onion (*Allium cepa* L.) as Influenced by Rate and Method of Phosphorus Placement. *Crop Res.* **2002a,** *23* (1), 105–107.

Turk, M. A.; Tawaha, A. M. Allelopathic Effect of Black Mustard (*Brassica nigra* L.) on Germination and Growth of Wild Oat (*Avena fatua* L.). *Crop Prot.* **2003a,** *22* (4), 673.

Turk, M. A.; Tawaha, A. M. Allelopathic Effect of Black Mustard (*Brassica nigra* L.) on Germination and Growth of Wild Barley (*Hordeum spontaneum*). *J. Agron. Crop Sci.* **2003b,** *189* (5), 298–303.

Turk, M. A.; Hmaeed, K. M.; Aqeel, A. M.; Tawaha, A. M. Nutritional Status of Durum Wheat Grown in Soil Supplemented with Olive Mill by-Products. *Agrochimica* **2003c,** *47* (5-6), 209–219.

Turk, M. A.; Lee, K. D.; Tawaha, A. M. Inhibitory Effects of Aqueous Extracts of Black Mustard on Germination and Growth of Radish. *Res. J. Agric. Biol. Sci.* **2005,** *1* (3), 227–231.

Turk, M. A.; Tawaha, A. M. Effect of Variable Sowing Ratios and Sowing Rates of Bitter Vetch on the Herbage Yield of Oat-Bitter Vetch Mixed Cropping. In *Proceedings* 7th *International Oat Conference*, Peltonen-Sainio, P., Topi-Hulmi, M., Eds.; MTT, 2004.

Turk, M. A.; Tawaha, A. M.; Samara, N. Effects of Seedling Rate and Date, and Phosphorous Application on Growth and Yield of Carbon Vetch (*Vicia narbonensis*). *Agronomie* **2003b,** *23*, 1–4.

Turk, M. A.; Tawaha, A. M.; Shatnawi, M. Lentil (*Lens culinaris* Medik) Response to Plant Density, Sowing Date, Phosphorous Fertilization and Ethephon Application in the Absence of Moisture Stress. *J. Agron. Crop Sci.* **2003a,** *189* (1), 1–6.

Tuzun, S.; Kloepper, J. W. Induced Systemic Resistance by Plant Growth-Promoting Rhizobacteria. In Ryder, et al., Eds.; CSIRO: Adelaide, Australia, 1994; pp 104–109.

Uddin, M. S. Cyanobacteria for Improving Fertility and Reclaiming Salinity of Saline Paddy Soils. M.Sc Thesis, Bangladesh Agricultural University, Mymensingh, 1996.

Urquiaga, S.; Cruz, K. H. S.; Boddey, R. M. Contribution of Nitrogen Fixation to Sugarcane: Nitrogen-15 and Nitrogen Balance Estimates. *Soil Sci. Soc. Am. J.* **1992,** *56*, 105–114.

Valente, E. F.; Ucha, A.; Quesada, A.; Leganes, F.; Carreres, R. Contribution of N_2 Fixing Cyanobacteria to Rice Production: Availability of Nitrogen from N Labelled Cyanobacterial and Ammonium Sulphate to Rice. *Plant Soil* **2000,** *221*, 107–112.

Van Kammen, A. Biological Nitrogen Fixation for Sustainable Agriculture. In *Biological Fixation of Nitrogen for Ecology and Sustainable Agriculture*; Legocki, A., Bothe, H., Puhler, A., Eds.; Springer-Verlag: Berlin, Germany, 1997; pp 177–178.

Vance, C. P. Enhanced Agriculture Sustainable through Biological Nitrogen Fixation. In *Biological Fixation of Nitrogen for Ecology and Sustainable Agriculture*; Legocki, A., Bothe, H., Puhler, A., Eds.; Springer-Verlag: Berlin, Germany, 1997; pp 179–186.

Vu, V. V.; Wong, F. S.; Kijne, H. W.; Planque, J. W.; Kraayenhof, K. Temperature, pH, and Bound Nitrogen on Photosynthesis and Nitrogen Fixation of *Azolla pinnata* and *Azolla filiculoides* Lam. *Photosyn.* **1986,** *20*, 67–73.

Wagner, G. M. *Azolla*: A Review of its Biology and Utilization. *Bot. Rev.* **1997,** *63*, 1–26.

Wani, S. P.; Rupela, O. P.; Lee, K. K. Sustainable Agriculture in the Semi-Arid Tropics thought Biological Nitrogen Fixation in Grain Legumes. *Plant Soil* **1995,** *174*, 29–49.

Watanabe, I. Use of Symbiotic and Free-Living Blue-Green Algae in Rice Culture. *Outlook Agriccult.* **1984,** *13*, 166–172.

Watanabe, I. Biological Nitrogen Fixation in Rice Soils. IRRI, 1985, pp 465–477.

Watanabe, I.; Liu, C. C. Improving Nitrogen Fixing Systems and Integrating them into Sustainable Rice Farming. *Plant Soil* **1992,** *141*, 57–67.

Yanni, Y. G.; Rizk, R. Y.; Corich, V.; Squartini, A.; Ninke, K.; Philip-Hollingsworth, G.; Orgambide, de Bruijn, F.; Stoltzfus, J.; Buckley, D.; Schmidt, T. M.; Matoes, P. F.; Ladha, J. K.; Dazzo, F. B. Natural Endophytic Association between *Rhizobium leguminosarum* bv. *Trifolii* and Rice Roots and Assessment of its Potential to Promote Rice Growth. *Plant Soil* **1997,** *194*, 99–114.

Yoshida, T.; Ancajas, R. R. Nitrogen Fixation by Bacteria in the Root Zone of Rice. *Soil Sci. Soc. Am. Proc.* **1971,** *35*, 156–157.

Young, J. P. W. Phylogenic Classification of Nitrogen-Fixing Organisms. In *Biological Nitrogen Fixation*; Stacey, G., Burris, R. H., Evans, H., Eds.; Chapman and Hall: New York, USA, 1992; pp 43–86.

Young, J. P. W.; Haukka, K. E. Diversity and Phylogeny of Rhizobia. *New Phytol.* **1996,** *133*, 87–94.

INDEX

N

O

P

R